Computer Science and Data Analysis Series

Semisupervised Learning for Computational Linguistics

Chapman & Hall/CRC
Computer Science and Data Analysis Series

The interface between the computer and statistical sciences is increasing, as each discipline seeks to harness the power and resources of the other. This series aims to foster the integration between the computer sciences and statistical, numerical, and probabilistic methods by publishing a broad range of reference works, textbooks, and handbooks.

Proposals for the series should be sent directly to one of the series editors above, or submitted to:

Chapman & Hall/CRC
23-25 Blades Court
London SW15 2NU
UK

Published Titles

Bayesian Artificial Intelligence
Kevin B. Korb and Ann E. Nicholson

Pattern Recognition Algorithms for Data Mining
Sankar K. Pal and Pabitra Mitra

Exploratory Data Analysis with MATLAB®
Wendy L. Martinez and Angel R. Martinez

Clustering for Data Mining: A Data Recovery Approach
Boris Mirkin

Correspondence Analysis and Data Coding with Java and R
Fionn Murtagh

R Graphics
Paul Murrell

Design and Modeling for Computer Experiments
Kai-Tai Fang, Runze Li, and Agus Sudjianto

Semisupervised Learning for Computational Linguistics
Steven Abney

Computer Science and Data Analysis Series

Semisupervised Learning for Computational Linguistics

Steven Abney

University of Michigan

Ann Arbor, U.S.A.

CRC Press is an imprint of the Taylor & Francis Group, an **informa** business

A CHAPMAN & HALL BOOK

CRC Press
Taylor & Francis Group
6000 Broken Sound Parkway NW, Suite 300
Boca Raton, FL 33487-2742

First issued in paperback 2019

ISBN-13: 978-1-58488-559-7 (hbk)
ISBN-13: 978-0-367-38863-8 (pbk)

Library of Congress Cataloging-in-Publication Data

Abney, Steven P.
Semisupervised learning in computational linguistics / Steven Abney.
p. cm. -- (Computer science and data analysis series)
ISBN 978-1-58488-559-7 (alk. paper)
1. Computational linguistics--Study and teaching (Higher) I. Title. II. Series.

P98.3.A26 2007
410.285--dc22 2007022858

Visit the Taylor & Francis Web site at
http://www.taylorandfrancis.com

and the CRC Press Web site at
http://www.crcpress.com

Steven Abney

Semisupervised Learning for Computational Linguistics

CRC PRESS
Boca Raton   Ann Arbor   London   Tokyo

Preamble

The primary audience for this book is students, researchers, and developers in computational linguistics who are interested in applying or advancing our understanding of semisupervised learning methods for natural language processing. The problem of semisupervised learning arose almost immediately when computational linguists began exploring statistical and machine learning methods seriously in the late 1980s and early 1990s. In fact, language applications – particularly text classification and information extraction – have provided a major impetus for interest in semisupervised learning in the machine learning community.

Statistical methods that combine labeled and unlabeled data go back at least to the 1960s, and theoretical understanding has advanced quickly over the last few years; but the rate of advancements has made it difficult for non-specialists to keep abreast of them. Those computational linguists whose interest in semisupervised learning is more practical and empirical would benefit from an accessible presentation of the state of the theory. For students, the need for an accessible presentation is urgent.

The purpose of the book is to provide students and researchers a broad and accessible presentation of what is currently known about semisupervised learning, including both the theory and linguistic applications. The background assumed is what can be reasonably expected of any graduate student (or even an advanced undergraduate) who has taken introductory courses in natural language processing that include statistical methods – concretely, the material contained in Jurafsky & Martin [119] and Manning & Schütze [141].

It is desirable that the book be self-contained. Consequently, its coverage will overlap somewhat with standard texts in machine learning. This is unavoidable, given that the target audience is not assumed to have background in machine learning beyond what is contained in the texts just mentioned, and given that semisupervised learning cannot be seriously tackled without understanding the methods for supervised and unsupervised learning that it builds on. My approach has been to treat only those topics in supervised and unsupervised learning that are necessary for understanding semisupervised methods, and to aim for intuitive understanding rather than rigor and completeness – again, except insomuch as a rigorous treatment is required for understanding the main topic, the semisupervised case. In short, the book does cover a number of topics that are found in general introductions to machine learning; but if viewed as a general introduction, it will seem eclectic in coverage and intuitive in treatment. I do not see this necessarily as a flaw.

I find that my own interests often run beyond the areas where I have solid foundations, and an intuitive overview gives me motivation to go back and fill in those foundations. I hope that students of computational linguistics who come with an interest in semisupervised problems but without a general training in machine learning will, above all, find the main topic accessible, but will also acquire a framework and motivation for more systematic study of machine learning.

Although the book is written with computational linguists in mind, I hope that it will also be of interest to students of machine learning. Simple text classification tasks are now familiar in the machine learning literature, but fewer machine learning researchers are aware of the variety of other linguistic applications. Moreover, linguistic applications have characteristic properties that differ in interesting ways from applications that have been the traditional focus in machine learning, and can suggest new questions for theoretical study. For example, natural language problems often have attributes with large sets of discrete values with highly skewed distributions (that is, word-valued attributes), or large sparse spaces of real-valued attributes (numeric attributes indexed by words), or learning targets that are neither discrete classes nor real values, but rather structures (text spans or parse trees).

Perhaps a few readers from even further afield will find the book useful. I have benefited from a book on clustering written for chemists [145], and I would be pleased if researchers from areas well outside of natural language processing find something useful here. Semisupervised learning is a topic of broad applicability. It has already been applied to image processing [98], bioinformatics, and security assessment [177], to name a few examples. Further applications are limited only by imagination.

Acknowledgments

First and foremost, I would like to thank Mark Abney for producing the cover and most of the illustrations. He has done a terrific job; without him the book would have taken much longer and been a good deal less attractive.

I would also like to thank my students and colleagues for reading an earlier draft and giving me indispensable feedback, especially Güneş Erkan, Jessica Hullman, Kevin McGowan, Terrence Szymanski, Richmond Thomason, Li Yang, and an anonymous reviewer.

Finally, I would like to thank my family, Marie Carmen, Anneliese, and Nina, for their patience and support while this book has been in the making.

Contents

1 Introduction

1.1 A brief history

1.1.1 Probabilistic methods in computational linguistics

Computational linguistics seeks to describe methods for natural language processing, that is, for processing human languages by automatic means. Since the advent of electronic computers in the late 1940s, human language processing has been an area of active research; machine translation in particular attracted early interest. Indeed, the inspiration for computing machines was the creation of a thinking automaton, a *machina sapiens*, and language is perhaps the most distinctively human cognitive capacity.

In early work on artificial intelligence, there was something of a competition between discrete, "symbolic" reasoning and stochastic systems, particularly neural nets. But the indispensability of a firm probabilistic basis for dealing with uncertainty was soon recognized. In computational linguistics, by contrast, the presumption of the sufficiency of grammatical and logical constraints, supplemented perhaps by ad hoc heuristics, was much more tenacious.

When the field recognized the need for probabilistic methods, the shift was sudden and dramatic. It is probably fair to identify the birth of awareness with the appearance in 1988 of two papers on statistical part-of-speech tagging, one by Church [44] and one by DeRose [75]. These were not the first papers that proposed stochastic methods for part of speech disambiguation, but they were the first in prominent venues in computational linguistics, and it is no exaggeration to say that the field was reshaped within a decade.

The main barrier to progress in natural language processing at the time was the brittleness of manually constructed systems. The dominant issues were encapsulated under the rubrics of ambiguity resolution, portability, and robustness. The primary method for **ambiguity resolution** was the use of semantic constraints, but they were often either too loose, leaving a large number of viable analyses, or else too strict, ruling out the correct analysis. Well-founded and automatic means for softening constraints and resolving ambiguities were needed. **Portability** meant in particular automatic means for adapting to variability across application domains. **Robustness** covers both the fact that input to natural language systems is frequently errorful, and also the fact that, in Sapir's terms, "all grammars leak" [201]. No manually constructed description of language is complete.

Together, these issues point to the need for automatic learning methods, and explain why the penetration of probabilistic methods, and machine learning in particular, was so rapid. Computational linguistics has now become inseparable from machine learning.

1.1.2 Supervised and unsupervised training

The probabilistic models used by Church and DeRose in the papers just cited were Hidden Markov Models (HMMs), imported from the speech recognition community. An HMM describes a probabilistic process or automaton that generates sequences of states and parallel sequences of output symbols. Commonly, a sequence of output symbols represents a sentence of English or of some other natural language. An HMM, or any model, that defines probabilities of word sequences (that is, sentences) of a natural language is known as a **language model**.

The probabilistic automaton defined by an HMM may be in some number of distinct **states**. The automaton begins by choosing a state at random. Then it chooses a symbol to emit, the choice being sensitive to the state. Next it chooses a new state, emits a symbol from that state, and the process repeats. Each choice is stochastic – that is, probabilistic. At each step, the automaton makes its choice at random from a distribution over output symbols or next states, as the case may be. Which distribution it uses at any point is completely determined by the kind of choice, either **emission** of an output symbol or **transition** to a new state, and the identity of the current state. The actual **model** consists in a collection of numeric values, one for each possible transition or emission, representing the probability that the automaton chooses that particular transition or emission when making one of its stochastic choices.

Learning an HMM is straightforward if one is provided with **labeled data**, meaning state sequences paired with output sequences. Each sequence pair is a record of the stochastic choices made by the automaton. To estimate the probability that the automaton will choose a particular value x when faced with a stochastic choice of type T, one can simply count how often the automaton actually chose x when making a choice of type T in the record of previous computations, that is, in the labeled data. If sufficient labeled data is available, the model can be estimated accurately in this way.

Church and DeRose applied HMMs to the problem of part-of-speech tagging by identifying the states of the automaton with parts of speech. The automaton generates a sequence of parts of speech, and emits a word for each part of speech. The result is a **tagged text**, which is a text in which each word is annotated with its part of speech.

Supervised learning of an HMM for part-of-speech tagging is quite effective; HMM taggers for English generally have an error rate of 3.5 to 4 percent.

Their effectiveness was what brought probabilistic models to the attention of computational linguists, as already mentioned.

1.1.3 Semisupervised learning

Creating sufficient labeled data can be very time-consuming. Obtaining the output sequences is not difficult: English texts are available in great quantity. What is time-consuming is creating the state sequences. One must essentially annotate each word of the texts with the state of the HMM from which it was emitted. For this reason, one would like to have a method for learning a model from **unlabeled data**, which, in this case, consists of simple English texts without state annotations. A learning method that uses unlabeled data is known as an **unsupervised** learning method, in contrast to **supervised** learning methods, which use labeled data.

An unsupervised learning method for HMMs has long been known, called the forward-backward algorithm. It is a special case of the Expectation-Maximization (EM) algorithm. It is widely used and effective in speech recognition. For example, it is used to estimate acoustic models from unlabeled data. For an acoustic model, unlabeled data consists of text paired with the speech signal resulting from reading the text, but without any annotation regarding the sequence of states that the model passes through while generating the speech from the text.

However, unsupervised learning turned out not to be very effective for part-of-speech tagging. The forward-backward algorithm is an iterative algorithm that begins with some initial model and improves it by repeated passes through the unlabeled data. The question of the effectiveness of unsupervised learning was posed by Elworthy [83] in the following form. If one uses labeled data to obtain the initial model for forward-backward training on unlabeled data, how many iterations of forward-backward training should one do for a given amount of labeled seed data? The answer was essentially zero iterations if one had more than a tiny amount of labeled data. (Merialdo [153] came to similar conclusions.) Intuitively, the states that are learned by an automaton trained on unlabeled data do not correspond at all well to linguistically motivated parts of speech.

The solution that emerged in the case of part-of-speech tagging was to use additional constraints. Specifically, if one uses unlabeled data and forward-backward training, but restricts words to take on only the parts of speech that are allowed for them by a dictionary, the results are comparable to using labeled data [62].

But despite its ineffectiveness for part-of-speech tagging, the idea of learning from a mixture of labeled and unlabeled data remained potent, and it has come to constitute the canonical setting for **semisupervised learning**. (One can view the dictionary constraints that proved more effective for tagging as providing partially labeled data, hence a variant of semisupervised learning. This will be discussed in more detail later.)

Subsequent work in computational linguistics led to development of alternative algorithms for semisupervised learning, the algorithm of Yarowsky [239] being a prominent example. These algorithms were developed specifically for the sorts of problems that arise frequently in computational linguistics: problems in which there is a linguistically correct answer, and large amounts of unlabeled data, but very little labeled data. Unlike in the example of acoustic modeling, classic unsupervised learning is inappropriate, because not just any way of assigning classes will do. The learning method is largely unsupervised, because most of the data is unlabeled, but the labeled data is indispensable, because it provides the only characterization of the linguistically correct classes.

The algorithms just mentioned turn out to be very similar to an older learning method known as **self-training** that was unknown in computational linguistics at the time. For this reason, it is more accurate to say that they were rediscovered, rather than invented, by computational linguists. Until very recently, most prior work on semisupervised learning has been little known even among researchers in the area of machine learning. One goal of the present volume is to make the prior and also the more recent work on semisupervised learning more accessible to computational linguists.

Shortly after the rediscovery of self-training in computational linguistics, a method called **co-training** was invented by Blum and Mitchell [21], machine-learning researchers working on text classification. Self-training and co-training have become popular and widely employed in computational linguistics; together they account for all but a fraction of the work on semisupervised learning in the field. We will discuss them in the next chapter. In the remainder of this chapter, we give a broader perspective on semisupervised learning, and lay out the plan of the rest of the book.

1.2 Semisupervised learning

1.2.1 Major varieties of learning problem

There are five types of learning problem that have received the preponderance of attention in machine learning. The first four are all cases of **function estimation**, grouped along two dimensions: whether the learning task is supervised or unsupervised, and whether the variable to be predicted is nominal or real-valued.

Classification involves supervised learning of a function $f(x)$ whose value is nominal, that is, drawn from a finite set of possible values. The learned function is called a classifier. It is given **instances** x of one or another class, and it must determine which class each instance belongs to; the value $f(x)$ is the classifier's prediction regarding the class of the instance. For example, an instance might be a particular word in context, and the classification task is to determine its part of speech. The learner is given labeled data consisting of a collection of instances along with the correct answer, that is, the correct class label, for each instance.

The unsupervised counterpart to classification is **clustering**. The goal in clustering is also to assign instances to classes, but the clustering algorithm is given only the instances, not the correct answers for any of them. (In clustering, the instances are usually called **data points** and the classes are called **clusters**.) The primary difference between classification and clustering is not the task to be performed, but the sort of data that is given to the learner as input; in particular, whether the data is labeled or not.

The remaining two function estimation tasks involve estimation of a function that takes on real values, instead of values from a finite range. The supervised version is called **regression**; it differs from classification only in that the function to be learned takes on real values. Unsupervised learning of a real-valued function can be viewed as **density estimation**. The learner is given an unlabeled set of training data, consisting of a finite sample of data points from a multi-dimensional space, and the goal is to learn a function $f(x)$ assigning a real value to every point in the space; the function is interpreted as (proportional to) a probability density.

Finally, we mentioned a fifth setting that does not fall under function estimation. This fifth setting is **reinforcement learning**. In reinforcement learning, the learner receives a stream of data from sensors, and its "answers" consist in actions, in the form of commands sent to actuators. There is, additionally, a reward signal that is to be maximized (over the long run). There are at least two significant ways this differs from the four function estimation settings. First is the sequential nature of the inputs. Even if we assume discrete time, there are temporal dependencies that cannot be ignored: in particular, actions have time-delayed effects on sensors and reward. Second is the indirect nature of the supervision. The reward signal provides information about the relative value of different actions, but it is much less direct than simply providing the correct answer, as in classification.

Semisupervised learning generalizes supervised and unsupervised learning. The generalization is easiest to see with classification and clustering. As already mentioned, classification and clustering involve essentially the same task and the same inputs; they differ primarily in whether the training data is labeled or not. (They also differ in the way they are evaluated, but the difference in evaluation is a consequence of the difference in the kind of training data – more on that later.) The obvious generalization is to give the learner labels for *some* of the training data. At one extreme, all of the data is labeled, and the task is classification, and at the other extreme, none of the data is labeled, and the task is clustering. The mixed labeled/unlabeled setting is indeed the canonical case for semisupervised learning, and it will be our main interest.

At the same time, a mix of labeled and unlabeled information is only one way of providing a learner with partial information about the labels for training data. Many semisupervised learning methods work with alternate kinds of partial information, such as a handful of reliable rules for labeling instances, or constraints limiting the candidate labels for particular instances. We will also consider these extensions of the canonical setting. In principle, the kind of indirect information about labels found in reinforcement learning qualify it as a kind of semisupervised learning, but the indirect-information aspect of reinforcement learning is difficult to disentangle from the temporal dependencies, and the connection between reinforcement learning and other semisupervised approaches remains obscure; it lies beyond the scope of the present work.

1.2.2 Motivation

For most learning tasks of interest, it is easy to obtain samples of unlabeled data. For many language learning tasks, for example, the World Wide Web can be seen as a large collection of unlabeled data. By contrast, in most cases, the only practical way to obtain labeled data is to have subject-matter experts manually annotate the data, an expensive and time-consuming process.

The great advantage of unsupervised learning, such as clustering, is that it requires no labeled training data. The disadvantage has already been mentioned: under the best of circumstances, one might hope that the learner would recover the correct clusters, but hardly that it could correctly label the clusters. In many cases, even the correct clusters are too much to hope for. To say it another way, unsupervised learning methods rarely perform well if evaluated by the same yardstick used for supervised learners. If we expect a clustering algorithm to predict the labels in a labeled test set, without the advantage of labeled training data, we are sure to be disappointed.

The advantage of supervised learning algorithms is that they do well at the harder task: predicting the true labels for test data. The disadvantage is that they only do well if they are given enough labeled training data, but producing sufficient quantities of labeled data can be very expensive in manual effort.

The aim of semisupervised learning is to have our cake and eat it, too. Semisupervised learners take as input unlabeled data and a limited source of label information, and, if successful, achieve performance comparable to that of supervised learners at significantly reduced cost in manual production of training data.

We intentionally used the vague phrase "a limited source of label information." One source of label information is obviously labeled data, but there are alternatives. We will consider at least the following sources of label information:

- labeled data
- a seed classifier

- limiting the possible labels for instances without determining a unique label
- constraining pairs of instances to have the same, but unknown, label (co-training)
- intrinsic label definitions
- a budget for labeling instances selected by the learner (active learning)

One of the grand aims of computational linguistics is unsupervised learning of natural language. From a psychological perspective, it is widely accepted that explicit instruction plays little part in human language learning, and from a technological perspective, a completely autonomous system is more useful than one that requires manual guidance. Yet, in contradiction to the characterization sometimes given of the goal of unsupervised learning, the goal of unsupervised language learning is not the recovery of arbitrary "interesting" structure, but rather the acquisition of the correct target language. On the face of it, learning a target classification – much less an entire natural language – without labeled data hardly seems possible.

Semisupervised learning may provide the beginning of an account. If a kernel of labeled data can be acquired through unsupervised learning, semisupervised learning might be used to extend it to a complete solution. Something along these lines appears to characterize human language acquisition: in the psycholinguistic literature, *bootstrapping* refers to the process by which an initial kernel of language is acquired by explicit instruction, in the form, for example, of naming an object while drawing a child's attention to it. The processes by which that kernel is extended to the entirety of the language are thought to be different; distributional regularities of linguistic forms, rather than direct connections to the physical world, seem to play a large role. Semisupervised learning methods provide possible characterizations of the process of extending the initial kernel.

1.2.3 Evaluation

With regard to evaluation, semisupervised algorithms are like supervised algorithms. The basic measure of success is classification performance on an unseen test set, used as an estimate of generalization error.

But in addition to measuring absolute performance, one would also like to measure the benefit obtained by the addition of unlabeled data. The most general way to pose the question is the level of performance as a function of human effort. More concretely, one considers prediction rule quality as a function of the number of labeled instances and the number of unlabeled instances. Two questions are of particular interest: (1) for a fixed number of labeled instances (i.e., a fixed annotation budget), how much improvement is obtainable as the number of unlabeled instances grows without bound; and

(2) for a fixed target level of performance, what is the minimum number of labeled instances needed to achieve it, as the number of unlabeled instances grows without bound.

1.2.4 Active learning

One way of characterizing the overarching goal of semisupervised learning is to use unlabeled data to amplify the information gained from a manually created seed. We focus almost exclusively on "batch" learning, in which the seed and a population of unlabeled data are given in advance. A natural next question is whether a better effort-performance curve can be obtained in an interactive setting, for example, by selecting instances to be labeled, and interleaving learning with labeling. Interactive semisupervised learning is called **active learning**. It lies beyond the scope of the current work.

1.3 Organization and assumptions

1.3.1 Leading ideas

Semisupervised learning methods have sprung up independently in several different areas, usually as modifications of existing algorithms. For example, if one's interest is classification, it is natural to ask how to modify a classifier-learning algorithm to make use of unlabeled data. Conversely, if one's interest is clustering, it is natural to ask how to make use of manually labeled examples, either to assign names to otherwise anonymous clusters, or to constrain the algorithm to produce clusters that are consistent with the manual labels.

We organize semisupervised algorithms by the leading idea that each is based on. These are the leading ideas, in our view:

- **Self-training** (chapters 2–3 and 8). If one comes from the perspective of supervised learning, and asks how unlabeled instances might be put to use in addition to labeled instances, a natural idea is to train a classifier on the labeled instances, apply it to the unlabeled instances, and take its predictions at face value, at least in those cases where its predictions are most confident. A new classifier is trained on the extended set of labeled instances, and the process repeats. This approach is known as **self-training**.

- **Cluster and label** (chapter 7). Coming to semisupervised learning from the perspective of unsupervised learning, a natural idea is to apply a clustering algorithm to the unlabeled data (one can also strip the labels from the labeled data and throw it in as well), and then use the labeled data to "name" the clusters. A cluster is associated with whichever label occurs most frequently on labeled instances in the cluster, and the prediction for an unlabeled instance is determined by the cluster that it is assigned to.

- **Application of "missing values" techniques** (chapter 8). The problem of missing values in a data set is very familiar in statistics. It is natural to think of unlabeled data as data with missing values for the dependent variable (that is, the class label), and apply a method for filling in missing information using a **generative model**. The canonical example is the Expectation-Maximization (EM) algorithm. The earliest literature on semisupervised learning falls under this rubric.

- **Label propagation in graphs** (chapter 10). Clustering algorithms are typically based on a similarity metric; clusters are defined to be groups of similar instances. A postulate sometimes called the **cluster hypothesis** is that similar instances have similar labels, or in geometric terms, that proximate instances have similar labels. A similarity function can be represented as a weighted graph, in which instances are nodes and edges are weighted by the similarity of the instances they connect. Two nodes connected by a heavily weighted edge should have the same label. An algorithmic correlate is to propagate labels along heavily weighted edges. Geometrically, one can view the graph as a fabric whose interior consists of unlabeled instances and whose boundary consists of labeled instances. The elevation of the fabric at a given point represents the label of that point, and the effect of propagation is to interpolate from the fixed boundary across the interior of the graph.

- **Boundaries in low density regions** (chapters 5–6). The contrapositive of the cluster hypothesis is what we might call the **separation hypothesis**: the idea that different labels imply distant data points, which is to say, that boundaries between classes lie where the data is sparse. **Transductive** maximum-margin methods, such as transductive Support Vector Machines (SVMs) and transductive boosting, can be understood in those terms. The goal is to find a linear inter-class boundary with a large *margin*, which is to say, a large distance to the nearest data points. Instead of looking for natural clusters, one looks for natural boundaries, but otherwise the approach is very similar to cluster-and-label. Natural boundaries are found without regard to labels, and the labeled instances are used to determine labels for the resulting regions.

- **Constraint- and agreement-driven learning** (chapters 2–3 and 9). In a sense, all semisupervised learning is driven by constraints. Sufficiently restrictive constraints can be almost as good as labels for unlabeled data – and in some cases even better, inasmuch as a constraint applies to the entire population of instances, whereas a label on an instance applies only to that instance. We have mentioned how graph methods translate similarities into soft constraints. A particularly salient class of constraints is agreement constraints. For example, in **co-training**, the learner is given two independent "views" of the data, and constructs one classifier for each view, under the constraint that the classifiers agree in their predictions on the unlabeled data. Effectively, instances come in pairs that are constrained to have the same label. The learner does not know what the label is, but does know that it is the same for both members of the pair. A non-algorithmic way of enforcing agreement is via a **random field** that penalizes disagreement.

- **Spectral methods** (chapters 11–12). One can build on the idea of interpolation that emerges from label propagation by using a "standing wave" to interpolate across the graph. Pursuing this idea leads to deep connections among apparently disparate ideas, including the cluster hypothesis and label propagation, "mincut" boundary-oriented methods, and random fields.

The plan of the book more or less follows the list of leading ideas just given, with a couple of rearrangements for the sake of a smoother line of development. We begin, in chapters 2 and 3, with a discussion of the semisupervised methods that are already well known in computational linguistics, namely, self-training and co-training. We turn then to methods that come from the machine learning literature, beginning in chapter 4 with an introduction to classification, including some detail on the topic of decision boundaries. The discussion of boundary-oriented methods follows naturally at that point (chapters 5–6). Then we turn to clustering in chapter 7, followed by discussion of the EM algorithm and related generative methods in chapter 8. There are connections between co-training and the generative methods of chapter 8, so the chapter on agreement methods is placed next (chapter 9). Finally, the strand of graph-based methods, begun in the chapter on clustering, is picked up in chapter 10, which concerns label propagation, and in chapters 11 and 12 on spectral methods.

1.3.2 Mathematical background

As stated in the preface, my goal is to bring the current state of the art in semisupervised learning within the reach of a student or researcher in computational linguistics who has mastered the standard textbooks, in particular, Manning and Schütze, and has acquired a certain familiarity with machine learning through references in the computational linguistics literature, but does not necessarily have a general background in machine learning. This goal is more than a little quixotic. To do things properly, we should lay a foundation of linear algebra, multivariate calculus, optimization theory, probability and statistics, and even a bit of physics (e.g., simple harmonic motion), and on that build a proper treatment of classification and clustering, before tackling the actual topic of interest, semisupervised learning. But doing so would involve replicating many volumes of material that has been well covered elsewhere. A reader who has already mastered all the background material just mentioned is in an excellent position to tackle the primary literature on semisupervised learning, and will probably not find this book particularly useful. On the other hand, readers who have not mastered all the necessary background material will rightfully feel daunted by the enormity of the task, and would under most circumstances decide that, however interested they may be in semisupervised learning, the cost of entry is simply too great to pay. Those are the readers for whom this book is intended.

My strategy has been to blaze a long thin trail, filling in just the background that is needed to give a reasonably detailed account of the selected semisupervised learning techniques. Two chapters provide an introduction to machine learning: one on classification (chapter 4) and one on clustering (chapter 7). They do not attempt to give a balanced overview of the field, but only to treat topics specifically needed for semisupervised learning. As for more general mathematical background, I have chosen not to collect it into a single chapter – the result would have been a disconnected collection of topics, and the reasons for their inclusion would only have become clear much later. Instead, these topics have been introduced "just in time." The cost is a rather lengthy run-up to the semisupervised techniques involved, especially SVM-based and spectral methods, but that seemed the lesser of the two evils.

1.3.3 Notation

I have collected here notational conventions that I use that are nonstandard or may not be familiar to all readers.

$[\![\Phi]\!]$	semantic value: 1 if Φ is true and 0 otherwise
$\sum_x [\![x \in S]\!] w(x)$	equivalent to: $\sum_{x \in S} w(x)$
$\lVert\mathbf{x}\rVert$	vector norm: $\sqrt{\sum_i x_i^2}$
$\lvert A\rvert$	cardinality of a set or absolute value of a number
$p[\phi]$	the expectation of ϕ under distribution p
$\tilde{p}(x)$	empirical distribution: relative frequency in sample
$f(x) = \bot$	$f(x)$ is undefined
$\equiv$	is defined as
$F \Rightarrow y$	rule: if the instance has feature F, predict class y
$x \leftarrow x + 1$	set the value of x (in an algorithm)
$\mathbf{D}_\mathbf{x}$	derivative with respect to a vector; see section 5.2.3

2

Self-training and Co-training

In the previous chapter we introduced self-training and co-training as the most widely used semisupervised learning algorithms in computational linguistics. We present them in more detail in this chapter.

Both algorithms are "naive" in the sense that they did not derive from a theory of semisupervised learning; rather, they are embodiments of simple algorithmic ideas that can be grasped without much background. This is not to say that they lack theoretical justification. Co-training and self-training can both be justified on the basis of a conditional independence assumption that is closely related to the independence assumption underlying the Naive Bayes classifier. And forms of the algorithms *have* arisen on the basis of a theory of semisupervised learning – particular examples are McLachlan's version of self-training (discussed in chapter 8) and de Sa's version of co-training (discussed in chapter 9). These theoretically motivated algorithms even predate the versions that are well known in computational linguistics. But we focus here on the familiar versions, whose primary attraction is simplicity and intuitiveness, rather than theoretical underpinnings.

2.1 Classification

Self-training and co-training are both classifier-learning algorithms. We introduced classification in the previous chapter as function estimation in which the function is nominal-valued and the learning is supervised. With self-training and co-training, we extend the notion of classification to include the semisupervised case. Ultimately, we will view semisupervised classification as a generalization of classification and clustering.

We begin with some grounding in classification.

2.1.1 The standard setting

An example of a classification problem which we have already mentioned is part-of-speech tagging. This is a well-studied task in computational linguistics. The input to a tagger is a plain text, such as

instance	label
[a] black dog appeared	det
a [black] dog appeared	adj
a black [dog] appeared	noun
a black dog [appeared]	verb

FIGURE 2.1
Inputs and outputs for a part of speech classifier.

a black dog appeared

and the output is a sequence of parts of speech, perhaps:

det adj noun verb

One can reduce tagging to a classification problem by treating each word separately. Each word constitutes an instance, and the possible class labels are the possible parts of speech. An example is given in figure 2.1. The left column shows inputs that the classifier receives, and the right column shows the output of the classifier for each input. The box indicating the "current word" is an essential part of the input; in fact, the inputs differ only in the location of the box. The box indicates which word is the word of interest, and the rest of the text provides the context. We note that the context does not include information about the classifier's own previous decisions. Previous decisions could be included in the context. Doing so would mean that training instances are not independently drawn, violating a standard assumption about the training data, but we have in fact already violated that assumption by using running text. We cannot include both previous and future decisions, however. There are methods for choosing the best overall sequence of labels – Hidden Markov Models and Conditional Random Fields (CRFs) are two well-known examples – but they take us beyond classification.

A classifier *learner* is given a set of training data as input, and produces a classifier as output. Training data is labeled: it consists of a set of instances, representing sample inputs, along with the correct output for each. The goal of the learner is to produce a classifer that generalizes well to the entire population of instances, and particularly to new instances that were not encountered in training.

2.1.2 Features and rules

For effective learning, instances must be broken down into **features**. A feature may be anything, but it is often the pairing of an **attribute** with a particular value. For example, we might adopt the attributes `string` for the current word as an ASCII string, `prev` for the previous word, and `next` for next word. The value for each attribute is an actual word, for example, "`prev=black`."

instance	label
{prev=null, string=a, next=black}	det
{prev=a, string=black, next=dog}	adj
{prev=black, string=dog, next=appeared}	noun
{prev=dog, string=appeared, next=null}	verb

FIGURE 2.2
Instances represented as sets of features.

There are two interchangeable ways of viewing the expression "prev=black." It may be viewed as the assignment of the value "black" to the attribute "prev," or, in its entirety, it represents a single feature, distinct from other features like "string=black" or "next=black" or "prev=dog." An instance consists of a set of features. For example, the training text in figure 2.1 might be represented as in figure 2.2.

Generally, a feature corresponds to a predicate F. An instance x possesses the feature just in case $F(x)$ is true. An attribute corresponds to a function h. The pairing of an attribute h and value v corresponds to a predicate which is true just in case $h(x) = v$, and false otherwise. A feature may alternatively correspond to the pairing of a two-place relation R with a value v. The predicate thus represented is true just in case $R(x, v)$. For example, a word string has multiple suffixes, of varying lengths. Consider the two-place predicate hasSuffix, such that hasSuffix(x, y) is true just in case y is a suffix of x. The expressions hasSuffix(dogs, s) and hasSuffix(dogs, gs) are both true. We will adopt the convention of writing hasSuffix:s for the feature that is possessed by an instance x just in case hasSuffix(x, s) is true. In short, the difference between a feature like string=dogs and a feature like hasSuffix:s is that an instance may possess only one "string=" feature, but it may possess multiple "hasSuffix:" features.

Many authors use "feature" and "attribute" synonymously. I should emphasize that I use them distinctly. An attribute has a value; the combination of an attribute and a value is a feature.

We have seen that an instance may be represented as a set of features. An instance may also be represented as a vector of attribute values. If a feature is not explicitly represented as the pairing of an attribute and a value, we may always take the feature itself to be the attribute, and the value to be T (true) if the instance possesses the feature and F otherwise. Similarly, the combination of a relation and value, like hasSuffix:s, can be taken as an attribute whose value is either T or F.

If we fix the order of attributes, a unique vector of values is determined for each instance. For example, if we fix the order of attributes as (prev, string, next), the instances of figure 2.2 can equivalently be represented as the vectors of figure 2.3.

instance	label
(null, a, black)	det
(a, black, dog)	adj
(black, dog, appeared)	noun
(dog, appeared, null)	verb

FIGURE 2.3
Instances represented as vectors of attribute values.

2.1.3 Decision lists

Decision lists provide a simple concrete example of a classifier, and they are also the supervised learner on which a popular version of self-training is based.

A decision list consists of a sequence of **rules** of the form "if instance x possesses feature F, then predict label y," for example:

```
if string=dog then predict noun
```

The order in which the rules are listed is significant. The list may also include a default rule "`predict` y."

A rule **matches** an instance x if x possesses the feature in the antecedent of the rule. A default rule always matches. The prediction of a decision list on input x is determined by the first rule in the list that matches x. For example, the decision list

```
if prev=a then predict noun
if string=black then predict adj
predict verb
```

predicts "`noun`" on the input

{`prev=a`, `string=black`, `next=dog`}

but it predicts "`adj`" on the input

{`prev=the`, `string=black`, `next=dog`}

Henceforth we abbreviate "`if` F `then predict` y" as "$F \Rightarrow y$" and the default rule "`predict` y" as "$\Rightarrow y$."

The task of the learner is to decide which rules to include in the list, and how to sort them. An easy choice regarding which rules to include is to include all that are attested in the training data. If there is an instance (x, y) in the training data, and x has feature F, then include the rule "$F \Rightarrow y$." Also, for each label y, include the default rule "$\Rightarrow y$." For example, in the little training set of figure 2.2, there are twelve features attested, and each only ever occurs with one label, yielding twelve rules, plus four default rules.

It may seem odd to include default rules for all four labels. Only one of them will ever be used in the end – whichever one is ordered first in the decision list will always match, and the others will never be used. Borrowing a term from phonology, we can say the first one **bleeds** the others. But we include them all in the initial list, and decide which one will will bleed the others when we sort the rules. If desired, the others could be pruned after the list is sorted. Similarly, wherever there are multiple rules $F \Rightarrow y$ and $F \Rightarrow y'$ with the same feature but different labels, the first one on the list will bleed the others, and all but the first can be pruned.

As for sorting, the main issue is how to sort rules that have different features. It is possible for more than one rule to match an instance x, each rule matching a different feature of x. The simplest way is to sort by the probability that the rule is right, given that it matches. That is, the learner sorts the list by descending conditional accuracy, breaking ties arbitrarily, where **conditional accuracy** of the rule is defined as

$$\text{CA}(F \Rightarrow y) = \Pr[\text{true label is } y | \text{rule fires}].$$

In most cases, we are concerned with conditional accuracy as a sample statistic, for either the training or test set, in which case

$$\text{CA}(F \Rightarrow y) = \frac{\sum_i [\![F \in x_i, y_i = y]\!]}{\sum_i [\![F \in x_i]\!]}.$$

where i ranges over the (indices of the) instances in the sample. Here we have borrowed from semantics the notation $[\![\phi]\!]$:

$$[\![\phi]\!] \equiv \begin{cases} 1 \text{ if } \phi \text{ is true} \\ 0 \text{ otherwise} \end{cases}.$$

Conditional accuracy is related to a perhaps more familiar quantity called **precision**, but it is not the same; the connection is discussed in chapter 4.

We have already mentioned the bleeding that occurs when two rules are conditioned on the same feature. We also note that a default rule bleeds *all* rules that follow it, since a default rule always matches. For example, suppose that the most-frequent label occurs on half the training instances, and suppose that a particular feature F occurs three times, each time with a different label. Then each of the three rules conditioned on F has conditional accuracy 1/3, but the default rule has conditional accuracy 1/2. Hence the default rule precedes all of the rules conditioned on F, and none of them ever matches an instance. They can be deleted without affecting the predictions of the classifier.

Instead of assuming an ordered list, we can interpret the conditional accuracy of each rule as a **score** or **weight**. Among the rules that match a given instance, the decision list's prediction is determined by the one that has the highest score.

Note that a decision list actually provides more information than just a prediction concerning the label for a given input instance. The score of the "winning" rule can be interpreted as a measure of confidence: the higher the score, the more confident the classifier is that its prediction is correct. A classifier such as this, that produces both a label prediction and also a measure of confidence in that prediction, is said to be a **confidence-rated** classifier. A measure of confidence often proves useful in adapting classifiers to a semisupervised setting.

Finally, we should note that decision lists may fail to make a prediction, that is, they may **abstain**. Abstention does not occur if the decision list contains a default rule, but if there is no default rule, and a given input instance possesses no feature that is mentioned in the decision list, then the decision list makes no prediction.

The conditional accuracy measure extends naturally from single rules to entire decision lists, or for that matter, to any classifier f that may abstain. Let us write $f(x) = \perp$ to represent the case in which classifier f abstains on input x, the symbol "$\perp$" representing "undefined." The conditional accuracy of f on a sample is

$$\mathrm{CA}(f) = \frac{\sum_i [\![f(x_i) = y_i \neq \perp]\!]}{\sum_i [\![f(x_i) \neq \perp]\!]}.$$

A complementary measure is the **firing rate** of the classifier,

$$\mathrm{FR}(f) = \frac{1}{n} \sum_i [\![f(x_i) \neq \perp]\!]$$

where n is the sample size. If we use the notation "$\tilde{p}[\phi]$" for the sample probability that event ϕ occurs, then we can write these quantities more perspicuously:

$$\mathrm{CA}(f) = \tilde{p}[\, f(X) = Y \mid f(X) \neq \perp]$$
$$\mathrm{FR}(f) = \tilde{p}[\, f(X) \neq \perp]$$

where we have introduced X and Y as random variables ranging over instances and their associated labels in the sample. The overall accuracy of the classifier is the product of these two values:

$$\begin{aligned} \mathrm{Acc}(f) &\equiv \tilde{p}[\, f(X) = Y] \\ &= \mathrm{FR}(f) \cdot \mathrm{CA}(f). \end{aligned}$$

2.2 Self-training

We turn now to self-training. The version of the algorithm that we describe is that of Yarowsky [239].

2.2.1 The algorithm

The algorithm is simple, and that is in fact a large part of its appeal. Use a seed set of labeled data to construct a confidence-rated classifier (such as a decision list), apply the classifier to unlabeled data, and take the predictions of the classifier to be correct for those instances where it is most confident. Expand the labeled data by the addition of the classifier-labeled data, and train a new classifier. The process may be iterated: continue labeling new data and retraining the classifier until a stopping condition is met.

Though Yarowsky's paper is the most widely cited, the key idea – taking an initial classifier's high-confidence predictions at face value and adding them to the training data to train a new classifier – was proposed in at least two earlier papers, one by Hindle & Rooth [111] and one by Hearst [109].

Hindle & Rooth address the problem of prepositional phrase (PP) attachment. Their seed consists in a set of manually constructed, deterministic decision rules that are treated as inerrant. Applying the deterministic rules to unlabeled data yields an initial labeled set, from which statistics are estimated for a probabilistic decision rule. The probabilistic decision rule is then applied to the remaining unlabeled data, and its high-confidence predictions are accepted at face value and used to augment the statistics.

Hearst, like Yarowsky, addresses the problem of word-sense disambiguation. She defines a decision rule based on certain statistics computed from labeled data, and then augments the statistics using unlabeled instances where the predictions of the decision rule are sufficiently confident. Neither Hearst nor Hindle & Rooth iterate the process of labeling and retraining.

None of these authors treats semisupervised learning as a paradigm in its own right, though Hearst comes closest. She explicitly describes her algorithm as a combination of supervised and unsupervised learning: "A period of supervised training, ... is required as a bootstrap before unsupervised training or testing can begin" (p. 5). Yarowsky also makes an oblique mention of bootstrapping, though he considers his algorithm to be unsupervised learning.

With hindsight, we will treat the three papers just mentioned as examples of **self-training**. (I have taken the term from Nigam & Ghani [166], though it may well be older.) The basic form of self-training is given in figure 2.4. The algorithm produces a sequence of classifiers, which are understood to be confidence-rated classifiers. A confidence-rated classifier takes an instance as input and produces as output a pair (y, s) consisting of a label y and confidence score s. The classifier may also abstain, in which case its output is $\perp$.

The function `label` in figure 2.4 applies the classifier c to each instance in the unlabeled data set U, producing one pair for each instance. The resulting list of (y, s) pairs is a **confidence-weighted labeling**. To be precise, the output of `label` is a confidence-labeled data set, consisting of both the confidence-weighted labeling and the instances to which the labeling applies.

The function `select` takes a confidence-labeled data set and selects only those instances where the confidence is "sufficiently high." The output is a

```
procedure selfTrain (L0, U)
1  L0 is labeled data, U is unlabeled data
2  c ← train(L0)
3  loop until stopping criterion is met
4      L ← L0 + select(label(U, c))
5      c ← train(L)
6  end loop
7  return c
```

FIGURE 2.4
Self-training, basic form.

(smaller) labeled data set (without confidences).

The function `train` represents a supervised classifier learning algorithm, which we call the **base learner**. It takes labeled data and produces a confidence-rated classifier.

2.2.2 Parameters and variants

The schema contains several open parameters, and there are a number of additional points where variations have been explored. What is essential to self-training is:

- Using an intermediate classifier to produce *hard-labeled* data
- Using the classifier-labeled data to train an improved classifier

By "hard-labeled" we mean that unlabeled instances are assigned definite labels, as opposed to probability distributions over labels. One can certainly imagine varying even that aspect of the algorithm, but the result is probably best seen as falling into a different class of algorithm. Training a classifier on the distributions predicted by a previous classifier is characteristic of the Expectation-Maximization (EM) algorithm, which we discuss in chapter 8.

The open choices in the schema include the choice of supervised learning algorithm (the base learner), the manner of thresholding confidence, and the stopping criterion. There are also points at which variations have been proposed.

Base learner. Virtually any supervised learning algorithm can be used to implement the "`train`" step. It is assumed that the classifier makes confidence-weighted predictions. Yarowsky uses a decision list algorithm. Hearst constructs a vector, indexed by features, for each word sense; the decision rule predicts the class whose vector is most similar to the feature vector of the instance to be classified. Hindle & Rooth make independence assumptions to permit a straightforward estimate of conditional probability of classes given the features of the instance.

Confidence measure. The function `select` in figure 2.4 preserves labeled instances on which the classifier is sufficiently confident, and discards the remaining instances. There are several ways of measuring confidence. A confidence-rated classifier is often based on an estimate of, or approximation to, the conditional probability of the predicted label, that is, $p(y|x)$, where x is the input and y is the classifier's output. For example, in the decision list algorithm (section 2.1.3), the classifier's prediction is determined by a single rule, the "winning rule," and the conditional accuracy of the winning rule provides the classifier's measure of confidence. Writing $h(x)$ for the prediction of the winning rule, with $h(x) = \bot$ if the rule does not match, the conditional accuracy of the winning rule is an estimate of $p[Y = y|h \neq \bot]$, and it can be seen as an approximation to $p[Y = y|X = x]$, under the assumption that the rule h is conditioned on the "most reliable" feature of x.

Another possibility is to consider not the absolute value $p(y|x)$, but rather the ratio

$$\frac{p(y_1|x)}{p(y_2|x)} \tag{2.1}$$

where y_1 is the most-probable label and y_2 is the second most probable label. The ratio (2.1) is maximized when the conditional distribution $p(y|x)$ is a point distribution with all its mass on one label (defining the value to be $+\infty$ if $p(y_2|x) = 0$), and it is minimized when the two most-likely labels are equally probable. In the two-label case, it is monotonically related to the entropy of the distribution $p(y|x)$. That entropy provides a third possible measure of confidence.

Stopping criterion. Our schema does not specify when to stop looping. One simple method is to run for a fixed, arbitrary number of rounds. A method that is almost as simple is to keep running until convergence, that is, until the labeled data and classifier stop changing.

A third alternative is to divide the training data into two parts. One part is passed to the learner as training data, and the other part is set aside as **validation data**. After each iteration, the performance of the classifier is measured on the validation data, and a stopping point is chosen based on performance on the validation data. A refinement is **cross-validation**. One divides the data into n segments, usually five or ten, and one uses each segment in turn as validation data, with the remaining segments supplying training data. Performance on the validation data is used to estimate the optimal number of iterations through the loop; by letting each segment play the role of validation data in turn, one obtains n different estimates. Averaging them gives a final estimate T of the optimal number of iterations. Then the entire data set is used for training, stopping after T iterations.

Seed. The schema assumes that the seed with which the "bootstrapping" process begins consists of labeled data (L_0). An alternative is to provide an initial classifer instead of labeled data. This is actually a point of difference between Hearst and Yarowsky: Hearst uses labeled data as a seed, whereas Yarowsky uses an initial classifier, in the form of a small list of highly reliable rules.

It should be noted that neither labeled data nor an initial classifier is generally reducible to the other. If one uses a seed classifier to label a data set and then one induces a new classifier from the data set, one does not generally obtain the seed classifier again. Conversely, if one begins with labeled data, induces a classifier, and relabels the data, the result is not generally identical to the original labeling. Labeled data has the advantage of being error-free (one hopes), but a seed classifier has the advantage of applying to the entire population of instances.

Other forms of seed are possible, for example, a set of constraints that restricts the allowable labels for each instance based on some feature or features of the instance. Such constraints have been used in part-of-speech tagging – we have already mentioned that a dictionary can be used to specify a limited set of possible parts of speech for each word. Such seed constraints produce a partial labeling of the data, in two senses of partial. The dataset is partially labeled in the sense that the constraints are usually sufficiently restrictive to assign unique labels to at least some instances; and, in addition, many instances are "partially labeled" in the sense that the constraints eliminate some labels from consideration, providing partial information about the correct label.

Indelibility. The basic version of the schema relabels the unlabeled data from scratch with every iteration. In a variation, labels, once assigned, are never recomputed; they are *indelible*. Indelibility is obtained by replacing the "L_0" in line 4 of figure 2.4 with "L". Indelibility is natural when the seed is labeled data and the classifier is recomputed from scratch in each iteration. The benefit of label indelibility is that it provides stability and prevents the learner from wandering too far from the original seed.

A weaker form of indelibility is what we might call **persistence**. Once an instance is labeled, it remains labeled, but the label may change.

Many classifiers can be represented as a list or set of simple rules, in which case indelibility may also apply to the classifier. If rules are indelible, the classifier is not retrained from scratch, but rather is grown monotonically, by adding new rules without deleting old ones.

Throttling. A variation on the behavior of `select` is the following. Instead of accepting all instances whose confidence exceeds a threshold, a limit k is placed on the number of instances accepted into the labeled set. This prevents a large influx of newly labeled instances from overwhelming the influence of the previously labeled instances in early iterations.

Balancing. Another modification to selection, often coupled with throttling, is to select the same number of instances for each class, even if it necessitates having different thresholds for the different classes. Classifiers often have a predeliction for one class over another, which is problematic because any under- or oversampling of classes tends to get amplified in subsequent iterations. Balancing counteracts that tendency. It can be seen as an additional weak constraint, providing the learner with some information about the target, namely, the (approximate) relative frequency of labels. As a practical matter, balancing is usually indispensable for good performance.

Preselection. A third variation on selection is to do two passes of selection. Instead of selecting the highest-confidence predictions from the entire set of unlabeled data, it can be beneficial to apply the classifier to a smaller sample of the unlabeled instances, sometimes called a **pool**. The use of a pool is motivated by the desire to keep newly labeled instances from being too homogeneous. It prevents the classifier from choosing a set of very similar instances that it particularly "likes," and instead forces it to choose a more heterogeneous set to add to the labeled set.

One way of selecting the members of the pool is random sampling. Random sampling is used in the original co-training paper [21], and has subsequently been applied to self-training as well. An alternative is to use one measure of confidence (or more generally, quality) for preselection, and a second measure for selection.

2.2.3 Evaluation

As should be apparent from the description just given, self-training is very much an algorithmic and heuristic approach. It is predicated on certain intuitions, but it has developed for the most part through empirical fixes to empirical problems, rather than algorithms based on mathematical considerations. Nonetheless, it is worthwhile to examine the intuitions briefly.

The basic reasoning is something like the following. A classifier performs better as you increase the amount of labeled data it is trained on, provided that the proportion of labeling errors is negligible, which we may take to mean less than some rate ϵ. Suppose that classifier confidence $c(y|x)$, computed on the labeled data, provides a good estimate of the true distribution $p(y|x)$ on the unlabeled data. Let X be the set of unlabeled instances where classifier confidence exceeds $1-\epsilon$, that is, where $\hat{y}$ maximizes $c(y|x)$ and $c(\hat{y}|x) > 1-\epsilon$. If c being a good estimate of p means that $p(\hat{y}|x) > 1-\epsilon$ for instances in X, then the proportion of erroneous predictions in X is less than ϵ, and we can take the predictions of the classifier on X at face value and increase the number of labeled instances for training, while keeping the proportion of labeling errors negligible.

This reasoning makes several assumptions that are not necessarily true. It is certainly true that training a classifier on a larger sample drawn from the

```
string=New_York → LOC
string=U.S. → LOC
string=California → LOC
hasSubstring:Mr. → PER
hasSubstring:Incorporated → ORG
string=Microsoft → Microsoft
string=I.B.M. → ORG
```

FIGURE 2.5
The seed classifier of Collins & Singer.

true instance distribution $D(x)$ is better than training it on a smaller sample. But the way we select new labeled instances for inclusion, selecting those where the current classifer is most confident, means that the newly sampled instances are almost certainly not drawn from anything like the distribution $D(x)$. (Incidentally, preselection can be seen as a way of forcing the distribution of the new sample to be more like $D(x)$ than it would be otherwise.) Similarly, a fixed error rate may be negligible if the errors are uncorrelated, but the selection process is likely to cause errors to be correlated. Finally, any biases may be amplified by the feedback loop of iterative retraining.

Nonetheless, despite the lack of guarantees that its assumptions and approximations hold, self-training does work on at least some data sets. When it works, it appears to behave in the way that the above reasoning would lead us to expect. A sequence of abstaining classifiers is produced. The algorithm is designed to keep conditional accuracy high, while gradually increasing the firing rate (that is, decreasing the abstention rate).

Figure 2.6 illustrates a typical trajectory. The task is that described by Collins & Singer [50]; noun phrases are to be classified as person, location, or organization. There are 89,000 unlabeled training instances and 1,000 labeled test instances. The solid line in the figure shows the conditional accuracy after each iteration of labeling and training. The base learner is a decision list, and the seed consists of seven rules, given in figure 2.5. As shown in figure 2.6, the seed classifier has a firing rate of 11% and a conditional accuracy of 100%. The dotted lines are contours of accuracy. A default prediction of "person" is assumed in cases where the classifier abstains, and the default prediction has an accuracy of 46%. Hence the only contour where accuracy is unaffected by firing rate, and the contour is a horizontal line, is the contour of accuracy equal to 46%. That contour is not shown, but it lies about halfway between the lowest contour that curves upward, representing an accuracy of 50%, and the highest contour that curves downward, representing 40%. The accuracy of the seed classifier is represented by the heavy dotted contour; its numeric value is 52%:

$$(0.11)(1.0) + (0.89)(0.46) = 0.52.$$

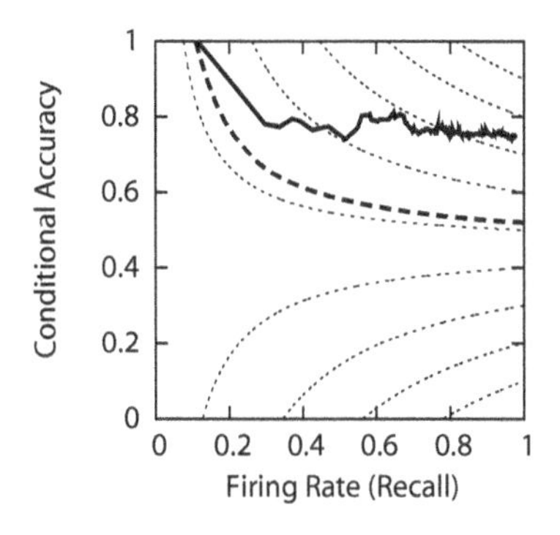

FIGURE 2.6
A typical trajectory for self-training. The dotted lines are contours of equal accuracy. The accuracy of the default classifier is 0.46; it is represented by the heavy dotted line.

After a drop in conditional accuracy in the very first round, conditional accuracy remains essentially constant, but firing rate increases. As a result, accuracy increases steadily. Learning was stopped after 1,000 iterations, at which point the firing rate was 98% and conditional accuracy and accuracy were both 75%. We note that it is possible to obtain considerably higher performance on this data set; performance varies significantly with different choices in the options discussed above.

2.2.4 Symmetry of features and instances

The large jump in firing rate, and corresponding drop in conditional accuracy, in the first iteration in figure 2.6 occurs because no pruning was done to the decision list. Collins & Singer [50] obtain considerably higher performance by doing preselection, throttling, and balancing *on the rules forming the decision list*, rather than on the labeled instances.

This points out a symmetry between features and instances that arises when the base learner is a decision list or other simple combination of **simple rules**, by which we mean rules that are conditioned on single features. A simple rule $F \Rightarrow y$ can be viewed as a **labeled feature**, where F is the feature and y is the label. A decision list is a set of labeled features. No feature appears twice in the list – or, at least, if any feature appears multiple times, all but one occurrence can be eliminated without affecting the predictions of the decision list, because the one with the highest score bleeds the others. Not all features necessarily appear in the decision list; the ones that do not appear are unlabeled features. In short, features are analogous to instances; simple rules ("labeled features") are analogous to labeled instances; and decision lists

procedure `selfTrain2` (C_0, U)
1 C_0 is a seed decision list, U is unlabeled data
2 $c \leftarrow C_0$
3 **loop until** stopping criterion is met
4 $L \leftarrow$ `label`(U, c)
5 $c \leftarrow C_0 +$ `select`(`train`(L))
6 **end loop**
7 **return** c

FIGURE 2.7
Self-training, "dual" form.

("labeled feature sets") are analogous to labeled sets of instances.

The symmetry between features and instances suggests a "dual" version of self-training in which confidence-based selection is applied to the rules, not to the instances (figure 2.7). All of the variations discussed in section 2.2.2 apply to the dual case as well. We already mentioned that an alternative form of seed is an initial classifier instead of a set of labeled instances; an initial classifier is the natural form of seed for the dual version. Throttling, balancing, and preselection can all be included in the `select` function in figure 2.7. The algorithm can be modified to introduce persistence or indelibility of feature labels. In the indelible version, once a rule $F \Rightarrow y$ is added to the classifier, it is never deleted (and no contradictory rule is added). In the persistent version, once a rule $F \Rightarrow y$ is added to the classifier, it may be replaced by a rule associating a different label with F, but there must always be some rule associating a label with F.

The Collins & Singer versions of self-training (the algorithms they call "Yarowsky-95" and "Yarowsky-cautious") are of the dual form. The better-performing one is "Yarowsky-cautious," which uses throttling and balancing. It uses a rule's conditional accuracy for preselection – specifically, only rules whose conditional accuracy exceeds 95%, as measured on the labeled data, are included in the pool. Selection is done using firing rate. For each label y, of the rules that predict y, the k rules with the highest firing rate are selected. The value of k starts at 5 and increases by 5 at each iteration. Rules are not persistent, but the growth of the classifier is like that of a persistent classifier that grows by k rules at each iteration.

Interestingly, the rule score used to determine the classifier's prediction on new instances is different from either of the measures used in selection or preselection. The score for rule $F \Rightarrow y$ when used for prediction is smoothed conditional accuracy:

$$\frac{\sum_i [\![F \in x_i, y = y_i]\!] + \epsilon}{\sum_i [\![F \in x_i]\!] + m\epsilon}$$

where m is the number of labels. The use of three different scores seems rather ad hoc, but it does emphasize the difference between rule selection (the `train` step) and prediction (the `label` step).

There is also a body of work on induction of "semantic lexica" that uses the dual form of self-training. By "semantic lexicon" is meant a list of entities of a particular semantic class – such as a list of names of people, or names of cities, or names of vehicles – or several such lists. A semantic lexicon is also called a **gazetteer**.

In traditional terminology, the elements of a semantic-class list are entity **types**, and their occurrences in running text are entity **tokens**. A type can be seen as a particular sort of feature, one whose value is the entire phrase (usually a noun phrase) but excluding context. This is a commonly used attribute; for example, it is the attribute called "`string`" in the list of seed rules of figure 2.5. A list of entities belonging to a particular class is equivalent to a list of rules "`string=`$s \Rightarrow y$" where s is the entity type and y is the semantic class that the list exemplifies. A collection of such lists defines a classifier.

Entity tokens in running text are instances. Based on statistics of tokens, new types are added to the lists; this corresponds to the `learn` step in figure 2.7. In the `label` step, additional tokens of the types in the lists are marked.

Typically, the seed consists of list items – that is, it is a classifier. Selection, including throttling and balancing, is typically applied in the `learn` step, not the `label` step. In both of these respects, the algorithm instantiates the dual form of self-training (figure 2.7). Semantic lexicon induction will be discussed in more detail in section 3.4.3.

2.2.5 Related algorithms

Given the simplicity of self-training, it is not surprising that it has arisen in many variants in different contexts. The idea of self-training will come up repeatedly throughout the book; we give a brief roadmap here.

- **k-means clustering** is a commonly used clustering algorithm that is quite similar to self-training. One can view self-training as an adaptation of k-means clustering to the semisupervised setting, with generalized labeling and training steps. K-means clustering is discussed in section 7.4.1.

- **Pseudo relevance feedback.** In information retrieval, documents are classified as either relevant or not relevant to a given query. A retrieval algorithm essentially constructs a binary classifier given a query and applies it to a body of (unlabeled) documents. Those where the classifier is most confident are returned to the user. In relevance feedback, the user manually labels some of the documents, and an improved classifier is constructed. In pseudo relevance feedback, the initial classifier's predictions are taken at face value where the classifier is most confident, and a new classifier is constructed from the labeled data. The process may be iterated. It is discussed in section 7.4.2.

- **Label propagation.** Yarowsky visualizes self-training as propagating labels through a graph. We will discuss label propagation in considerable detail, for example in chapter 10. The formulation of self-training as label propagation is discussed in section 7.6.2.

- **Decision-directed approximation** is an unsupervised learning algorithm that alternates training and labeling in a way that is very similar to self-training. One begins with an arbitrary assignment of labels, and one uses a supervised learning algorithm that maximizes likelihood to construct a classifier. Next, one relabels the data by choosing, for each instance, the highest-probability label according to the classifier. Then the process repeats, alternating training and labeling. In the semisupervised version, labels for some instances are given, and relabeling applies only to the initially unlabeled data. Decision-directed approximation is discussed in section 8.1.3.

- **McLachlan's algorithm** can be viewed as a special case of decision-directed approximation in which the learned model is a mixture of Gaussians and a nearest-neighbor rule is used for relabeling – though it would be historically more accurate to say that decision-directed approximation is a generalization of McLachlan's algorithm. McLachlan's algorithm appears to be the oldest version of self-training. It is discussed in section 8.1.4.

- **Game self-teaching.** There are two distinct learning approaches that have been called "self-teaching." One arises in the context of learning to play a game; it is the idea of having a system teach itself to play a game by playing against itself. It is discussed in section 6.2, in connection with the perceptron algorithm.

- **Agreement-based self-teaching.** The second setting that has been called self-teaching involves two separate modalities, such as vision and hearing, that provide information about a single learning task. It is closely related to co-training, and we will discuss it in detail in section 9.2, when we turn to the theory of agreement-based algorithms.

2.3 Co-Training

Co-training was proposed by Blum and Mitchell [21]. It is predicated on having two independent **views** of each instance. The example used in the

```
procedure cotrain (L, U)
1  L is labeled data, U is unlabeled data
2  P ← random selection from U
3  loop until stopping criterion is met
4      f1 ← train(view1(L))
5      f2 ← train(view2(L))
6      L ← L + select(label(P, f1)) + select(label(P, f2))
7      Remove the labeled instances from P and replenish P from U
8  end loop
```

FIGURE 2.8
Co-training.

original paper is classification of computer science web pages; specifically, the task is binary classification of web pages as being course home pages or not. An instance is a web page, and the two views are (1) the contents of the web page itself, and (2) the contents of hyperlinks pointing to the web page. Another example in which multiple views arise is in multi-modal settings, for example, simultaneous audio and video streams. The idea is to construct separate classifiers for each view, and to have the classifiers teach each other by labeling instances where they are able. If the two views are independently sufficient to learn the target function, and they are not merely redundant, then each can learn from the other.

This suggests the following algorithm (see figure 2.8). Start with some labeled data. Train one classifier using the first view, and a separate classifier using the second view. For classifier training, Blum & Mitchell use the Naive Bayes algorithm, which we will discuss in section 4.1.1, but any classifier learning algorithm can be used. Use the two classifiers to label some new data. For each, keep the labeled instances where it is most confident, and add them to the labeled data. Iterate until a stopping criterion is met; the discussion of stopping criteria in section 2.2.2 is applicable without modification.

There are strong similarities between co-training (figure 2.8) and self-training (figure 2.4). If we view the classifier pair (f_1, f_2) as being itself a classifier with interesting internal structure, then co-training can be seen as a special case of self-training. The version shown in figure 2.8 is persistent (line 6 has "L" rather than "L_0" after the assignment operator) and uses preselection (lines 2 and 7), but those are variations that we already discussed in connection with self-training. The other dimensions of variation discussed for self-training can also be applied to co-training. This is not to minimize the importance of the assumption of two views; the introduction of two views is a significant modification, and plays a central role in the theory of co-training.

Co-training will be discussed in more detail later in the book:

- **Label propagation.** Like self-training, co-training can be formulated as label propagation. This was noted in the original Blum & Mitchell paper [21], and will be discussed in section 7.6.3.
- **Agreement-based methods,** of which co-training is the prime example, are discussed in chapter 9.
- **Co-boosting.** An agreement component can be added to the objective function for boosting, a boundary-oriented method. The resulting "coboosting" method is discussed in section 6.3.3.

We introduce the co-training algorithm at this point because it is widely used, along with self-training, in current computational-linguistic applications of semisupervised learning. These applications are the topic of the next chapter.

3

Applications of Self-Training and Co-Training

Currently, most applications of semisupervised learning to problems in computational linguistics have involved self-training and co-training. For that reason, it is natural to discuss the applications at this point. It is hoped that the presentation of less familiar, but often mathematically sophisticated, methods in the remainder of the book will lead to their wider application in computational linguistics in the future.

3.1 Part-of-speech tagging

Part-of-speech tagging has already been introduced. Instances are word occurrences in running text. An instance is represented as a collection of features. For part-of-speech tagging, the preceding word and following word are adequate for good performance. The classes are parts of speech: noun, verb, adjective, and so on. There are a number of different sets of parts of speech ("tagsets") in use. The most widely used is the Penn tagset, which distinguishes 36 parts of speech, or 48 if one includes punctuation tags [142].

Part-of-speech tagging, as just characterized, can be viewed as a straightforward classification task. A complication arises, however, in that classification decisions for different instances are not independent. There are sequential dependencies among parts of speech assigned within a sentence. For this reason, part-of-speech tagging is not usually approached as a simple classification problem; rather generative models such as Hidden Markov Models (HMMs) or Conditional Random Fields (CRFs) are used to select an entire sequence of parts of speech at once. Generative models will be discussed in chapter 8, but explicitly sequential models such as HMMs and CRFs will not be covered. The reader is referred to Manning & Schütze [141] for a treatment of HMMs.

As previously mentioned, part-of-speech tagging was the first task in computational linguistics to which semisupervised techniques were applied. (It was also the first task to which probabilistic techniques in general were applied.) The Expectation-Maximization (EM) algorithm was well established as a standard means of estimating an HMM from unlabeled data even before

HMMs were applied to part of speech tagging. When HMMs were applied to part-of-speech tagging, labeled data (namely, the Brown corpus) was used, and impressively good results were obtained.

It was a natural next step to use the EM algorithm for unsupervised learning of a part-of-speech tagger, but the results were poor. EM applied to unlabeled data is essentially a clustering algorithm, and the word clusters induced had little correspondence to the target parts of speech.

The idea of applying EM to a combination of labeled data and unlabeled data was the first application of semisupervised learning in computational linguistics. The results were again disappointing. In a well-known pair of studies, Merialdo [153] and Elworthy [83] trained HMMs from labeled data, and used the trained models as initial models for EM estimation using unlabeled data. The EM algorithm is an iterative algorithm that constructs a sequence of models with increasing likelihood; likelihood measures how well the model predicts the training data. The algorithm requires some model as a starting point, and the quality of the model it ends with can be strongly influenced by the choice of starting point. (The algorithm will be discussed in more detail in section 8.2.) Merialdo and Elworthy used models trained on labeled data as starting points.

The conclusions of Merialdo's and Elworthy's experiments are that incorporating unlabeled data using the EM algorithm actually hurts tagger performance. An exception is when only very small amounts of training data are available. To be precise, unlabeled data helped when the supervised model was trained on 100 sentences (about 2370 words), but did not help when the model was training on 2000 sentences (47,400 words) or more. It should be noted that 2370 training instances is actually a moderately large training set in many contexts. Be that as it may, the degradation one observes with EM training when larger amounts of labeled data are available is consistent with the poor performance of EM on purely unlabeled data. It is also consistent with theoretical results that show that the use of unlabeled data can hurt when the assumptions of the model are incorrect [53, 54].

The negative results of Merialdo and Elworthy also contrast with successful attempts at semisupervised learning by Cutting et al. [62] and Kupiec [131]. These latter approaches used a dictionary giving the possible parts of speech for each word, but not their probabilities. A significant proportion of words in English are unambiguous, so a dictionary implies a labeling for part of the data. For example, in the Penn Treebank, 85% of words are unambiguous. (This includes a large number of words that only appear once in the corpus.) Moreover, the average number of parts of speech that an ambiguous word has is small. In the Penn Treebank, an ambiguous word has on average 2.24 parts of speech out of the 36 parts of speech in the tagset. A dictionary simplifies the learning problem by greatly reducing the number of labels available for each instance. It can be seen as providing a partial labeling in two senses: it assigns a label to a significant part of the data, and it assigns "partial labels" to the remaining instances, where a "partial label" is a restricted set of labels

Y. An unlabeled instance is one whose label is restricted to $\mathcal{L}$, the full set of labels. A labeled instance is one whose label is restricted to $\{y\}$, a singleton set. A partially labeled instance is one whose label is restricted to a set Y that is neither a singleton set nor the full set of labels. (Recall the discussion of constraints as seeds for self-training in section 2.2.2.)

Cutting et al. and Kupiec used the EM algorithm for semisupervised learning of a part-of-speech tagger. Subsequently, other forms of semisupervised learning have been applied. For example, Brill [25] uses a variant of self-training, on the basis of a supervised algorithm known as transformation-based learning. A dictionary is assumed that provides an initial partial labeling. Rules are considered of the form $f : Y \Rightarrow y$ where f is a feature, Y is a partial label (that is, a set of labels), and $y \in Y$ is a label. The rule matches any instance with feature f, and it predicts label y. However, the rule only *applies to* a partially labeled instance that has feature f and partial label Y. The algorithm is iterative. In each iteration, rules are scored according to their conditional accuracy on the (unambiguously) labeled instances that they match, and the best rule is added to the classifier. The classifier is an ordered list of rules; new rules are added at the end of the list. The classifier is then applied to the partially labeled instances, labeling some of them. Then the process is repeated, selecting a new rule and applying it to the partially labeled instances. The resulting algorithm is a version of self-training in which both the classifier and labels are persistent, and in which only one rule is selected in each iteration. The use of partial labels is novel, however.

3.2 Information extraction

The aim in information extraction is to populate a database from unstructured text. The typical approach involves at least four steps: entity recognition, relation recognition, coreference resolution, and slot filling. Of these, entity recognition and relation recognition have received the most attention with respect to semisupervised learning.

Information extraction typically involves a small number of types of semantic entities and relations, corresponding to the schema of the database to be populated. The classic example is the "acquisitions and mergers" domain of the Message Understanding Conference (MUC), which assumes entity types such as `person`, `place`, and `company`, and relations such as *person* `is-CEO-of` *company*, *company* `acquires` *company*, etc.

Entity recognition is the identification of phrases (usually noun phrases) referring to semantic entities. It is often divided into two steps, segmentation and classification. In segmentation, one determines the boundaries of likely entity occurrences, and in classification, one determines the semantic type of the occurrences. Segmentation is typically reduced to a tagging problem. In the simplest form, one tags the sentence with three tags, B, I, and O, where "B" marks the first word in an entity reference, "I" marks a word inside an entity occurrence that is not the first, and "O" marks words that are outside of any entity occurrence. For example, the sequence

On	March	4,	1992,	Ronald	Arbiter	founded	Arbutus
O	B	I	I	B	I	O	B

is an encoding for the entity segmentation

`On [March 4, 1992,] [Ronald Arbiter] founded [Arbutus].`

Hence the application of semisupervised learning methods to entity segmentation largely reduces to their application to the part-of-speech tagging problem.

Entity classification is the assignment of semantic types to segmented entities. For instance, in the example just given, it remains to be determined that the three entities have types `date`, `person`, and `company`, respectively. This is a straightforward classification problem. The instances are segmented entity references. Features are often divided into "spelling" features, meaning features derived from the words inside the entity reference, and "context" features, meaning features derived from the surrounding context of the entity reference. Spelling features and context features can be used as two views for co-training. The labels for the classification problem are the semantic entity types, plus "none of the above." Collins & Singer [50] show the effectiveness of both self-training and co-training at this task, and they also develop a method called co-boosting, which will be discussed later (section 6.3.3).

A third information-extraction subproblem to which semisupervised learning has been applied is relation recognition. An early example here is a method proposed by Brin [26] for filling a database of book titles and their authors. The idea, in a word, is to begin with a small list of author-title pairs, find web pages in which the known pairs occur, extract features that are indicative of author-title pairs, and use those features to find new pairs. Then the process repeats. This can be seen again as a variety of self-training, in which the instances are potential author-title pairs, the labels are positive and negative, and "features indicative of author-title pairs" means rules predicting the positive class. One begins with a small number of (positively) labeled instances, uses them to construct a classifier (a list of features indicative of positive instances), then one uses the classifier to label new instances, and the cycle repeats. Distinctive properties of this particular application of self-training are the very large space of instances, and the focus on the positive class – there are no negatively labeled instances and no rules predicting the negative class. These two properties are actually related. The very large space of instances means that almost all instances are negative, so "unlabeled" is almost equivalent to "negative." Only exceptional (positive) instances are explicitly marked.

A final aspect of information extraction to which semisupervised learning has been applied is the induction of lexical resources, particularly gazetteers, or lists of known examples of particular semantic types. This is discussed below, under taxonomic inference.

3.3 Parsing

Beyond information extraction, in complexity and generality of structures detected, is full parsing. There has been some pioneering work on applying semisupervised learning to parsing. A specialization of the EM algorithm for parsing, called the Inside-Outside algorithm, was developed early on, but the structures it induces have little resemblance to standard parse trees. Pereira & Schabes [180] were more successful by beginning with partial bracketings.

Sarkar [202] has applied co-training to parsing. Parsing is a true structure-building task; unlike part-of-speech tagging, one cannot reduce parsing even approximately to independent classification tasks for each word. Parsing is the assignment of a tree to an entire sentence. Unlike labels in a classification task, the number of trees is infinite, and trees have rich internal structure that is essential to the task.

A typical stochastic parsing model consists of three things: a representation scheme, a scoring function, and a search algorithm. The representation scheme defines the way that a parse tree is decomposed into small pieces of structure that play the role of features. The scoring function assigns scores to structure-pieces based on their frequency in a labeled training set. "Labeled" here means that the correct parse tree is given for each sentence. "Training" a model means computing the statistics to which the scoring function refers. The score for a complete parse tree is the sum of scores of the structure-pieces that make it up. The job of the search algorithm is to assemble structure-pieces in such a way as to arrive at the parse tree with the maximal score, or close to the maximal score, among all valid parse trees for the given sentence.

Instead of constructing two classifiers based on separate views of instances, Sarkar constructs two different stochastic parsers, based on two different models. Each has its own distinct representation scheme, scoring function, and search algorithm. The process begins with a seed set of labeled data, as well as a dictionary of allowable structure-pieces for individual words, along the lines of the tag dictionary used in semisupervised part-of-speech tagging. Each model is trained separately on the labeled data, and then each model is applied to a randomly selected pool of sentences. Applying the model to a sentence yields not only a best parse, but also a score for that parse. The score is interpreted like a confidence measure. Where the model is most confident, its predicted parse is taken at face value and added to the labeled data.

This is done for each model separately. Then the models are retrained on the extended labeled set, the pool is replenished, and the process is repeated. The algorithm is conceived as an instance of co-training, using stochastic parsing models in the place of classifiers, sentences in place of instances, and entire parse trees in place of labels.

3.4 Word senses

We begin with a brief description of what has become the standard word-sense taxonomy for English, WordNet. Then we discuss word sense disambiguation and taxonomic inference.

3.4.1 WordNet

A taxonomy, or "is-a" hierarchy, is an acyclic directed graph whose nodes are **concepts** and whose edges represent specialization. The first large taxonomic lexical database was WordNet, and it is by far the most widely used taxonomy in computational linguistics [87]. It is nearly a tree: nodes with multiple parents do exist, but they are rare.

A significant problem for semantic relations is identification of the basic units. A multitude of efforts to identify "concepts," much less the relations among them, have disappeared into a swamp of subjective judgments. WordNet obtains traction by focussing on relationships among words. Instead of attempting to define concepts in the abstract, WordNet equates concepts with **word senses**. WordNet takes the stance that every concept is lexicalizable, or at least it limits its attention to lexicalized concepts, which is to say, concepts that are expressible as a word or compound word.

The relation between words and concepts is many-many. A given word may have more than one sense, and a given concept may be expressible as more than one word. Rather than using arbitrary designators for concepts, such as "`PLANE-1`" versus "`PLANE-2`," WordNet observes that two words are synonymous just in case they share a sense, implying that the synonyms of a word vary according to the sense of the word that is of interest. Hence a word sense can be identified by giving the list of synonyms that share the sense. Instead of "`PLANE-1`" and "`PLANE-2`," we may use "airplane/aeroplane/plane" for the flying machine and "plane/sheet" for the mathematical plane. In WordNet terms, these are **synsets**. For WordNet, the terms "synset," "word sense," and "concept" are all equivalent. Both examples just given are senses of the word "plane." The synset "airplane/aeroplane/plane" is a shared sense of the words "airplane," "aeroplane," and "plane," and "plane/sheet" is a shared sense of the words "plane" and "sheet." The word "sheet" also has the sense "sheet/bed sheet," which is not shared with "plane."

Ideally, no two synsets should be identical as sets of words, though this ideal is not maintained in practice. There do exist pairs of synsets that contain exactly the same words, but are nonetheless distinct; the difference is reflected in a gloss giving a more detailed description of the sense.

An important basic problem in computational linguistics is **word sense disambiguation**: the problem of determining the applicable word sense for a given word in context. It is also one of the problems for which self-training was originally developed.

WordNet treats taxonomy as a relation among synsets. The WordNet term is *hypernymy*, and its inverse is *hyponymy*:

- A **hypernym** is a generic word, naming a class or natural kind. X is a hypernym of Y if X is a generalization of Y. For example, *dog* is a hypernym of *golden retriever.*

- A **hyponym** is a specific term, naming a member or specialization of a natural kind. Hyponymy is the lexical relationship corresponding to the "is-a" relation. For example, *golden retriever* is a hyponym of *dog.*

Before WordNet, a few thesauri were available, most notably Roget's [197], the 1911 edition of which is out of copyright and has been available for many years in electronic form. There was also considerable research on extracting information, including taxonomic information, from machine-readable dictionaries. Interest in processing machine-readable dictionaries waned with the availability of WordNet and other large, manually redacted electronic resources, but one of the issues that it raised is as relevant as ever, namely, how to supplement dictionary information from text.

Linguistics has long distinguished between synchronic and diachronic accounts of language. Outside of the subdiscipline of historical linguistics, a strictly synchronic point of view – an idealization to a static linguistic system at a single point in time – is widely adopted. The synchronic idealization is so pervasive in linguistics that it becomes natural to consider any deviations (for example, coinages and borrowings) to be negligible if one limits attention to language within a relatively small timeframe, say, a single generation.

But the new prominence of corpora in computational linguistics has led, essentially, to an abandonment of the idealization to synchronicity. It is now a truism that the lexicon is dynamic. Each new text brings new vocabulary. For example, a fundamental issue in speech recognition is the out-of-vocabulary (OOV) rate. A fundamental issue in any language processing step is how to deal with unknown words. One now takes as given that a non-negligible percent of the words in a test set will never have been encountered in training. In addition to proper nouns, coinages, borrowings, and rare words that simply had not been previously encountered, each technical area has its own vocabulary, and technical vocabularies change rapidly.

`used troops and tanks to quash a pro democracy`	weapon
`tools and equipment, pumps, hoses, water tanks`	container
`an enlisted man in the royal air force and tank corps`	weapon
`case of fighter aircraft the engines and fuel tanks`	container

FIGURE 3.1
Four labeled instances in word-sense disambiguation.

These considerations lead to a second important problem, **taxonomic inference**. Taxonomic inference in the narrow sense is the question of how to place the new terms one encounters into an existing taxonomy. A more fundamental question is how to induce the entire hierarchy, such as for entirely new domains or new languages. Taxonomic inference was also an area to which semisupervised learning techniques were applied early on, and it continues to be an area of active research.

3.4.2 Word-sense disambiguation

Word-sense disambiguation is the problem of choosing the contextually appropriate sense for words in text. It can be viewed as a classification problem in which instances are word tokens, and labels are word senses. However, it is a classification problem with special structure. The list of word senses is very long – for example, WordNet 2.1 contains more than 100,000 word senses (synsets). But for any given word, the effective size of the label set is much smaller. A large majority (83%) of word types are unambiguous. Limiting attention to ambiguous words, there are fewer than four senses on average for verbs, and fewer than three senses on average for nouns, adjectives, and adverbs [234].

Word-sense disambiguation was one of the first tasks to which self-training was applied. Typically, an instance corresponds to a particular occurrence of an ambiguous word in context, and the classes are the different senses of the word. Essentially, each ambiguous word represents a separate learning problem. For example, Hearst [109] uses the word *tank* to define one learning problem. Two classes are the "container" sense and "weapon" sense of the word, and four labeled instances are given in figure 3.1. As usual, instances are represented as sets of features or, equivalently, as vectors of attribute values. The features most commonly used are represented by words in the proximity of the target word, with various definitions of proximity. In some cases, parse information is used, for example, the identity of the head that governs the target.

Identifying the classes, which is to say, deciding how many senses a word has and how to (manually) distinguish among them, is often difficult. For English, the easiest course is to adopt the WordNet senses, though they do tend to be relatively fine-grained, and hence pose a difficult discrimination task. Traditional print dictionaries, particularly ones available in machine-readable form, have also been used in the past, though they are generally less convenient than WordNet.

A disadvantage of WordNet, or any dictionary, is the degree of subjectivity in the discrimination of word senses. This is reflected in the variation among dictionaries in their decisions regarding word senses.

An alternative that ameliorates this particular shortcoming is to define word senses by translations. Given an English text and its translation into another language, one can define the sense of a given word occurrence to be its translation in the other language. Different occurrences of a word typically are translated in different ways, corresponding, at least to a first approximation, to the different senses of the word. For example, *interest* can be translated into German either as *Interesse* (intellectual interest) or as *Zinsen* (interest paid on an account balance). This way of defining word senses does have the advantage of circumventing the need for manual annotation, when a translation is available, but it does not actually eliminate the element of subjectivity – the translator faces a subjective choice among possible translations. It is also less than definitive in that the choice of target language can have a significant effect on the way that word occurrences are divided into "senses."

The word-sense disambiguation task makes some special constraints available that can be used to improve learning. It has been observed [91] that a given word occurring multiple times in a single discourse (for example, in the same newspaper article) is almost always translated the same way each time it occurs, even though multiple translations are possible in principle. Yarowsky incorporated this constraint into self-training for word-sense disambiguation.

An important empirical question is whether semisupervised learning is effective, that is, whether it affords improvements over using just the labeled data. Although Yarowsky [239] and Collins & Singer [50] report that semisupervised learning gives substantial improvement over the performance of the seed classifier they use, they intentionally use a very weak seed classifier. Putting even moderate effort into supervised classification can make the comparison less clear-cut. For example, Mihalcea [154] does an empirical comparison of self-training and co-training on word sense disambiguation. She examines variation in performance with respect to a few parameters: the number of iterations of self- or co-training, the pool size used in `select`, and the number of instances selected in each iteration. If the ideal parameter settings are known in advance, good performance gains (25% reduction in error rate) are achievable. However, if one is unlucky in guessing the appropriate parameter settings, the use of unlabeled data hurts performance.

Another open question is whether one can do better than treating each ambiguous word as an independent learning task. Mihalcea observes that choosing a single set of parameter settings to optimize performance for all words gives better average performance than optimizing for each word separately. One would like to be able to transfer a good deal more information across the tasks.

3.4.3 Taxonomic inference

A more ambitious task than word-sense disambiguation is the automatic construction of a word-sense hierarchy. A task intermediate between word-sense disambiguation and hierarchy construction is the identification of the appropriate parent node for a new concept in an existing, but incomplete, hierarchy. The construction of a taxonomy is usually based on the detection of hyponomous word pairs (Y is an X), which correspond to child-parent links in the taxonomy, and similarly distributed word pairs, which may correspond to siblings in the taxonomy – what are called **coordinates** in WordNet. Constructing a taxonomy from scratch is most naturally formulated as a clustering problem, but the more limited problem of extending a taxonomy is naturally formulated as a classification problem. This is particularly true in limited domains, such as are common in information extraction. For example, one might address the limited question of classifying nouns as subtypes of `person`, `place`, `organization`, or none of the above.

Unlike word-sense disambiguation, taxonomy induction is not generally concerned with word **tokens**, that is, individual occurrences of words or phrases, but with word **types**. By classifying *president* as `person`, one is making a statement about the meaning of the word *president* in general, not about any particular occurrence of the word in text. To be precise, one is asserting that one of the senses of *president* is a subtype of `person`. Usually, then, one takes instances to be word types, not word tokens.

The features of a word type are constructed from features of word tokens. Features of word tokens are typically instantiations of templates of various types. For example, "previous word" and "following word" are examples of templates in the intended sense. Applied to a token in context, they yield concrete features, such as "`previousWord=powerful`." Features of types are usually created by aggregating token features. One natural approach is to view a token feature as a type attribute whose value is a count or relative frequency. Continuing with our example, the type *explosion* might have the value 4 for the attribute `previousWord=powerful`, meaning that there are four occurrences of *explosion* that are immediately preceded by the word "powerful," or it might have the value 0.002, meaning that 0.2% of its occurrences are immediately preceded by the word "powerful." Real-valued attributes of this sort are natural for clustering, but are less immediately useful for some classification algorithms, such as decision lists. The commonest expedient is to impose a threshold: *explosion* is defined to have the feature `previousWord=powerful` if there are at least n occurrences of *explosion* immediately preceded by *powerful.*

In addition to features that are generally useful for constructing a classifier, Hearst [108] proposed several templates for predicting both hyponymy and coordinacy that continue to be widely used. Here are the patterns that she proposes. The variables match noun phrases.

- x ,? such as y_1 (, y_2)* (,? (and|or) y_3)? $\rightarrow$ y_i isa x

- such x as y_1 (, y_2)* (,? (and|or) y_3)? $\rightarrow$ y_i isa x
- y_1 (, y_2)* ,? (and|or) other x $\rightarrow$ y_i isa x
- x ,? including y_1 (, y_2)* (,? (and|or) y_3)? $\rightarrow$ y_i isa x
- x ,? especially y_1 (, y_2)* (,? (and|or) y_3)? $\rightarrow$ y_i isa x

If an instantiation of one of the templates occurs in text, she postulates, then each pair (y_i, x) is an instance of hyponomy, and each pair (y_i, y_j) is a sibling pair.

When the classification task involves identification of the hypernym (that is, parent) of an instance, the Hearst templates can be interpreted not just yielding features when instantiated, but as yielding labeled instances. In this way, they play the role of a seed classifier. An instantiation of one of the templates is obtained by replacing the variables with particular noun phrases, for example:

```
siege engines, such as the ballista and onager
```

Assuming that `siege engine` is one of the allowable hypernyms, this yields two labeled instances:

Instance	Label
`ballista`	`siege engine`
`onager`	`siege engine`

The Hearst templates can also be used simply to generate features, the attributes being likely hypernyms and likely coordinates. For example:

Instance	Features
`ballista`	`likelyHypernym=siege_engine, likelyCoordinate=onager`
`onager`	`likelyHypernym=siege_engine, likelyCoordinate=ballista`

With word types as instances, and features of the sort defined, self-training can be, and often has been, applied to the task of extending a taxonomy. Self-training is most often applied to a task with a small number of candidate hypernyms serving as labels; an additional label "`other`" representing "none of the above" is also added. As already mentioned, a task of this sort arises particularly in the context of information extraction, where it is often couched in terms of fleshing out a gazetteer. A gazetteer is a collection of lists, each list representing members (that is hyponyms) of a given semantic class. Finding list members is equivalent to labeling instances with one of the classes.

Typically, one begins with a small number of examples of each class, which means a small number of labeled instances. One constructs a variant of a decision list: one identifies rules $F \Rightarrow y$ such that feature F occurs frequently with instances that share label y; the classifier consists of a list of such rules.

To be sure, the process is usually couched in terms of collecting a list of "patterns" for each class, but a "pattern" is simply a template instantiation, which is to say, a feature, and associating a feature F with a class y is equivalent to adding rule $F \Rightarrow y$ to the classifier.

Approaches along these general lines have been proposed by Riloff & Shepherd [192, 193]. Roark & Charniak [195] propose a refinement of Riloff & Shepherd's approach, using Charniak's parser to identify lists, conjunctions, and appositives, and considering each pair of terms in such constructions to represent likely coordinate pairs. Other work along these lines includes Riloff & Jones [194], Widdows & Dorrow [231], Thelen & Riloff [219], and Phillips & Riloff [182].

4

Classification

In the previous chapters we have considered the semisupervised learning algorithms that are most familiar in computational linguistics. In the rest of the book, we turn to methods from the machine learning literature. We do not assume that the reader's primary area of research is machine learning, nor that the reader is necessarily familiar with all the mathematical techniques that are taken for granted in the primary literature of machine learning. For that reason, we include chapters that do not treat semisupervised learning per se, but provide necessary background material, beginning with the present chapter, on classification.

4.1 Two simple classifiers

We begin with some concrete examples of classifiers, and then turn to the more abstract setting.

4.1.1 Naive Bayes

The Naive Bayes algorithm is a particularly simple classifier; it was mentioned earlier as the classifier used by Blum & Mitchell in co-training. For the purpose of the Naive Bayes algorithm, an instance is viewed as a vector of attribute values: $\mathbf{x} = (x_1, \ldots, x_m)$. By the chain rule of conditional probability, the probability $p(\mathbf{x}, y)$ of an instance $\mathbf{x}$ with label y can be expressed as

$$p(\mathbf{x}, y) = p(y)p(x_1|y) \ldots p(x_m|y, x_1, \ldots, x_{m-1}).$$

The naiveté of the Naive Bayes algorithm rests in an independence assumption it makes, namely, that the choice of value for any attribute is conditionally independent of the choice of value for any other attribute, given the label. That is

$$p(\mathbf{x}, y) = p(y) \prod_j p(x_j|y).$$

Though almost certainly false, the assumption greatly simplifies computation, and as a practical matter, the algorithm performs respectably well in a wide range of learning problems.

Len	*Wid*	*Label*	*Len*	*Wid*	*Label*
small	small	−	small	large	+
small	med	−	med	large	+
small	med	−	med	large	+
med	small	−	large	small	+
large	small	−	large	large	+
small	small	−	large	med	+
small	large	−			
small	med	−			

FIGURE 4.1
Example of training data for Naive Bayes.

The algorithm estimates $p(y)$ and $p(x_j|y)$ by relative frequency in the training data. Given a new instance $\mathbf{x}$, the prediction of the classifier is the label that maximizes the conditional probability

$$p(y|\mathbf{x}) = \frac{p(\mathbf{x}, y)}{p(\mathbf{x})}.$$

Since the denominator is constant across choices of label, we can ignore it. Hence the prediction of the classifier is

$$\arg\max_y p(y) \prod_j p(x_j|y).$$

For example, suppose we are given the training data shown in figure 4.1. We have already sorted it into negative instances and positive instances. Counting the positives and negatives gives the following estimates for $p(y)$:

y	$p(y)$
−	8/14
+	6/14

(4.1)

Counting the relative frequencies of each attribute value within the positive and negative groups gives the estimates for $p(v|y)$, where v is a value for a given attribute. Note that the expression $p(v|y)$ is ambiguous because it leaves the attribute implicit; it can be written more explicitly as either $\Pr[\text{Len} = v|Y = y]$ or $\Pr[\text{Wid} = v|Y = y]$, depending on which attribute is intended.

	$y = -$		$y = +$	
	Len	Wid	Len	Wid
v	$p(v\|y)$	$p(v\|y)$	$p(v\|y)$	$p(v\|y)$
small	6/8	4/8	1/6	1/6
med	1/8	3/8	2/6	1/6
large	1/8	1/8	3/6	4/6

(4.2)

Suppose that we are given a new input: (med, med). Given the parameter values in the two tables above, we have:

$$\begin{aligned} p(\mathbf{x}, -) &= p(-)\Pr[\text{Len} = \text{med}|Y = -]\Pr[\text{Wid} = \text{med}|Y = -] \\ &= (8/14)(1/8)(3/8) \\ &= .027 \end{aligned}$$

$$\begin{aligned} p(\mathbf{x}, +) &= p(+)\Pr[\text{Len} = \text{med}|Y = +]\Pr[\text{Wid} = \text{med}|Y = +] \\ &= (6/14)(2/6)(1/6) \\ &= .024. \end{aligned}$$

Accordingly, the classifier predicts negative, since $p(\mathbf{x}, -) > p(\mathbf{x}, +)$.

4.1.2 k-nearest-neighbor classifier

Another very simple classifier, but one that is in a certain sense at the other end of the spectrum from Naive Bayes, is the k-nearest-neighbor classifier. The k-nearest-neighbor learner is essentially trivial: it simply memorizes the training data. All the work is done by the classifier at "run time." Given a new instance $\mathbf{x}$ to be classified, the classifier finds the k training examples that are most similar to $\mathbf{x}$, and looks at their labels. Whichever label occurs most frequently among the k nearest neighbors is chosen as the predicted label for $\mathbf{x}$.

To specify the algorithm fully, one must choose a value for k and a measure of similarity between instances. There is a simple geometric interpretation of the k-nearest-neighbor classifier in the case when $k = 1$ (the 1-nearest-neighbor classifier) and similarity is Euclidean distance. To see it, we must think about instances as points in **feature space**, which is to say, the space whose axes are defined by the attributes. We represent instances as vectors of values, and view them as points in an m-dimensional space, where m is the number of attributes, hence the number of components in the vectors.

For example, consider the following little training set.

Length	Width	Label
10	7	+
13	8	+
17	8	+
12	5	−
15	6	−

In figure 4.2, the instances are plotted as points in a space whose axes are Length and Width. In the figure, we have also marked the boundaries of the decision regions. If we consider a new point $\mathbf{x}$ anywhere in the space, its (predicted) label is determined by the nearest training point. Consider any

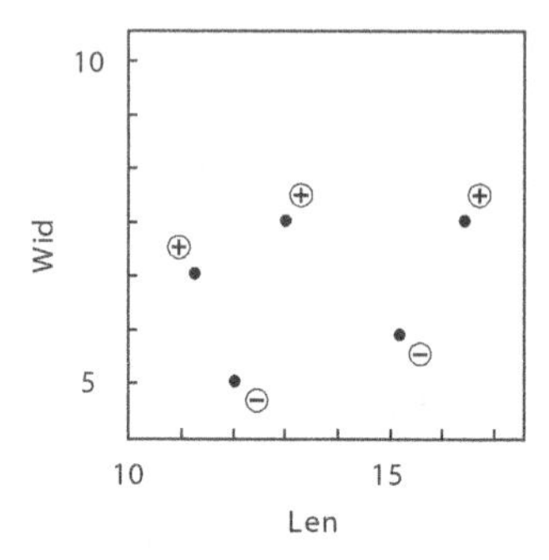

FIGURE 4.2
Instances as points in feature space.

two training points, and an arbitrary point **x** in their vicinity (figure 4.3). By definition, **x** "goes with" (its label is determined by) the nearer of the two training points. Clearly, the boundary between the two regions is the perpendicular bisector of the line connecting the two training points.

Each training point P is surrounded by a region consisting of the points closer to P than to any other training point. The boundary of the region is piecewise linear, consisting of perpendicular bisectors between P and nearby training points. The result is a tiling of the feature space known as a **Voronoi tesselation**, shown in figure 4.4. The decision boundary between the positive and negative examples is the reinforced line.

The nearest-neighbor algorithm usually produces a complicated, zig-zagging decision boundary. As long as any duplicate examples in the training data are consistently labeled, the 1-nearest-neighbor algorithm obviously makes the correct prediction for every training example.

In general, there is a trade-off between the **complexity** of a model and **fit** to the training data. The nearest-neighbor algorithm represents one extreme in that trade-off: it produces very complicated models, since it does not "digest" the training data at all, but it gets very good fit to the data. The apparent goodness of the predictions can be deceptive, however. By modeling the training data so closely, the nearest-neighbor algorithm captures not only the "signal" in the data, but also the "noise." The Naive Bayes algorithm is closer to the other end of the spectrum. The "naive" conditional independence assumption permits a simple model, and, as we will see below, in section 5.1.4, the decision boundary in the case of Naive Bayes is a single straight line, no matter how many training points there are. As a consequence, though, the classifier may be unable to make the correct prediction for all the training points: there may be positives mixed among the negatives or vice versa, so that no straight line is able to separate them.

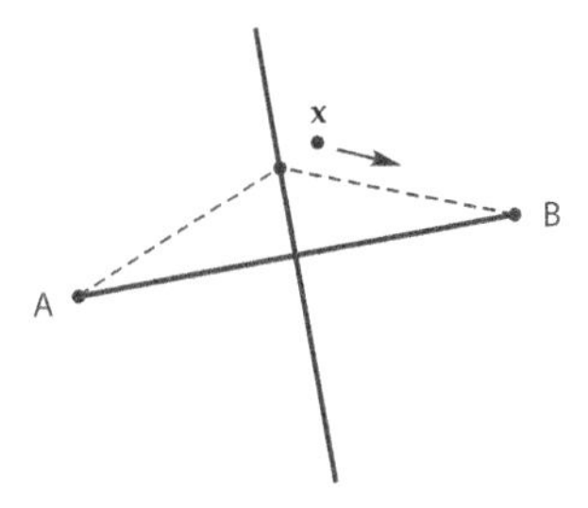

FIGURE 4.3
A point **x** in the vicinity of two training points A and B. The points equidistant from A and B constitute the perpendicular bisector of the line segment connecting A and B.

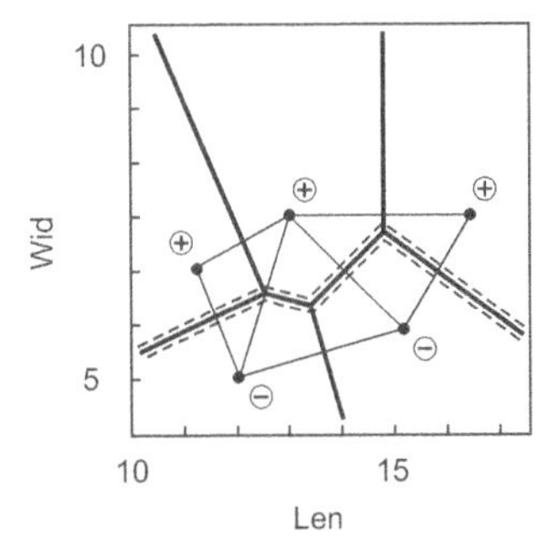

FIGURE 4.4
The decision regions; they constitute a Voronoi tesselation of the space.

Ideally, we would like a learning algorithm that captures all of the general features of the training data (the signal) without being distracted by its quirks (the noise). The "general features" are those that generalize to test data, and the quirks are the accidents of the training data that do not generalize. To capture every detail of the training data, one requires a complicated model. Simple models cannot capture all the details, but one hopes that one captures the features that generalize, leaving out only the quirks. If one chooses too simple a model, however, one loses essential detail, and generalization again suffers. A model that captures too much detail is said to **overfit** the training data, and a model that is too simple, and captures too little detail, is said to **underfit**. The goal is to find the model that, in the words of Goldilocks, is "just right."

4.2 Abstract setting

With these concrete examples in mind, let us define classification, and the learner's goal, more abstractly.

4.2.1 Function approximation

As was mentioned earlier (section 1.2.1), classification is one example of function approximation, three other examples being clustering, regression, and density estimation. In classification and clustering, the function to be learned, or **target function**, is nominal-valued, and in regression and density estimation, it is real-valued. In all cases, the target function takes *instances* or *data points* or *sample points* (the three terms are synonymous) as inputs. Instances are drawn from a sample space Ω and are distributed according to a fixed but unknown probability distribution D. If Ω is a discrete space, then D is a probability mass function, otherwise it is a probability density function. A random variable is by definition any function whose domain is the sample space Ω, so the functions to be learned are, formally, random variables.

The learner is given a training sample $(\mathbf{x}_1, \mathbf{x}_2, \ldots, \mathbf{x}_n)$ of instances drawn from the population in accordance with distribution D. The instances are assumed to be drawn independently, so that the probability of the sample is

$$\prod_{i=1}^{n} D(\mathbf{x}_i).$$

That is, the training sample is i.i.d.: the individual data points are *independent* and *identically distributed.*

As a practical matter, the assumption of independence is often violated. A typical training set for a natural language problem is constructed from running texts, with the consequence that there are significant statistical dependencies among instances. Little analytic traction can be gained, however, unless one assumes independently drawn instances. Fortunately, violating the independence assumption does not usually render the methods ineffective for practical purposes.

What has been said so far applies to all function approximation methods. It suffices for density estimation. The goal in density estimation is to use the (finite) training sample to construct an estimate $\hat{D}$ that is a good approximation to D over the entire (possibly infinite) population of instances.

The remaining function approximation settings differ from density estimation in that they involve a second variable – the label – in addition to the instance $\mathbf{x}$. Conventionally, we write "y" for the label. Unlabeled data is a list of instances $(\mathbf{x}_1, \mathbf{x}_2, \ldots, \mathbf{x}_n)$. In the supervised settings, classification and regression, the learner receives in addition a second list $(y_1, y_2, \ldots, y_n)$ containing labels. Equivalently, we may view labeled data as a list of pairs $((\mathbf{x}_1, y_1), (\mathbf{x}_2, y_2), \ldots, (\mathbf{x}_n, y_n))$. Let us write Y for the target function, that is, the function to be learned. The correct label for a given instance $\mathbf{x}$ is the output $y = Y(\mathbf{x})$. In particular, $y_i = Y(\mathbf{x}_i)$ for all pairs $(\mathbf{x}_i, y_i)$ in the labeled training data. A central question for learning is how to predict the value of Y for instances that are not in the training sample.

In classification and regression, the learner is given labeled data as input, and the goal is to construct an estimate or **hypothesis** $\hat{Y}$ that is a good approximation to Y over the whole population of instances. The difference between classification and regression lies only in the kind of values that Y produces as output. In the case of classification, Y maps an instance to one of a finite set of classes. In the case of regression, Y maps an instance to a real number.

Clustering differs from classification with respect to supervision. In both cases, there is a fixed but unknown function Y that maps instances $\mathbf{x}$ to classes y. The goal of clustering, as for classification, is to construct an approximation $\hat{Y}$. However, what constitutes a good approximation differs in the two cases.

For classification, a good approximation $\hat{Y}$ is one for which

$$\hat{Y}(\mathbf{x}) = Y(\mathbf{x})$$

for most instances $\mathbf{x}$ in the population. For clustering, this requirement is too stringent. Given only unlabeled training data, we cannot expect a clustering algorithm to use the same labels y as the unknown function Y uses – it has no way of knowing what those labels are. Rather, we require only that the clustering function $\hat{Y}$ divide up the data into clusters in a way that corresponds well to the true classes determined by Y.

Namely, the values y of the target function can be viewed as arbitrary labels for classes. The class itself can be viewed as a set of instances:

$$\{\mathbf{x} \in \Omega | Y(\mathbf{x}) = y\}. \tag{4.3}$$

In this way, Y defines a partitioning of the population of instances. The clustering function $\hat{Y}$ similarly defines a partitioning of the instances. Each value z of $\hat{Y}$ represents an arbitrary label for a cluster. The cluster itself is a set of instances:

$$\{\mathbf{x} \in \Omega | \hat{Y}(\mathbf{x}) = z\}. \tag{4.4}$$

We emphasize that the cluster labels z are arbitrary designators, and need not be related to the unknown labels y produced by the target function Y. We require only that the cluster function partition the instances in (approximately) the same way as the target function. The cluster function $\hat{Y}$ is a good approximation to the target function Y if the clusters (4.4) correspond well to the classes (4.3). This is the same as saying that a cluster function is good if there exists a one-one correspondence μ mapping cluster labels to class labels such that

$$\mu(\hat{Y}(\mathbf{x})) = Y(\mathbf{x})$$

for most instances $\mathbf{x}$.

We note in passing that one often encounters characterizations of the goal of clustering that are rather less stringent than ours. Clustering is often presented without the assumption of a target function determining "true classes." Rather, the goal may be characterized subjectively, as finding "interesting" structure in the data, or indirectly, as finding clusters that are useful for some other task. Clustering is certainly useful for exploratory data analysis or for preprocessing, but our primary interest will be in settings in which a target function does exist.

In this book, we restrict our attention almost exclusively to classification and clustering. We are particularly interested in them as instances of a more general function-approximation problem in which the target function is discrete-valued, the task is to approximate it, and the learner receives a mixture of labeled and unlabeled training data. We may call this generalized setting **semisupervised classification**. Classification is the special case in which all the data is labeled, and clustering, when treated as *unsupervised classification*, is the special cases in which all the data is unlabeled.

4.2.2 Defining success

A classifier is a function $\hat{Y}(\mathbf{x})$ that takes an unlabeled instance as input and produces a label prediction as output. The performance of a classifier is measured by how well it approximates the target function Y that assigns the correct label to each instance in the population. Specifically, the measure of classifier performance is **generalization error**, which is the probability that the classifier's prediction differs from the correct label:

$$\epsilon = \sum_{\mathbf{x} \in \Omega} [\![\hat{Y}(\mathbf{x}) \neq Y(\mathbf{x})]\!] D(\mathbf{x}) \tag{4.5}$$

where $[\![\phi]\!]$ has value 1 if ϕ is true and 0 otherwise. If the population of instances is not a discrete space, the sum in (4.5) must be replaced with an integral.

In the case of regression, Y and $\hat{Y}$ are real-valued functions, so the natural way to compare them is different. Demanding that a prediction be exactly correct is overly stringent; it is sufficient that it be approximately correct. The natural measure of error for a given instance is the absolute difference $|Y(\mathbf{x}) - \hat{Y}(\mathbf{x})|$, though the squared difference $[Y(\mathbf{x}) - \hat{Y}(\mathbf{x})]^2$ is generally more convenient mathematically. The measure of overall discrepancy between $\hat{Y}$ and Y is the average discrepancy:

$$\epsilon = \sum_{\mathbf{x} \in \Omega} |Y(\mathbf{x}) - \hat{Y}(\mathbf{x})| D(\mathbf{x}) \tag{4.6}$$

or

$$\epsilon = \sum_{\mathbf{x} \in \Omega} [Y(\mathbf{x}) - \hat{Y}(\mathbf{x})]^2 D(\mathbf{x}). \tag{4.7}$$

All three of these measures of error, (4.5), (4.6), (4.7), have the same general form. At each data point $\mathbf{x}$, there is a measure of the discrepancy between $\hat{Y}$ and Y, measured variously as

$$L_{\hat{Y},Y}(\mathbf{x}) = [\![\hat{Y}(x) \neq Y(x)]\!]$$

$$L_{\hat{Y},Y}(\mathbf{x}) = \hat{Y}(x) - Y(x)|$$

or

$$L_{\hat{Y},Y}(\mathbf{x}) = [\hat{Y}(x) - Y(x)]^2.$$

A measure $L_{\hat{Y},Y}(\mathbf{x})$ of the discrepancy between $\hat{Y}$ and Y at each data point $\mathbf{x}$ is called a **loss** function.

The overall measure of error is the expected loss over the sample space:

$$R(\hat{Y}) = D[L_{\hat{Y},Y}]$$

where $D[\phi]$ is the expectation of the quantity ϕ under distribution D. The expected loss $R(\hat{Y})$ is the **risk** of the hypothesis $\hat{Y}$. The lower the risk, the better the hypothesis. A learning algorithm is successful if it finds a hypothesis that minimizes risk, or nearly so.

In the case of classification, the natural loss function is that of (4.5). It has value 1 (complete loss) if the prediction is wrong and 0 (no loss) if it is right, and is commonly called **0-1 loss**. The corresponding risk function, the expectation of 0-1 loss, is called **generalization error**. It represents the error rate within the population.

In most practical contexts, generalization error cannot be measured exactly. It is usually estimated by the **test error**, in which the loss (specifically, 0-1 loss) is averaged over a freshly drawn sample of instances, the **test data**. That is, the test error is the proportion of test instances on which the classifier and target function disagree:

$$\hat{\epsilon} = \frac{1}{n} \sum_{i=1}^{n} [\![\hat{Y}(\mathbf{x}_i) \neq y_i]\!] = \tilde{p}[\hat{Y} \neq Y]$$

where $(\mathbf{x}_1, \ldots, \mathbf{x}_n)$ here represents the test sample and $\tilde{p}$ is the empirical distribution with respect to the test sample.

The test error is a binomial variable whose bias is the generalization error ϵ. Its distribution is approximately normal if the sample size is sufficiently large, where "sufficiently large" is usually taken to mean that

$$n\hat{\epsilon}(1 - \hat{\epsilon}) \geq 5$$

It follows that, with high probability:

$$\epsilon = \hat{\epsilon} \pm z\sqrt{\frac{\hat{\epsilon}(1 - \hat{\epsilon})}{n}} \tag{4.8}$$

where z is a free parameter that controls the stringency of the qualifier "with high probability." The larger z is, the higher the probability that (4.8) will be true for a test sample drawn i.i.d. from the instance distribution. Specifically, the probability that (4.8) holds is the area under the standard normal curve excluding both tails that lie outside the interval $[-z, +z]$.

To summarize, we have mentioned four different quantities:

- The test error is the proportion of erroneous predictions on the test sample.
- The generalization error is the expected proportion of errors over the entire population of instances. The test error is an estimate of the generalization error, and will be an accurate estimate if the test sample is large enough.
- The loss is the measure of discrepancy between an individual prediction $\hat{y}$ and the correct label y. Test error and generalization error assume 0-1 loss: the prediction is either completely right or completely wrong.
- The risk is the expected loss over the entire instance population. Generalization error is the special case of risk using 0-1 loss.

4.2.3 Fit and simplicity

Because test error provides an estimate of generalization error, and generalization error is the usual measure of classifier quality, it may seem that one should draw a sample, the **training sample**, and choose the hypothesis $\hat{Y}$ that minimizes error on the training sample. However, that strategy is not always best. Error rate directly measures the **fit** of a hypothesis to the training data. If one is allowed to choose from a rich enough set of functions – that is, if one is allowed to construct models of arbitrary complexity – one can find a hypothesis that fits a given training sample arbitrarily well. But one does not thereby necessarily improve the fit to unseen data. As discussed earlier, beyond a certain point, one begins modeling accidental features of the training data, resulting in *worse* performance on test data, a phenomenon known as **overfitting**.

For this reason, many learning algorithms do not simply minimize risk on the training data, but rather optimize some other **objective function** on the training data. Ideally, there should be a proven connection between the hypothesis that minimizes the objective function on the training data, and the hypothesis that minimizes risk over the entire instance space. Establishing such connections is a major goal of theoretical research in machine learning.

Either implicitly or explicitly, learning algorithms trade off fit against simplicity. An explicit trade-off often takes the form of **regularization**, which is the optimization of an objective function that is a weighted average of fit and simplicity. Most commonly, one maximizes fit and simplicity by minimizing training-sample loss and complexity, that is, one chooses $\hat{Y}$ to minimize

$$\tilde{p}[L_{\hat{Y},Y} + \gamma|\hat{Y}|]$$

where $\tilde{p}$ is here the empirical distribution of the training sample, $|\hat{Y}|$ represents a measure of model complexity, and γ is a coefficient controlling the relative importance of fit and simplicity.

4.3 Evaluating detectors and classifiers that abstain

4.3.1 Confidence-rated classifiers

When we discussed decision lists, in section 2.1.3, we noted that classifiers often provide more information than just a prediction regarding the label: a **confidence-rated** classifier provides not only a prediction but also a score that can be interpreted as a measure of confidence. For example, the prediction of a decision list is determined by the highest-scoring rule that matches the input instance, and the score of that rule is interpretable as a measure of confidence.

To give another example, the prediction of the Naive Bayes classifier is the label that maximizes $p(y|\mathbf{x})$, where p is the distribution defined by Naive Bayes estimation. The probability $p(\hat{y}|\mathbf{x})$ for the most probable label $\hat{y}$ might be taken as a measure of confidence. At least in the case of binary classification, however, there is a related measure of confidence that turns out to have more interesting properties, namely, the log likelihood ratio

$$f(\mathbf{x}) = \log \frac{p(+|\mathbf{x})}{p(-|\mathbf{x})} \tag{4.9}$$

The classifier predicts positive just in case $p(+|\mathbf{x}) > p(-|\mathbf{x})$, which is to say, just in case $f(\mathbf{x}) > 0$. That is, the sign of $f(\mathbf{x})$ is equal to the predicted label, and the absolute value $|f(\mathbf{x})|$ is interpretable as confidence. The log likelihood (4.9) is a very common form for confidence-rated binary classifiers.

Incidentally, a decision list can readily be recast in the form (4.9). In the binary case, we have

$$p(-|\mathbf{x}) = 1 - p(+|\mathbf{x})$$

Hence the ratio

$$\frac{p(+|\mathbf{x})}{p(-|\mathbf{x})}$$

varies monotonically with $p(+|\mathbf{x})$, and since log is monotone, the log likelihood ratio (4.9) also varies monotonically with $p(+|\mathbf{x})$. As a consequence, we can define an alternate version of a decision list that uses (4.9) as a confidence measure, but makes exactly the same predictions as the original decision list. Instead of scoring rules $F \Rightarrow y$, we assign a score to each feature, namely,

$$f(F) = \log \frac{p(+|F)}{p(-|F)}$$

For a given input $\mathbf{x}$, the "winning feature" is the one with highest confidence $|f(F)|$, and the prediction is the sign of $f(F)$. The winning feature is guaranteed to be the feature that appears in the winning rule $F \Rightarrow y$ of the original classifier, which maximizes $p(y|F)$, and the prediction, $\text{sign}(f(F))$, will be equal to y.

4.3.2 Measures for detection

When a confidence-rated classifier has confidence 0, it effectively makes no prediction for the instance in question; it **abstains**. For the purpose of computing test error, an abstention is an error, but, as a consequence, test error is not necessarily the best measure of performance for classifiers that abstain on a significant proportion of instances.

There does not appear to be a standard approach for evaluating multiclass classifiers that abstain. Below, we will treat a multiclass classifier with abstention as the combination of a detector and a classifier without abstention. A **detector** is usually treated as a binary classifier, but we prefer to view it as a unary classifier that abstains.

Let us write "$\hat{Y} = \bot$" to mean that the classifier abstains, that is, that the value $\hat{Y}(\mathbf{x})$ is undefined. For the sake of symmetry, we also consider the possibility that the target function is partial, that is, that we sometimes have

$Y = \bot$. A detection problem is one in which there is a single class, but only some instances belong to that class. The remaining instances have $Y = \bot$. This is equivalent to a binary classification problem in which $\bot$ is interpreted as the negative label, rather than the absence of a label.

The **rate of occurrence** is the probability that $Y \neq \bot$. If it is small, then small test error can be obtained by the "know-nothing" detector, which always abstains: the error rate of the "know-nothing" detector is equal to the rate of occurrence. In such a case, it can be more informative to consider the performance of the detector on (true) positive instances and (true) negative instances separately. The ability of the detector to detect positive instances, measured as $\Pr[\hat{Y} \neq \bot | Y \neq \bot]$, is known as **sensitivity**, and its ability to reject negative instances, measured as $\Pr[\hat{Y} = \bot | Y = \bot]$, is **specificity**. **Accuracy** is $\Pr[\hat{Y} = Y]$, which is equal to the weighted average

$$\text{accuracy} = \Pr[Y \neq \bot] \cdot \text{sensitivity} + \Pr[Y = \bot] \cdot \text{specificity} \tag{4.10}$$

Sensitivity is weighted by the rate of occurrence, and specificity by the rate of non-occurrence. Note that accuracy is one minus the test error.

A number of synonyms exist for sensitivity and specificity. Sensitivity is also called the *true positive rate*, *hit rate*, or **recall**. One minus the specificity is also known as the *false positive rate*, *false alarm rate*, or *fallout*.

An alternative to specificity that is commonly used in information retrieval and computational linguistics is **precision**. It is the probability that the detector is correct, given that it fires: $\Pr[Y \neq \bot | \hat{Y} \neq \bot]$.

All the quantities we have mentioned we have described as probabilities, which is to say, population proportions. We are almost always more interested in the corresponding test-sample proportions. We will not bother distinguishing between "sensitivity" and "estimated sensitivity," and so on. We will omit the qualifier "estimated"; in almost all cases the quantities of interest will be the estimates.

The relations among the quantities mentioned are readily seen in the contingency table (figure 4.5a). The numbers are assumed to be proportions of total instances; that is, all the counts are assumed to have been divided by the total number of instances. We have used "T" for the positive class label. Note that the "positive" in "true positive" and "false positive" refers to the prediction $\hat{Y}$, not to the target value Y. Figures 4.5b and 4.5c give graphical depictions of recall (R), specificity (S), and precision (P), as well as rate of occurrence (r) and firing rate (t). The dark lines indicate the regions to which probabilities are conditionalized.

Rate of occurrence	$r = \Pr[Y \neq \bot]$	$= tp + fn$
Rate of firing	$t = \Pr[\hat{Y} \neq \bot]$	$= tp + fp$
Precision	$P = \Pr[Y \neq \bot \mid \hat{Y} \neq \bot]$	$= tp/t$
Sensitivity (recall)	$R = \Pr[\hat{Y} \neq \bot \mid Y \neq \bot]$	$= tp/r$
Specificity	$S = \Pr[\hat{Y} = \bot \mid Y = \bot]$	$= tn/(1-r)$
Accuracy	$A = \Pr[\hat{Y} = Y]$	$= tp + tn$

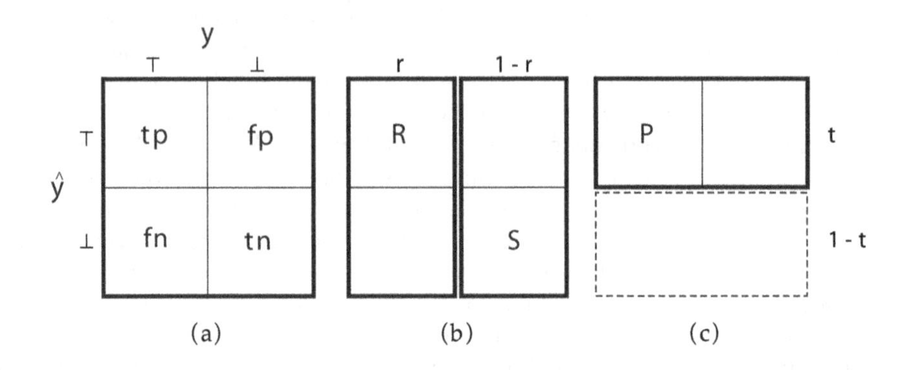

FIGURE 4.5
(a) Contingency table for detection; (b) recall (sensitivity) R and specificity S; (c) precision P.

Up to now we have assumed that the prediction of a confidence-rated classifier is determined simply by taking the class with the highest probability. In the case of detection, that implies that $\hat{Y}$ is positive just in case the confidence, measured as the log likelihood ratio (4.9), exceeds zero, and $\hat{Y}$ is negative otherwise. Writing $c(\mathbf{x})$ for the detector's confidence that instance $\mathbf{x}$ has the positive label, we have

$$\hat{Y} = +1 \quad \text{if and only if} \quad c(\mathbf{x}) > \theta$$

with $\theta = 0$.

Instead of taking a fixed threshold of $\theta = 0$, one can vary the threshold. Different choices of threshold yield different contingency tables, and a different trade-off of recall against precision or recall against specificity. The threshold is monotonically related to the rate of firing, t. As θ decreases, t increases.

If the detector simply guesses, then there is no correlation between c and Y, and recall (sensitivity) is equal to the firing rate:

$$R = \Pr[\hat{Y} \neq \perp | Y \neq \perp] = \frac{\Pr[\hat{Y} \neq \perp] \Pr[Y \neq \perp]}{\Pr[Y \neq \perp]} = t.$$

Similarly, if the detector simply guesses, specificity is equal to one minus the firing rate ($S = 1 - t = 1 - R$) and precision is equal to the rate of occurrence ($P = r$). If we plot precision against recall, the guessing detector's performance is a horizontal line whose level is the rate of occurrence (figure 4.6a). The points on the line represent different firing rates. If we plot recall against specificity, the detector's performance is the diagonal shown in figure 4.6b. In a recall-specificity plot, the curve for the guessing detector is the

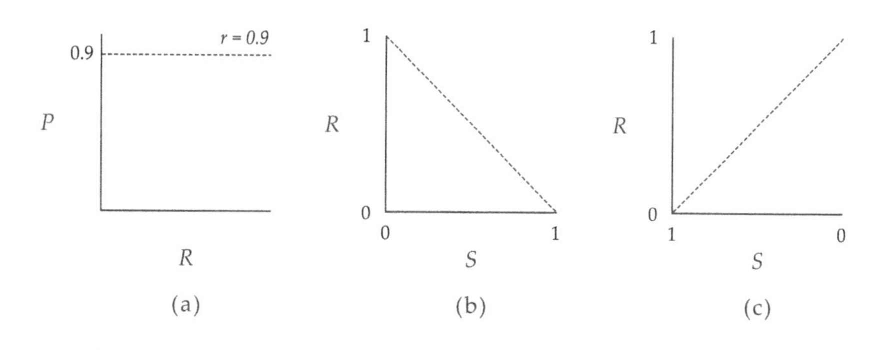

FIGURE 4.6
The performance of the guessing detector as a curve, plotted as (a) precision against recall, (b) specificity against recall, (c) ROC graph.

same regardless of the rate of occurrence. That, and the straightforward connection between sensitivity (that is, recall), specificity, and accuracy (4.10), make recall-specificity plots more attractive than precision-recall plots. Incidentally, when plotting sensitivity against specificity, it is conventional to plot specificity *decreasing* to the right (figure 4.6c). The resulting plot is called a Receiver Operating Characteristic (ROC) graph. In plots 4.6a and 4.6b, performance improves from lower left to upper right. In 4.6c, performance improves from lower right to upper left. A detailed discussion of ROC graphs can be found in Fawcett [86].

4.3.3 Idealized performance curves

If a single number is desired to measure performance, an alternative to accuracy is the area under the curve, either the area under the ROC curve (called AUC) or the area under the recall-precision curve (called average precision).

A different single-number measure often used in information retrieval and computational linguistics is the harmonic mean of recall and precision, called the **F-measure**:

$$F = \left(\frac{P^{-1} + R^{-1}}{2}\right)^{-1}$$

which simplifies to:

$$F = \frac{2PR}{P + R}.$$

Contours of equal F-measure can be viewed as idealized performance curves. See figure 4.7. Along the main diagonal, where $P = R$, both P and R are equal to F. This represents a kind of neutral point on the performance curve.

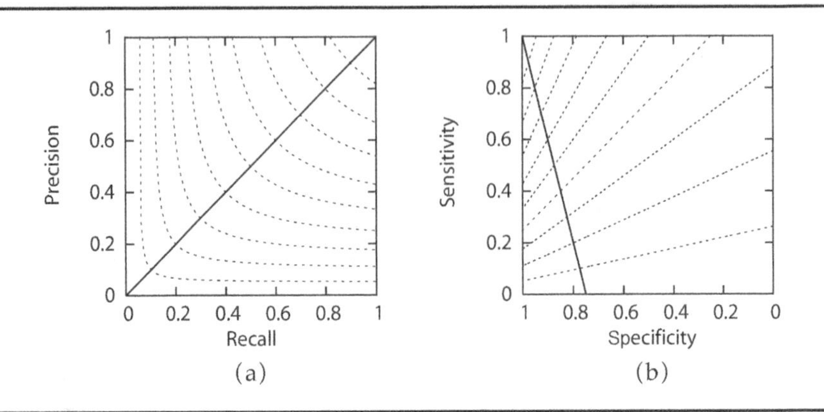

FIGURE 4.7
Contours of F ($r = 0.2$).

In figure 4.7a, F-measure contours and the line of equal recall and precision are plotted on a recall-precision graph, and in figure 4.7b, the same are plotted on an ROC graph, assuming a rate of occurrence of 10% positives. Interestingly, F-measure contours become linear in ROC space.*

It is difficult to justify F-measure probabilistically. A better-motivated set of idealized performance contours can be obtained by starting with a simple model of detector development. Suppose that we start with a guessing detector that fires (that is, guesses positive) with probability t_0. As we have already seen, the performance of such a detector lies on the line $R = S$, and more specifically at the point where $R = S = t_0$. Let the true rate of positives be r. Being a guessing detector means that $\hat{Y}$ and Y are uncorrelated, hence that the probability of true positives is rt_0, the probability of true negatives is $(1-r)(1-t_0)$, the probability of false positives is $(1-r)t_0$, and the probability of false negatives is $r(1-t_0)$.

To go from our initial detector to a perfect detector, we must reduce the rates of false positives and false negatives to zero. Suppose we do so stepwise, reducing the two error rates by, say, 10% at each step. After one step, the proportion of false positives is $\alpha \cdot fp_0$ where $\alpha = 0.9$. After two steps, it is

*Finding R as a function of F, S, and r is somewhat tedious, but merely an exercise in algebra. It is

$$R = \frac{F(1 - S(1 - r))}{r(2 - F)}.$$

If $P = R$, then both equal F. Substituting R for F in the previous equation yields the equation for the line $R = P$:

$$R = 2 - \frac{1}{r} + S\frac{1 - r}{r}.$$

$\alpha^2 \cdot fp_0$. The proportion of false negatives decreases similarly. In general:

$$fp_k = \alpha^k \cdot fp_0 = \alpha^k(1-r)t_0$$

$$fn_k = \alpha^k \cdot fn_0 = \alpha^k r(1-t_0).$$

Then we can compute sensitivity (recall) and specificity after k steps:

$$R_k = 1 - \frac{fn_k}{r} = 1 - \alpha^k(1-t_0) \tag{4.11}$$

$$S_k = 1 - \frac{fp_k}{1-r} = 1 - \alpha^k t_0. \tag{4.12}$$

We can see that both R_k and S_k are linear functions of α^k, and since $0 < \alpha < 1$, α^k ranges from 1 to 0 as k ranges from 0 to infinity. The detector with initial firing rate t_0 begins at point $(1-t_0, t_0)$ in (S, R) space, and moves along a line to $(1, 1)$. The trajectories for several initial firing rates are shown as solid lines in the graphs of figure 4.8. The trajectories are linear in (S, R) space, but curved in (R, P) space. If one guesses the proportion of positives correctly, and starts with $t_0 = r$, then the trajectory is along the diagonal of the recall-precision graph.

Solving (4.12) for t_0 and substituting into (4.11) yields

$$R_k = 2 - \alpha^k - S_k. \tag{4.13}$$

Taking the number of steps k as our measure of effort, the points (S_k, R_k) describe an equal-effort contour. Unlike F-measure contours, the equal error-reduction contours described by (4.13) are not sensitive to the rate of occurrence r. The contours of (4.13) are plotted as dotted lines in figure 4.8. The two plots in figure 4.8 should be contrasted with the corresponding plots in figure 4.7. Equal-effort contours are qualitatively similar to F-measure contours, and accomplish the same purpose of providing a single number to measure performance, but they have the advantage of a firm probabilistic basis and invariance under changes in the rate of occurrence.

In general, using specificity and sensitivity is preferable to using recall and precision. If one can directly sample from true positives and from instances on which the detector fires, but one cannot sample from the instance population as a whole and one has no information about either the rate of occurrence or rate of firing, then measuring precision and recall may be the only alternative. Such a case is rare, however; one usually acquires instances on which the detector fires by sampling from the population and applying the detector.

4.3.4 The multiclass case

As mentioned above, there is no standard generalization to the multiclass case with abstention. Abstention could be treated simply as an additional class

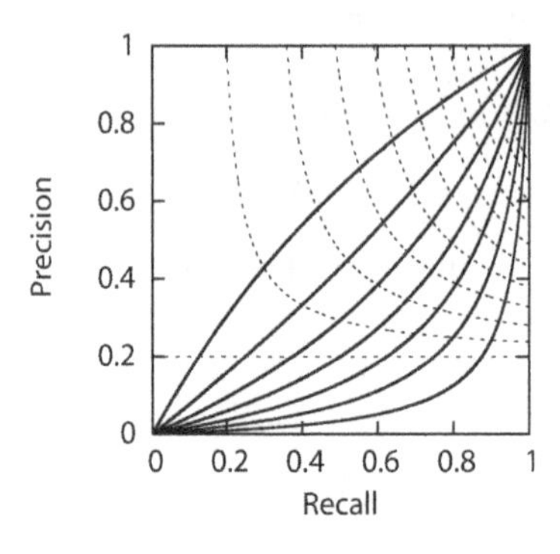

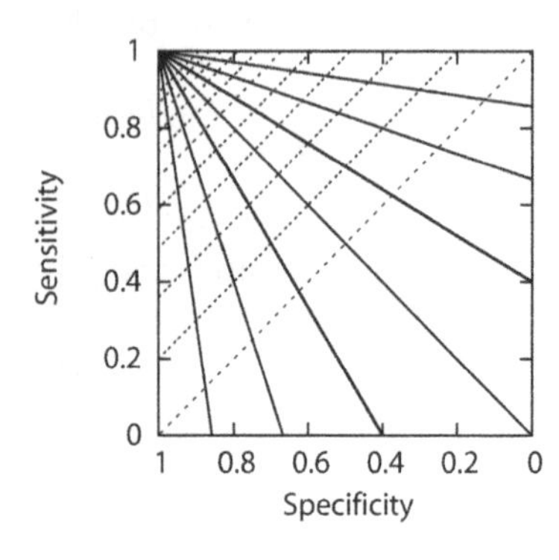

FIGURE 4.8
Equal error reduction contours. The solid lines are trajectories of ideal classifiers at $t_0 = 1/8, 2/8, \ldots, 7/8$. The dotted lines are contours of equal error reduction ($r = 0.2$).

(the "garbage" class), but doing so is less than satisfactory. An alternative is to treat the multiclass case with abstention as the combination of a detection problem and a classification problem. The detection problem concerns whether or not $\hat{Y} = \bot$ and $Y = \bot$, and performance is evaluated as we have just discussed. The classification problem is evaluated by **conditional accuracy**, that is, accuracy where attention is limited to the case that neither $\hat{Y} = \bot$ nor $Y = \bot$ (the upper left cell of the contingency table):

$$B = \Pr[\hat{Y} = Y | \hat{Y} \neq \bot, Y \neq \bot].$$

Graphical depictions of accuracy (A) and conditional accuracy (B) are given in figure 4.9. The overall accuracy A, which treats $\bot$ as a class rather than abstention, is conditional accuracy B times the true positive rate, plus the true negative rate. The true positive rate can be obtained either as the rate of occurrence times the recall, or as the firing rate times the precision. That is,

$$A = rRB + (1 - r)S = tPB + (1 - r)S.$$

If the test data is fully labeled, but the classifier may abstain, then $r = 1$, $P = 1$, $R = t$, and $A = tB$. In such a case, the natural performance curve is firing rate (which in this case equals recall) against conditional accuracy; this is what we used in the discussion of self-training in section 2.2.3. Figure 4.10a shows the contours of equal accuracy in a plot of firing rate against conditional accuracy. The similarity to F-measure is striking (see the left graph in figure 4.7), but in this case the measure is probabilistically well motivated.

Note that the accuracy contours in figure 4.10a treat abstentions as errors. As a practical matter, one would fall back to a default classifier instead.

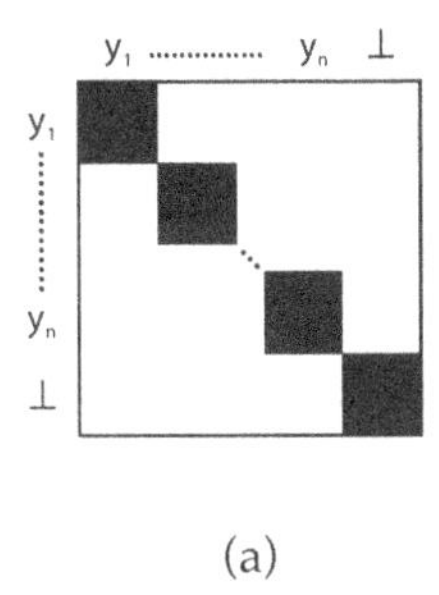

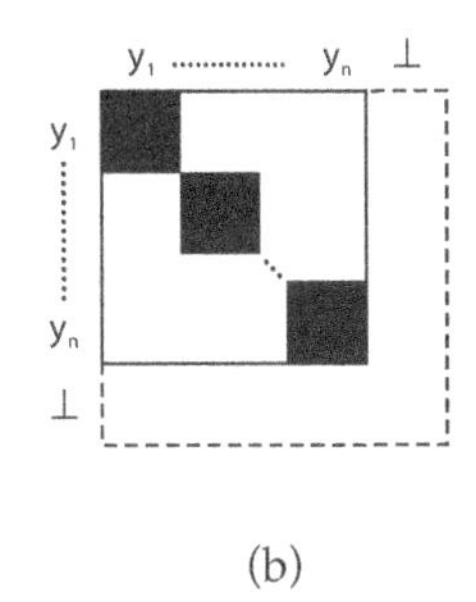

FIGURE 4.9
(a) Accuracy and (b) conditional accuracy in the multiclass case. In each case, the probability of interest is represented by the total area of the black regions as a proportion of the area of the encompassing solid square.

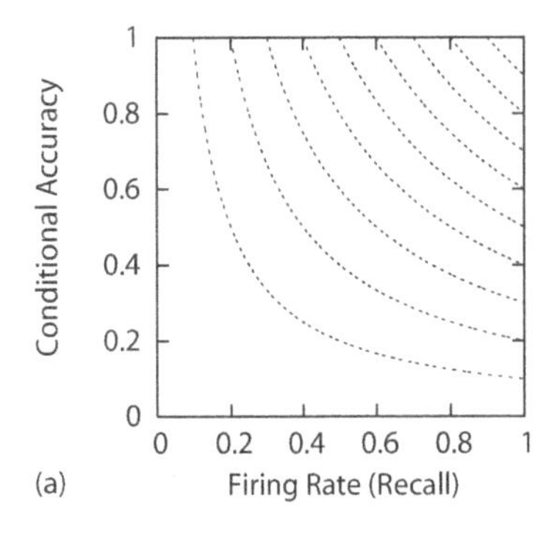

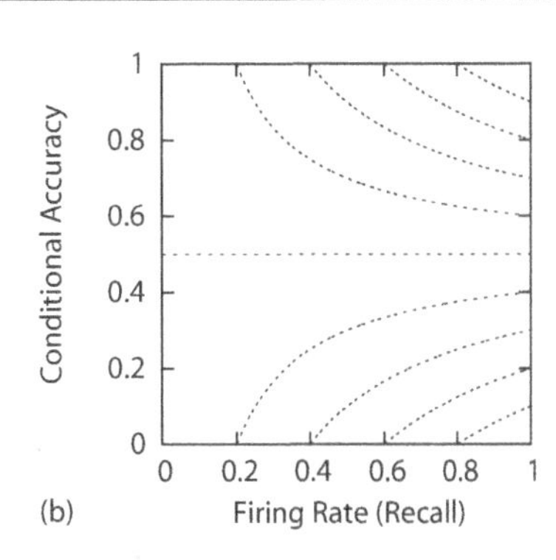

FIGURE 4.10
Contours of equal accuracy for classification with abstention. In the case of abstention, a fallback classifier is assumed with constant accuracy of (a) 0% or (b) 50%.

Figure 4.10b shows the accuracy contours if one falls back to a classifier with a constant 50% accuracy.

In the case of simple rules of form $F \Rightarrow y$, with $y \neq \perp$, there is a close relationship between precision and conditional accuracy. In a detection task, the natural measure for the quality of the rule is precision. There is only one target class, and precision is the probability that the rule's prediction is correct, when it fires. In a classification task with a total target function (Y is never $\perp$), the corresponding measure is conditional accuracy. In the detection case, the conditional accuracy of a simple rule is always 1: there is only one class (excluding $\perp$), so the rule is always correct when it fires and the target is not $\perp$. In the classification case, the precision of a simple rule is always 1: the target function Y never has value $\perp$, so certainly $Y \neq \perp$ on instances where the rule fires. Moreover, we can view the classification task as a collection of detection tasks, one for each label. The conditional accuracy of the rule $F \Rightarrow y$ on the classification task is equal to its precision on the detection task for label y.

4.4 Binary classifiers and ECOC

From this point, we will restrict our attention to binary classifiers, that is, classifiers whose output is one of two classes, represented by the values +1 and −1. Considering the binary case often simplifies matters considerably. And despite first appearances, limiting attention to binary classification does not entail any loss of generality. There are methods for addressing multiclass problems with binary classifiers.

A simple method is to train a separate binary classifier for each class. Suppose there are four classes, y_1, y_2, y_3, y_4. We construct a classifier c_1 that returns positive for instances labeled y_1, and negative for all other instances, and similar classifiers for each of the other labels. The following matrix represents this scheme. The entry in cell (i, j) is the label assigned to an instance in the training data for classifier c_i, if its original label is y_j. For example, for classifier c_2, instances whose original label is y_2 are treated as positives, and all other instances are treated as negatives.

	y_1	y_2	y_3	y_4
c_1	+1	−1	−1	−1
c_2	−1	+1	−1	−1
c_3	−1	−1	+1	−1
c_4	−1	−1	−1	+1

(4.14)

Each binary classifier can be viewed as discriminating between two sets of labels. For example, classifier c_2 distinguishes the set $\{y_2\}$ from the set

$\{y_1, y_3, y_4\}$.

The predictions of the individual classifiers must be combined to give a prediction for the composite, multiclass classifier. That is easy when one classifier says "yes" (output +1) and all others say "no" (output −1). But other cases are possible. The following illustrates several possible cases.

	(a)	(b)	(c)	(d)
c_1	-1	$+.6$	$-.5$	$-.5$
c_2	$+1$	$-.4$	$+.1$	$-.9$
c_3	-1	$-.2$	$+.4$	$-.3$
c_4	-1	$-.8$	$-.9$	$-.1$

(4.15)

Each column represents an output vector, with one component for each binary classifier. The question is which class the composite classifier should predict in each case. Case (a) is the easy one: the output should obviously be y_2. In case (b), the outputs are less crisp, but nonetheless, only c_1 says "yes" and all others say "no," so it seems clear that the output should be y_1. In case (c), both c_2 and c_3 say "yes." We interpret the absolute value of the classifier output as its confidence, so it is natural to "believe" the classifier that is most confident. In this case, c_3 is more confident than c_2, so we predict label y_3. Finally, in case (d), all of the classifiers say "no." In this case, it is natural to predict the label of the classifier that is least confident in saying "no": namely, we predict y_4. In short, in each case, we predict the label corresponding to the classifier with the greatest output value.

There is another way to think about the problem. We can view the columns of the matrix in (4.14) as codewords for the labels. The classifier outputs, which is to say, the columns of (4.15), may fail to match any of the codewords exactly. But we compare them to each codeword in turn, and output the label whose codeword is most similar to the classifier output vector. The measure of similarity most commonly used is Hamming distance, also known as L_1 distance; the Hamming distance between an output vector $\mathbf{z}$ and a codeword $\mathbf{w}$ is

$$L_1(\mathbf{z}, \mathbf{w}) \equiv \sum_i |z_i - w_i|.$$

For example, let $\mathbf{z}$ be the vector in (4.15c), namely $(-.5, +.1, +.4, -.9)$, and let $\mathbf{w}_i$ be the codeword for y_i. Then

$$\begin{aligned} L_1(\mathbf{z}, \mathbf{w}_1) &= 1.5 + 1.1 + 1.4 + .1 = 4.1 \\ L_1(\mathbf{z}, \mathbf{w}_2) &= .5 + .9 + 1.4 + .1 = 2.9 \\ L_1(\mathbf{z}, \mathbf{w}_3) &= .5 + 1.1 + .6 + .1 = 2.3 \\ L_1(\mathbf{z}, \mathbf{w}_4) &= .5 + 1.1 + 1.4 + 1.9 = 4.9. \end{aligned}$$

The codeword with the greater similarity to the output vector is $\mathbf{w}_3$, the codeword for y_3, so y_3 is the predicted label.

This technique works just as well if we add classifiers that distinguish between pairs of classes. Suppose we extend matrix (4.14) as follows:

$$\begin{array}{c|cccc|} & y_1 & y_2 & y_3 & y_4 \\ \hline c_1 & +1 & -1 & -1 & -1 \\ c_2 & -1 & +1 & -1 & -1 \\ c_3 & -1 & -1 & +1 & -1 \\ c_4 & -1 & -1 & -1 & +1 \\ c_5 & +1 & +1 & -1 & -1 \\ c_6 & +1 & -1 & +1 & -1 \\ c_7 & +1 & -1 & -1 & +1 \\ \hline \end{array} . \quad (4.16)$$

There are other rows that might be added that contain two positives and two negatives, but they are all equivalent to one of the existing rows. For example, the possible row $(-1,+1,-1,+1)$ is equivalent to the existing row $(+1,-1,+1,-1)$; both classifiers distinguish between the set $\{y_1, y_3\}$ and the set $\{y_2, y_4\}$.

Now consider the following output $\mathbf{z}$. Taking its Hamming distance with each of the codewords gives the results in the bottom row:

$$\begin{array}{ccccc} \mathbf{z} & \mathbf{w}_1 & \mathbf{w}_2 & \mathbf{w}_3 & \mathbf{w}_4 \\ \begin{bmatrix} -.3 \\ +.1 \\ +.2 \\ +.2 \\ -.6 \\ -.4 \\ +.1 \end{bmatrix} & \begin{bmatrix} +1 \\ -1 \\ -1 \\ -1 \\ +1 \\ +1 \\ +1 \end{bmatrix} & \begin{bmatrix} -1 \\ +1 \\ -1 \\ -1 \\ +1 \\ -1 \\ -1 \end{bmatrix} & \begin{bmatrix} -1 \\ -1 \\ +1 \\ -1 \\ -1 \\ +1 \\ -1 \end{bmatrix} & \begin{bmatrix} -1 \\ -1 \\ -1 \\ +1 \\ -1 \\ -1 \\ +1 \end{bmatrix} \\ & 8.7 & 7.3 & 6.7 & 5.7 \end{array} .$$

The most-similar codeword is $\mathbf{w}_4$, which is the codeword for y_4, so the prediction is y_4.

Incidentally, when the codewords consist of positive and negative ones, as here, the Hamming distance is closely related to the dot product between the two vectors. Consider the Hamming distance between $\mathbf{z}$ and $\mathbf{w}_3$. The first components are $-.3$ and -1. The signs agree, so the absolute difference is .7, which is the difference $1 - .3$. Likewise, the third components are $+.2$ and $+1$. The signs are positive this time, but they still agree, and the absolute difference is .8, which is again the difference $1 - .2$. On the other hand, the fourth components are $+.2$ and -1. In this case, the signs disagree, and the absolute difference is 1.2, which is the sum $1 + .2$. In general, we add if the signs disagree, and subtract if they agree. That is, the i-th components z_i and w_i contribute $1 - z_i w_i$ to the Hamming distance. Summing over all components, we have:

$$L_1(\mathbf{z}, \mathbf{w}) = n - \mathbf{z} \cdot \mathbf{w}$$

where n is the total number of components.

This leaves the question of how to choose the codewords for the classes. If the number of classes is small, we can enumerate all partitions of the classes, as in (4.16); each row represents a different partitioning of the classes into two sets. In general, however many classes there are, it is desirable to have codewords that are maximally distant from each other. That is, the columns of the matrix (4.16) should be maximally distant from one another, in Hamming distance. That makes the codewords maximally easy to distinguish from one another. But in order to have distance in the codeword space correspond to discriminability, the axes of the codeword space, represented by the individual binary classifiers, must be maximally separated, meaning that the rows of (4.16) must also be maximally distinct from each other. Intuitively, if multiple classifiers face the same or similar discrimination tasks, they will make correlated errors.

If classifiers make uncorrelated errors, and codewords are maximally distant from one another, the encoding method is robust to errors made by the individual classifiers. The encoding is called an Error-Correcting Output Code (ECOC); the idea of using ECOC for encoding multiclass problems as collections of binary classification problems was proposed by Dietterich & Bakiri [76], who discuss methods of obtaining good codewords when there is a large number of classes.

5

Mathematics for Boundary-Oriented Methods

In the discussion of the nearest-neighbor classifier (section 4.1.2), we viewed the classifier in terms of the decision boundary that it defines in the feature space, as, for example, in figure 4.4. This can be a very profitable way of viewing classifiers in general, and a number of methods focus explicitly on constructing good boundaries. We will discuss those methods in the next chapter. Before we can do so, however, we require some additional mathematical background.

This chapter treats three topics in linear algebra: the formal representation of a high-dimensional linear boundary, that is, a hyperplane; the gradient, along with rules for taking derivatives of linear-algebraic expressions; and constrained optimization.

5.1 Linear separators

The simplest variety of decision boundary is arguably a linear boundary, and many classifiers are based on the idea of constructing a linear decision boundary. In two dimensions, a linear boundary is a line; in three dimensions, it is a plane; in a general n-dimensional space it is a hyperplane of n minus one dimensions. A single linear boundary obviously divides the space of instances into only two classes, hence it suffices only for a binary classifier; but we have already seen how a multiclass problem can be reduced to a collection of binary classification problems.

5.1.1 Representing a hyperplane

The most convenient way to specify a hyperplane is to specify a normal vector $\mathbf{w}$, which gives the orientation of the hyperplane in space, and the distance r from the origin to the hyperplane along the direction of $\mathbf{w}$. Figure 5.1 illustrates in the case of a plane (a two-dimensional hyperplane) in a three-dimensional space. The vector $\mathbf{w}$ is normal (that is, perpendicular) to the plane, and serves to specify the orientation of the plane; and r is the distance from the origin. To be more precise, r is a "signed distance." A negative value for r indicates that the plane lies on the side of the origin opposite to

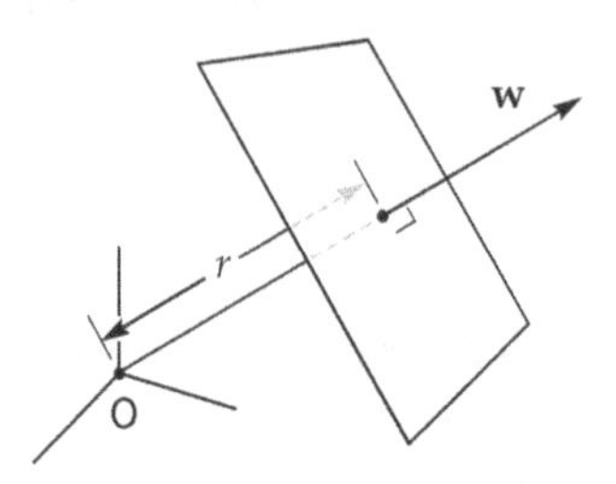

FIGURE 5.1
Specifying a hyperplane using a normal vector **w** and a distance r from the origin.

the direction of **w**.

Figure 5.2 shows a point **x** in the space. The prediction of the classifier on input **x** will be positive if **x** is on the "positive side" of the hyperplane, and negative otherwise. The positive side is defined to be the half-space that **w** points to. In figure 5.2, that is the side opposite the origin, but it need not be: as already mentioned, r may be negative, in which case **w** points toward the origin.

To determine the classifier's prediction for point **x**, we compute the signed distance d from the hyperplane to **x**. If **x** is on the positive side of the hyperplane, d is positive, and negative otherwise. The points that lie in the hyperplane are those for which $d = 0$. The absolute value of d is the distance from the hyperplane to **x**.

To compute d, we project **x** onto **w**. The projection of **x** onto **w** is the vector **v** in figure 5.2. Its direction is the same as that of **w**, but its length is the dot product of **x** with the unit vector in the direction of **w**:

$$\|\mathbf{v}\| = \mathbf{x} \cdot \frac{\mathbf{w}}{\|\mathbf{w}\|}.$$

This is the distance from the origin along the line of **w** to a point level with **x**. To obtain d, we subtract r:

$$d = \mathbf{x} \cdot \frac{\mathbf{w}}{\|\mathbf{w}\|} - r. \tag{5.1}$$

It will be convenient if we do not need to normalize **w**, so let us write this in

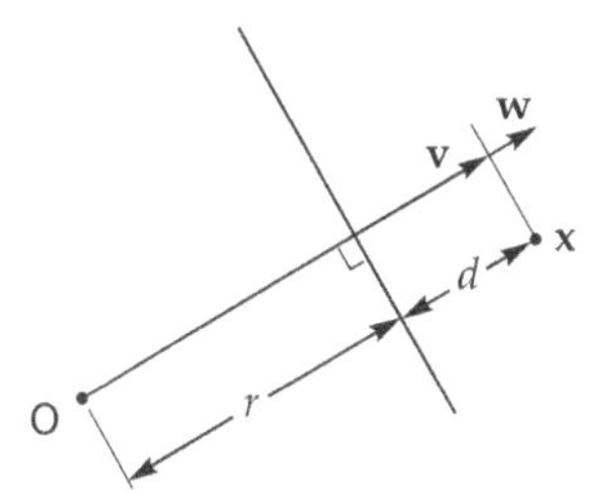

FIGURE 5.2
Computing the distance d from the hyperplane to a point $\mathbf{x}$. The vector $\mathbf{v}$ is the projection of $\mathbf{x}$ onto $\mathbf{w}$.

the equivalent form:

$$d\|\mathbf{w}\| = \mathbf{x} \cdot \mathbf{w} - r\|\mathbf{w}\|. \tag{5.2}$$

Thinking of $1/\|\mathbf{w}\|$ as one "$\mathbf{w}$-unit," this says that $\mathbf{x} \cdot \mathbf{w}$ less the threshold $r\|\mathbf{w}\|$ gives the distance from the separator in $\mathbf{w}$-units: $d\|\mathbf{w}\|$ times the unit size $1/\|\mathbf{w}\|$ equals d.

An important corollary is that scaling the vector $\mathbf{w}$ does not affect d. Define $\mathbf{w}' \equiv \beta\mathbf{w}$ to be a vector in the same direction as $\mathbf{w}$, but scaled by a factor β. Then

$$\begin{aligned}\frac{x \cdot \mathbf{w}'}{\|\mathbf{w}'\|} - r &= \frac{x \cdot \beta\mathbf{w}}{\|\beta\mathbf{w}\|} - r \\ &= \frac{\beta(x \cdot \mathbf{w})}{\beta\|\mathbf{w}\|} - r \\ &= d.\end{aligned}$$

5.1.2 Eliminating the threshold

The form (5.2) would be simpler if we could eliminate the threshold r. We can accomplish that by folding the threshold into the vector $\mathbf{w}$. The trick is to add an extra component to every data point $\mathbf{x}$, and add an extra component to the normal vector $\mathbf{w}$, as follows. If $\mathbf{w} = (w_1, \ldots, w_n)$ and $\mathbf{x} = (x_1, \ldots, x_n)$, we define the new representations

$$\mathbf{w}' = (w_1, \ldots, w_n, r\|\mathbf{w}\|)$$

$$\mathbf{x}' = (x_1, \ldots, x_n, -1)$$

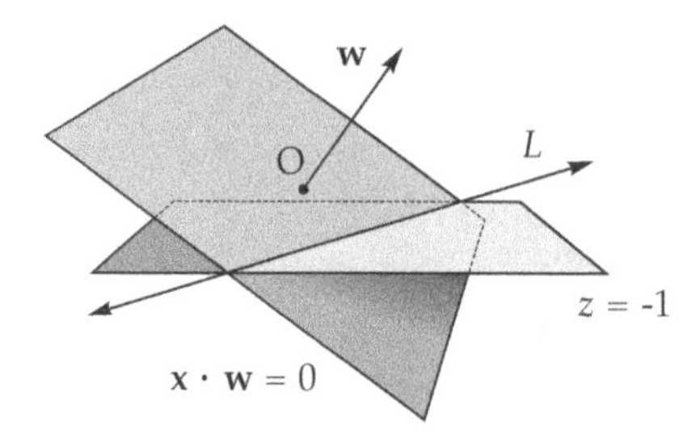

FIGURE 5.3
By adding an additional dimension z, with all data points living in the plane $z = -1$, we can specify an arbitrary linear separator L in the data plane by means of a hyperplane that passes through the origin (O) in the new space.

Then

$$\begin{aligned} \mathbf{w}' \cdot \mathbf{x}' &= (w_1, \ldots, w_n, r\|\mathbf{w}\|) \cdot (x_1, \ldots, x_n, -1) \\ &= \mathbf{w} \cdot \mathbf{x} - r\|\mathbf{w}\| \\ &= d\|\mathbf{w}\|. \end{aligned}$$

In the modified space, the distance (in $\mathbf{w}$-units) between $\mathbf{x}'$ and the hyperplane is simply the dot product $\mathbf{x}' \cdot \mathbf{w}'$.

Geometrically, the effect is the following. Imagine the data to lie in a two-dimensional space. We add an extra "z" axis perpendicular to the data plane, and place the data plane at level negative one on the z axis (figure 5.3). Now we can limit attention to separating hyperplanes that contain the origin; no threshold is necessary to specify distance from the origin. By rotating and tilting the hyperplane, with the origin as pivot, we can cut the data plane along any line L that we desire, as illustrated in the figure.

5.1.3 The point-normal form

Sometimes it is convenient to specify a hyperplane using a normal vector $\mathbf{w}$ as before, but using, instead of the distance r from the origin, a point $\mathbf{a}$ that lies on the hyperplane (figure 5.4). We again consider an arbitrary point $\mathbf{x}$, and we compute the signed distance d from the hyperplane to $\mathbf{x}$. The dotted vector pointing from $\mathbf{a}$ to $\mathbf{x}$ is the vector difference $\mathbf{x} - \mathbf{a}$. The length of its projection onto $\mathbf{w}$ is our desired signed distance d:

$$d = (\mathbf{x} - \mathbf{a}) \cdot \frac{\mathbf{w}}{\|\mathbf{w}\|}. \tag{5.3}$$

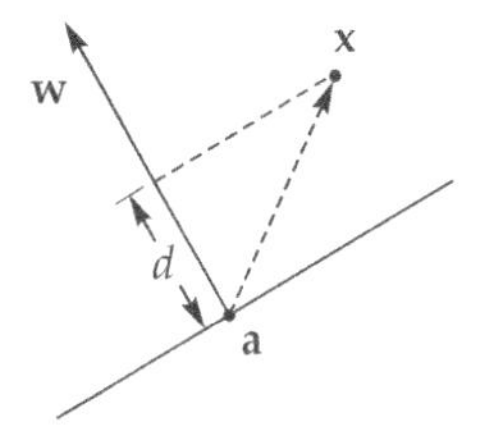

FIGURE 5.4
The point-normal form. The signed distance d from the hyperplane can be computed from the normal vector $\mathbf{w}$ and a point $\mathbf{a}$ lying in the hyperplane.

As before, d is positive for points in the half-space toward which $\mathbf{w}$ points, and negative in the other half-space. The points that lie in the hyperplane itself have $d = 0$.

The point-normal form (5.3) is easily translated to the form (5.1):

$$d = \mathbf{x} \cdot \frac{\mathbf{w}}{\|\mathbf{w}\|} - \mathbf{a} \cdot \frac{\mathbf{w}}{\|\mathbf{w}\|}.$$

This is equivalent to (5.1) with $r = \mathbf{a} \cdot \mathbf{w}/\|\mathbf{w}\|$.

A particular example in which the point-normal form arises involves a linear separator consisting of points equidistant between two designated points $\mathbf{m}_1$ and $\mathbf{m}_2$, as in figure 5.5. This situation arises in the 1-nearest-neighbor algorithm, for example, in which case the points $\mathbf{m}_1$ and $\mathbf{m}_2$ are training instances, and the linear separator between them is a boundary segment, such as we saw earlier in figure 4.4. A similar situation arises in k-means clustering (section 7.4.1), in which case the points are cluster centers.

The vector whose tail is at $\mathbf{m}_2$ and whose head is at $\mathbf{m}_1$ is normal to the separating hyperplane; call it $\mathbf{w}$. It is the vector difference of the centers:

$$\mathbf{w} = \mathbf{m}_1 - \mathbf{m}_2.$$

The point $\mathbf{a}$ is the midpoint of the connecting line segment:

$$\mathbf{a} = \frac{\mathbf{m}_1 + \mathbf{m}_2}{2}.$$

With these values of $\mathbf{w}$ and $\mathbf{a}$, the signed distance from the separating hyperplane to an arbitrary point $\mathbf{x}$ is given by (5.3), with the hyperplane itself consisting of those points for which $d = 0$.

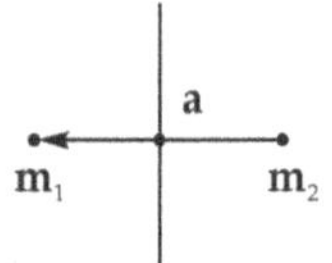

FIGURE 5.5
A linear separator equidistant between centers $\mathbf{m}_1$ and $\mathbf{m}_2$.

5.1.4 Naive Bayes decision boundary

It was mentioned in the previous chapter that the decision boundary for the Naive Bayes classifier is a linear separator. We are now in a position to see why that is true.

We continue to limit attention to binary classification. The Naive Bayes classifier predicts positive if $p(\mathbf{x}, +) > p(\mathbf{x}, -)$, which is to say if

$$f(\mathbf{x}) \equiv \log \frac{p(\mathbf{x}, +)}{p(\mathbf{x}, -)} > 0.$$

As mentioned in section 4.3.1, the value $f(\mathbf{x})$ is known as a **log likelihood ratio**.

Recall that, under the Naive Bayes independence assumption, $p(\mathbf{x}, y)$ is defined as

$$p(\mathbf{x}, y) = p(y) \prod_i p(x_i|y)$$

where $\mathbf{x} = (x_1, \ldots, x_m)$. Hence:

$$\begin{aligned} f(\mathbf{x}) &= \log \frac{p(+)\prod_i p(x_i|+)}{p(-)\prod_i p(x_i|-)} \\ &= \sum_i \log \frac{p(x_i|+)}{p(x_i|-)} + (-1) \log \frac{p(-)}{p(+)}. \end{aligned} \tag{5.4}$$

For example, consider the Naive Bayes model given in section 4.1.1, in (4.1) and (4.2). We have

$$\log \frac{p(-)}{p(+)} = \log \frac{8/14}{6/14} = .288$$

and for the attributes:

$$
\begin{array}{c|c|c|}
 & \text{Len} & \text{Wid} \\
x & \log \frac{p(x|+)}{p(x|-)} & \log \frac{p(x|+)}{p(x|-)} \\
\hline
\text{small} & \log \frac{1/6}{6/8} = -1.504 & \log \frac{1/6}{4/8} = -1.099 \\
\text{med} & \log \frac{2/6}{1/8} = 0.981 & \log \frac{1/6}{3/8} = -.811 \\
\text{large} & \log \frac{3/6}{1/8} = 1.386 & \log \frac{4/6}{1/8} = 1.674
\end{array}
. \tag{5.5}
$$

The value of $f(\mathbf{x})$ for $\mathbf{x} = (\text{med}, \text{med})$ is

$$f(\text{med}, \text{med}) = 0.981 - .811 - .288 = -.118. \tag{5.6}$$

The prediction is negative, with the absolute value (.118) interpretable as confidence.

Let us write w_{ij} for the number in the (i, j)-th cell of (5.5). For example, $w_{12} = -1.099$ is the number associated with Wid = small. In (5.6), the numbers 0.981 and $-.811$ are w_{21} and w_{22}, respectively. In general, the sum expanding $f(\mathbf{x})$ will contain the "bias" term $-.288$ and one weight w_{i1} for the Len attribute and one weight w_{i2} for the Wid attribute. We can write (5.6) in a more general form like this:

$$
\begin{aligned}
&f(\text{med}, \text{med}) = \\
&\quad (0)(-1.504) + (1)(0.981) + (0)(1.386) + \\
&\quad (0)(-1.099) + (1)(-.811) + (0)(1.674) + \\
&\quad (-1)(.288).
\end{aligned}
$$

The sum has coefficient one for the two weights corresponding to Len = med and Wid = med, and coefficient zero everywhere else, except for the bias term, which has coefficient negative one. A different value for $\mathbf{x}$ will have a sum that differs only in the choices of the ones and zeros. That is, let us represent the instance (med, med) by the vector:

$$\mathbf{x} = (0, 1, 0, 0, 1, 0, -1)$$

and let us define

$$\mathbf{w} = (-1.504, 0.981, 1.386, -1.099, -.811, 1.674, .288).$$

Then we have

$$f(\mathbf{x}) = \mathbf{x} \cdot \mathbf{w}.$$

In short, the Naive Bayes prediction $f(\mathbf{x})$ can be expressed as a dot product between a vector $\mathbf{w}$ representing the model, and a vector $\mathbf{x}$ representing the instance. Instances are represented as points in a space whose axes correspond to attribute-value pairs, plus a "z" axis for the bias. In that space, the Naive Bayes model defines a linear separator, which is the hyperplane $\mathbf{x} \cdot \mathbf{w} = 0$.

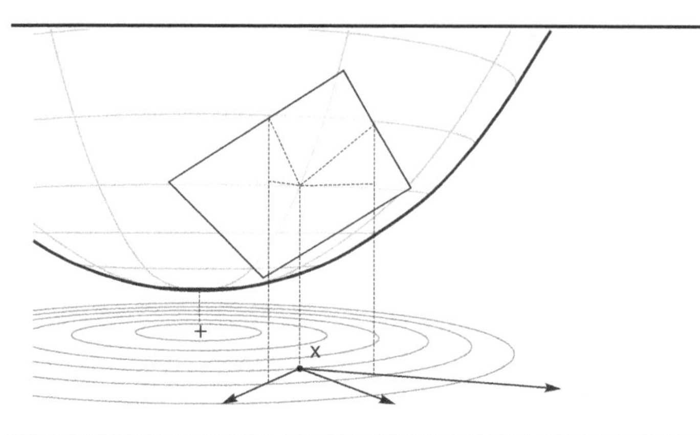

FIGURE 5.6
A function, tangent plane, contours.

5.2 The gradient

We change topics now to discuss optimization. One common method of minimizing a function is **gradient descent**. The gradient is a vector that points in the direction in which the function is increasing most rapidly, and by following the gradient "backward," one seeks to find a lowest point, that is, a minimum of the function.

5.2.1 Graphs and domains

Consider a function $f(\mathbf{x})$ whose domain is $\mathbf{R}^n$. The **graph** of f is a surface in $\mathbf{R}^{n+1}$, consisting of points $(x_1, \ldots, x_n, y)$ such that $y = f(\mathbf{x})$ and $\mathbf{x} = (x_1, \ldots, x_n)$. The additional dimension is the codomain of f. If D is the domain of f and C is its codomain, we call $D \times C$ the "graph space" of f. In our example, the domain is $\mathbf{R}^n$, the codomain is $\mathbf{R}$, and the graph space is $\mathbf{R}^{n+1}$.

A **level curve** of a function f is the set of points satisfying

$$f(\mathbf{x}) = c$$

for a particular value of c. A level curve resides in the *domain* of f, not in the graph space of f. A collection of level curves gives a "contour map" of the function, as in figure 5.6. The position of the cross marks the point at which f takes on its minimum value. Thinking of the values $f(\mathbf{x})$ as elevations, the cross marks the lowest point and the minimum is its elevation.

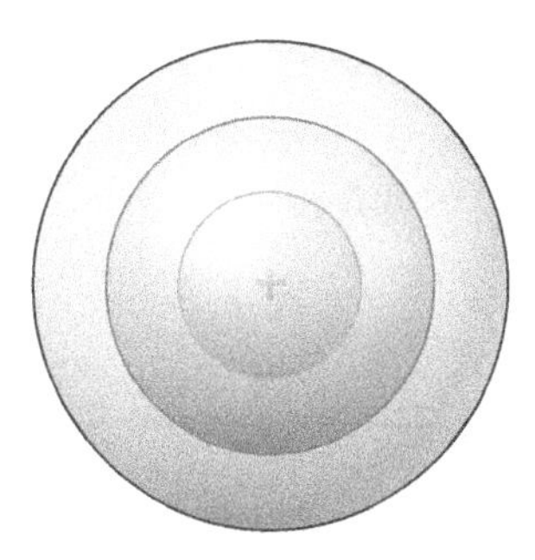

FIGURE 5.7
A three-dimensional domain.

The gradient of f is the vector of partial derivatives

$$\nabla f(\mathbf{x}) = \left(\frac{\partial}{\partial x_1} f(\mathbf{x}), \ldots, \frac{\partial}{\partial x_n} f(\mathbf{x}) \right).$$

It gives the rate of increase of the function at point $\mathbf{x}$ in each of the the axis-parallel directions. Note that ∇f is a function, and its value at point $\mathbf{x}$ is a vector.

The gradient can be interpreted geometrically as the direction of steepest increase in f. For example, consider the function f in figure 5.6, with $\partial f / \partial x_1 = \partial f / \partial x_2 = 1$ at a particular point x. The tangent plane to f at point x rises with slope $+1$ in the x_1 direction and with slope $+1$ in the x_2 direction. The direction directly "up" the tangent plane is at 45°, which is to say, in the direction of the vector

$$(1, 1) = \nabla f(\mathbf{x}).$$

Notice that figure 5.6 depicts the graph space of f, but the gradient, like the level curves, lives in the *domain,* not in the graph space. The "floor" in figure 5.6 is the domain.

Importantly, the gradient at point $\mathbf{x}$ is normal (that is, perpendicular) to the level curve passing through $\mathbf{x}$. Figure 5.6 illustrates the case where the domain is two-dimensional. In the case of a three-dimensional domain (figure 5.7), the graph space is four dimensional, hence it cannot easily be drawn. The level curves are surfaces within the domain, forming expanding "shells" around the minimizer. The gradient at point x is a vector that is perpendicular to the level surface, pointing in the direction away from the minimizer.

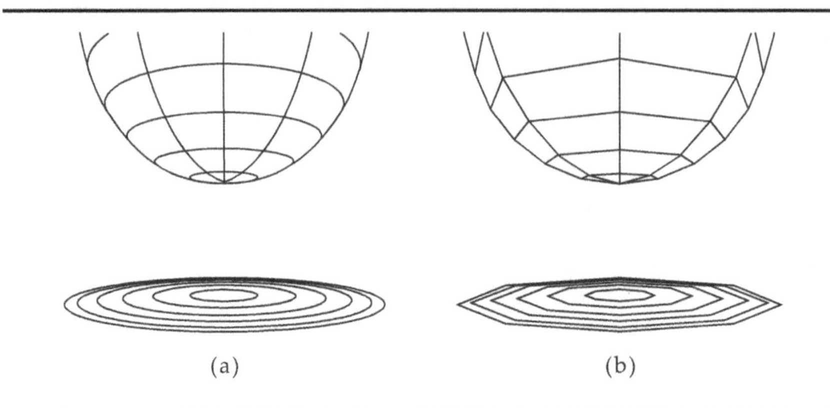

FIGURE 5.8
Convex functions. (a) Smooth, (b) piecewise linear.

5.2.2 Convexity

We will be particularly interested in a certain class of functions, the convex functions. A convex function is, intuitively, one shaped like a bowl. It may be smooth, as in figure 5.8a, or piecewise linear, as in figure 5.8b, or a combination of both. Consider any **chord**, that is, the the line segment connecting two points on the graph. A function is convex if it its graph lies on or below the chord, for every chord (figure 5.9). Algebraically, the line segment connecting points $\mathbf{x}$ and $\mathbf{y}$ is the set of points $\alpha\mathbf{x} + \beta\mathbf{y}$, for $\alpha, \beta \geq 0$ and $\alpha + \beta = 1$, and the z-value for the chord at each of those points is

$$\alpha f(\mathbf{x}) + \beta f(\mathbf{y}).$$

Hence a function is convex just in case:

$$f(\alpha\mathbf{x} + \beta\mathbf{y}) \leq \alpha f(\mathbf{x}) + \beta f(\mathbf{y})$$

for all points $\mathbf{x}$ and $\mathbf{y}$ in the domain of f and for all $0 \leq \alpha, \beta$ and $\alpha + \beta = 1$.

A linear combination of convex functions is itself convex. Suppose that the functions f_i are convex, and suppose there are real numbers c_i such that:

$$g(\mathbf{x}) = \sum_i c_i f_i(\mathbf{x}).$$

Then

$$\begin{aligned} g(\alpha\mathbf{x} + \beta\mathbf{y}) &= \sum_i c_i f_i(\alpha\mathbf{x} + \beta\mathbf{y}) \\ &\leq \sum_i c_i [\alpha f_i(\mathbf{x}) + \beta f_i(\mathbf{y})] \end{aligned}$$

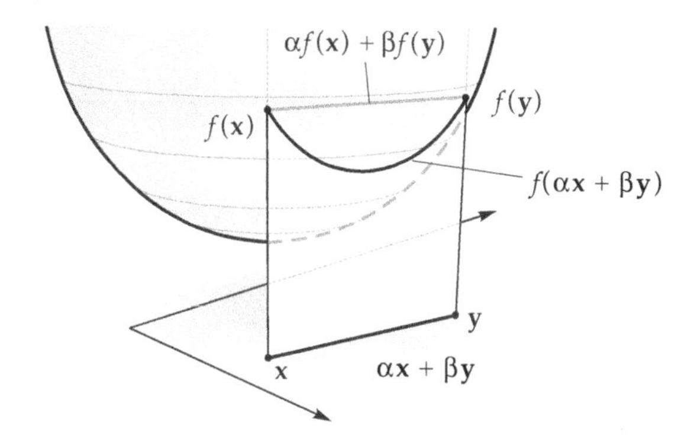

FIGURE 5.9
Definition of convexity.

$$
\begin{aligned}
&= \alpha \sum_i c_i f_i(\mathbf{x}) + \beta \sum_i c_i f_i(\mathbf{y}) \\
&= \alpha g(\mathbf{x}) + \beta g(\mathbf{y}).
\end{aligned}
$$

A related concept is that of a **convex set**, which is a set of points S that has the following property. If $\mathbf{x}$ and $\mathbf{y}$ belong to S, then so does every point

$$\alpha\mathbf{x} + \beta\mathbf{y}$$

on the line segment connecting $\mathbf{x}$ and $\mathbf{y}$, where $\alpha, \beta \geq 0$ and $\alpha + \beta = 1$. The connection between convex functions and convex sets is this:

> *A level curve of a convex function describes the boundary of a convex set.*

Recall that a level curve is the set of points satisfying $f(\mathbf{x}) = c$ for a fixed value c. The set of points satisfying $f(\mathbf{x}) \leq c$ is a convex set.

One more property of note is the following. Consider the parabola $f(x)$ in figure 5.10a. The derivative $f'(x)$ is the slope of the tangent line. It increases as one moves out from the minimum. This is true in general of convex functions. The reason should be clear intuitively: if the derivative ever decreased, the "bowl" would curve outward, and the resulting "lip" would provide a concavity where the graph of the function is higher than a chord (figure 5.10b).

To be more precise, the slope is nondecreasing as one moves away from the minimum. It is strictly increasing if the function is strictly convex.

In three and higher dimensions, the generalization of slope is the length of the gradient. Figure 5.11 shows the rise in the tangent plane in the direction of the gradient (which is the direction of maximum rise). For brevity, we

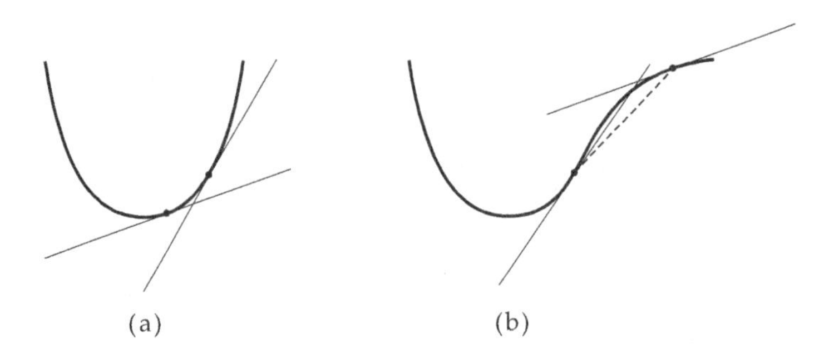

FIGURE 5.10
(a) Convex function, slopes of tangent lines always increase. (b) Slopes of tangent lines decrease in region of concavity, function lies above chord.

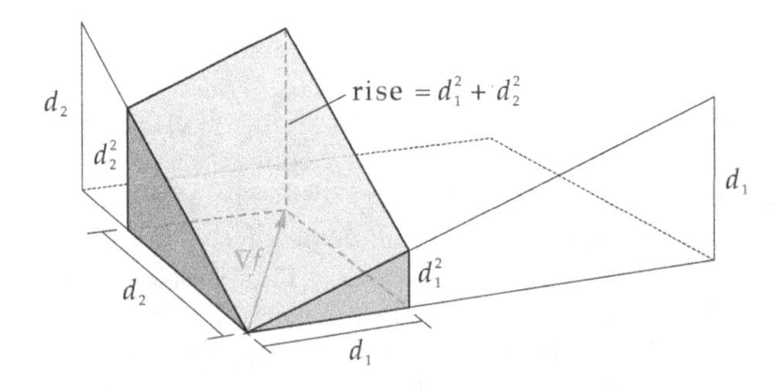

FIGURE 5.11
The slope of the tangent plane in the direction of the gradient.

have written d_1 for $\partial f/\partial x_1$ and d_2 for $\partial f/\partial x_2$. The figure depicts the case $d_1 = 1/3$ and $d_2 = 3/4$, but nothing hinges on those choices. The quantity d_1 is the rise in the tangent plane if we move one unit in the x_1 direction. If we only move d_1 units in the x_1 direction, then the rise is d_1^2: for example, here it is 1/3 of 1/3. The figure also makes clear that the rise at the point of the gradient is $d_1^2 + d_2^2$, which is equal to $\|\nabla f\|^2$. The "run" is simply the length of the gradient vector. The slope of the tangent plane is the rise over the run:

$$\frac{\|\nabla f\|^2}{\|\nabla f\|} = \|\nabla f\|.$$

We conclude that the length of the gradient of a convex function is monotonically related to the distance from the minimum point. It is strictly increasing if the function is strictly convex. This is extremely useful for optimization: the direction of the gradient tells us in which direction the minimum lies, and its magnitude decreases as we get closer, becoming zero when we reach the minimum.

5.2.3 Differentiation of vector and matrix expressions

The gradient is an example of a function that maps vectors to vectors; in fact, as we saw, it is the multidimensional counterpart of the derivative. It is often easy to set up an expression for a gradient, but tedious to simplify it using componentwise partial derivatives. Not only for the gradient, but generally for functions involving operations on vectors and matrices, more direct rules of differentiation are a great convenience.

The derivative of a vector with respect to a scalar is simply the vector in which the differentiation is done elementwise:

$$\frac{d}{dx}\mathbf{y} = \begin{bmatrix} \frac{d}{dx}y_1 \\ \vdots \\ \frac{d}{dx}y_m \end{bmatrix}.$$

The derivative of a vector $\mathbf{y}$ with respect to a vector $\mathbf{x}$ is a *matrix*, in which the j-th column is the derivative of $\mathbf{y}$ with respect to x_j:

$$\mathbf{D_x y} = \begin{bmatrix} \frac{\partial}{\partial x_1}\mathbf{y} \cdots \frac{\partial}{\partial x_n}\mathbf{y} \end{bmatrix} = \begin{bmatrix} \frac{\partial y_1}{\partial x_1} & \cdots & \frac{\partial y_1}{\partial x_n} \\ \vdots & \ddots & \vdots \\ \frac{\partial y_m}{\partial x_1} & \cdots & \frac{\partial y_m}{\partial x_n} \end{bmatrix}. \tag{5.7}$$

The derivative of a scalar with respect to a vector can be treated as the special case in which $\mathbf{y}$ contains only one component y:

$$\mathbf{D_x} y = \begin{bmatrix} \frac{\partial y}{\partial x_1} \cdots \frac{\partial y}{\partial x_n} \end{bmatrix}. \tag{5.8}$$

Note that this is in fact the definition of the gradient:

$$\nabla y = \mathbf{D_x} y$$

although to be quite precise ∇y is a vector whereas $\mathbf{D_x} y$ is a matrix containing a single row.

We do not consider differentiation of matrices or with respect to matrices. The differentiation operator $\mathbf{D}$ here is an operator on vectors. As such, it is immaterial whether the vector $\mathbf{y}$ or the vector $\mathbf{x}$ is considered as a row vector or a column vector; the derivative is the same.

Now we turn to the basic rules for taking derivatives. All the rules can be derived fairly straightforwardly from the definitions just given and the familiar rules for differentiation of real numbers.

For example, consider $\mathbf{D_x x}$. By definition, $\mathbf{D_x x}$ is a matrix whose (i, j)-th entry is $\partial x_i / \partial x_j$. For a given independent variable x_i, the other independent variables x_j are treated as constants; that is, $\partial x_i / \partial x_j$ is one if $i = j$ and zero if $i \neq j$. Hence, $\mathbf{D_x x}$ is the identity matrix:

$$\mathbf{D_x x} = I.$$

For another example, consider $\mathbf{D_x a^T y}$, where we understand $\mathbf{y}$ to be a vector-valued function of $\mathbf{x}$, but $\mathbf{a}$ to be a constant vector. Because $\mathbf{a^T y}$ is a scalar, we apply definition (5.8):

$$\begin{aligned}
\mathbf{D_x a^T y} &= \left[\cdots \frac{\partial}{\partial x_j} \mathbf{a^T y} \ldots\right] \\
&= \left[\cdots \frac{\partial}{\partial x_j} \textstyle\sum_i a_i y_i \cdots\right] \\
&= \left[\cdots \textstyle\sum_i a_i \frac{\partial}{\partial x_j} y_i \cdots\right] \\
&= \mathbf{a^T} \left[\cdots \frac{\partial}{\partial x_j} \mathbf{y} \cdots\right] \\
&= \mathbf{a^T D_x y}.
\end{aligned}$$

Combining the previous two results yields

$$\mathbf{D_x a^T x} = \mathbf{a^T D_x x} = \mathbf{a^T} I = \mathbf{a^T}.$$

What if we reverse $\mathbf{a}$ and $\mathbf{y}$? We have

$$\mathbf{D_x y^T a} = \left[\cdots \frac{\partial}{\partial x_j} \mathbf{y^T a} \ldots\right].$$

Since $\mathbf{y^T a} = \mathbf{a^T y}$, the remainder of the derivation is identical to the one given just above, and we conclude that

$$\mathbf{D_x y^T a} = \mathbf{a^T D_x y}.$$

To give one more example, let us consider $\mathbf{D_x y^T z}$ where both $\mathbf{y}$ and $\mathbf{z}$ are vector-valued functions of $\mathbf{x}$.

$$
\begin{aligned}
\mathbf{D_x y^T z} &= \mathbf{D_x} \sum_i y_i z_i \\
&= \left[\cdots \tfrac{\partial}{\partial x_j} \textstyle\sum_i y_i z_i \cdots \right] \\
&= \left[\cdots \textstyle\sum_i \left(z_i \frac{\partial}{\partial x_j} y_i + y_i \frac{\partial}{\partial x_j} z_i \right) \cdots \right] \\
&= \left[\cdots \textstyle\sum_i z_i \frac{\partial}{\partial x_j} y_i \cdots \right] + \left[\cdots \textstyle\sum_i y_i \frac{\partial}{\partial x_j} z_i \cdots \right] \\
&= \mathbf{z}^{\mathrm{T}} \left[\cdots \tfrac{\partial}{\partial x_j} \mathbf{y} \cdots \right] + \mathbf{y}^{\mathrm{T}} \left[\cdots \tfrac{\partial}{\partial x_j} \mathbf{z} \cdots \right] \\
&= \mathbf{z^T D_x y} + \mathbf{y^T D_x z}.
\end{aligned}
$$

The following is a useful list of derivation rules, including the ones we have just derived, but others as well whose proof is left as an exercise for the reader. In the following, $\mathbf{x}$ is always the independent variable (a vector) and the variables $\mathbf{y}$ and $\mathbf{z}$ represent vector-valued functions of the independent variable. The variable a represents a scalar constant, $\mathbf{a}$ represents a vector constant, and $\mathbf{A}$ represents a matrix constant.

$$
\begin{aligned}
\mathbf{D_x a} &= \mathbf{0} \\
\mathbf{D_x x} &= \mathbf{I} \\
\mathbf{D_x} a\mathbf{y} &= a\mathbf{D_x y} \\
\mathbf{D_x (y + z)} &= \mathbf{D_x y} + \mathbf{D_x z} \\
\mathbf{D_x a^T y} &= \mathbf{a^T D_x y} \\
\mathbf{D_x y^T a} &= \mathbf{a^T D_x y} \\
\mathbf{D_x A y} &= \mathbf{A D_x y} \\
\mathbf{D_x y^T A} &= \mathbf{A^T D_x y} \\
\mathbf{D_x y^T z} &= \mathbf{y^T D_x z} + \mathbf{z^T D_x y} \\
\mathbf{D_x y^T y} &= 2\mathbf{y^T D_x y} \\
\mathbf{D_x y^T A y} &= \mathbf{y^T (A + A^T) D_x y} \\
\mathbf{D_x y^T A y} &= 2\mathbf{y^T A D_x y} \qquad\qquad \text{if } \mathbf{A} \text{ is symmetric.}
\end{aligned}
$$

5.2.4 An example: linear regression

Linear regression provides a good example of the utility of our rules for computing the derivative. In linear regression, there is a real-valued target function $Y(\mathbf{x})$, and we construct a hypothesis of the form

$$\hat{Y}(\mathbf{x}) = \mathbf{w} \cdot \mathbf{x} + \alpha. \tag{5.9}$$

The question is how to choose the weight vector $\mathbf{w}$.

As discussed in section 5.1.2, we can fold α into $\mathbf{w}$. Specifically, α becomes a new final component in the vector $\mathbf{w}$, and every instance vector $\mathbf{x}$ is extended by the addition of a new constant component with value one. After this change of representation, (5.9) becomes

$$\hat{Y}(\mathbf{x}) = \mathbf{w} \cdot \mathbf{x}. \tag{5.10}$$

The learner is provided with a labeled training sample, which we represent as a matrix $\mathbf{X}$ whose rows are the training instances $\mathbf{x}_i$, and a vector $\mathbf{y}$ whose components are the corresponding labels y_i. Using (5.10), we obtain a vector $\hat{\mathbf{y}}$ containing predicted values for labels:

$$\hat{\mathbf{y}} = \mathbf{X}\mathbf{w}.$$

To choose the best weight vector $\mathbf{w}$, we minimize squared error. The error on the i-th instance is $y_i - \hat{y}_i$. The sum of squared errors can be represented as a dot product:

$$\sum_i (y_i - \hat{y}_i)^2 = (\mathbf{y} - \hat{\mathbf{y}}) \cdot (\mathbf{y} - \hat{\mathbf{y}}).$$

Let us define the error vector $\mathbf{e}$ as

$$\mathbf{e} \equiv \mathbf{y} - \hat{\mathbf{y}}.$$

We wish to minimize squared error $\mathbf{e}^{\mathrm{T}}\mathbf{e}$. To find the value of $\mathbf{w}$ that minimizes squared error, we take the derivative of $\mathbf{e}^{\mathrm{T}}\mathbf{e}$ with respect to $\mathbf{w}$, and set it to zero. This is the point at which the rules from the previous section show their worth:

$$\begin{aligned}
\mathbf{D_w}\mathbf{e}^{\mathrm{T}}\mathbf{e} &= 2\mathbf{e}^{\mathrm{T}}\mathbf{D_w}\mathbf{e} \\
&= 2\mathbf{e}^{\mathrm{T}}\mathbf{D_w}(\mathbf{y} - \mathbf{X}\mathbf{w}) \\
&= 2\mathbf{e}^{\mathrm{T}}(\mathbf{D_w}\mathbf{y} - \mathbf{D_w}\mathbf{X}\mathbf{w}) \\
&= -2\mathbf{e}^{\mathrm{T}}\mathbf{X}\mathbf{D_w}\mathbf{w} \\
&= -2\mathbf{e}^{\mathrm{T}}\mathbf{X}.
\end{aligned}$$

Substituting in the definition for $\mathbf{e}$, we obtain

$$\begin{aligned}
\mathbf{D_w}\mathbf{e}^{\mathrm{T}}\mathbf{e} &= -2(\mathbf{y} - \mathbf{X}\mathbf{w})^{\mathrm{T}}\mathbf{X} \\
&= -2\mathbf{y}^{\mathrm{T}}\mathbf{X} + 2\mathbf{w}^{\mathrm{T}}\mathbf{X}^{\mathrm{T}}\mathbf{X}.
\end{aligned}$$

Setting this to zero, we conclude

$$\mathbf{w}^{\mathrm{T}}\mathbf{X}^{\mathrm{T}}\mathbf{X} = \mathbf{y}^{\mathrm{T}}\mathbf{X}$$

$$\mathbf{w} = (\mathbf{X}^{\mathrm{T}}\mathbf{X})^{-1}\mathbf{X}^{\mathrm{T}}\mathbf{y}$$

provided that the inverse is defined.

5.3 Constrained optimization

5.3.1 Optimization

An optimization problem is the problem of finding a point that minimizes or maximizes a function $f(\mathbf{x})$, called the **objective function**. We will limit our attention to minimization. In the general case, this is not an onerous restriction, inasmuch as the problem of maximizing a function $f(\mathbf{x})$ can be expressed as the problem of minimizing $-f(\mathbf{x})$. To be sure, we also limit attention (for the most part) to convex functions, and if $f(\mathbf{x})$ is convex, then $-f(\mathbf{x})$ is not convex, it is concave; so the restriction to minimization is a limitation after all. But then again, with a few uninteresting exceptions, convex functions are unbounded – they have no maxima – so it is sufficient to be able to minimize convex functions or maximize concave functions (by minimizing their negatives).

The standard method for minimizing a function is to find the point at which its gradient is equal to 0, as we did in the previous section when minimizing squared error in linear regression. As we have seen, the gradient at $\mathbf{x}$ gives the rate of increase of the function from $\mathbf{x}$ in each of the axis directions. If the function is not increasing in any direction, one is either at a maximum or a minimum (or a saddle point: maximum in some directions and minimum in others).

In this section, we consider optimization under constraints. Constraints are of two sorts: equality constraints and inequality constraints. An equality constraint is an equation:

$$g(x) = c$$

which must be satisfied. We limit our attention to equality constraints of the form:

$$g(x) = 0.$$

This does not represent a real limitation, as any constraint $g(x) = c$ can be written:

$$g(x) - c = 0.$$

An inequality constraint takes the form:

$$h(x) \leq 0.$$

A constraint $h(x) \leq c$ can be written as $h(x) - c \leq 0$, and a constraint $h(x) \geq c$ can be written as $c - h(x) \leq 0$.

The goal in constrained optimization is not to find the global minimum of the objective function f, but rather to find the point *among those satisfying the constraints* where f is at a minimum. The points satisfying the constraints are known as the **feasible points**.

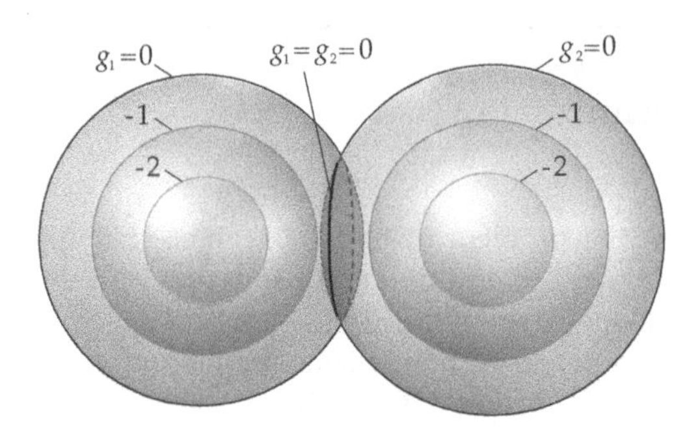

FIGURE 5.12
Intersection of two constraint surfaces.

5.3.2 Equality constraints

Let us consider a case with a three-dimensional domain and two equality constraints:

$$g_1(\mathbf{x}) = 0 \qquad g_2(\mathbf{x}) = 0 \tag{5.11}$$

with $\mathbf{x} \in \mathbf{R}^3$. The "level curves" of the function g_1 are actually level surfaces, typically forming nested shells of the sort that we already saw in figure 5.7, and similarly for g_2. The two sets of nested shells are shown in figure 5.12. (The minimum points for g_1 and g_2 are not shown, but each lies in the center of its set of nested shells.) Each constraint in (5.11) picks out one of the shells. That shell, which we call the **constraint surface**, is a 2-dimensional surface in 3-space. The feasible set is the intersection of the two constraint surfaces, a 1-dimensional curve. It is the dark circle in figure 5.12.

At each point $\mathbf{x}$ in the figure, there is a gradient $\nabla g_1(\mathbf{x})$ that points perpendicularly out from the surface of the g_1 shell at that point, and there is also a gradient $\nabla g_2(\mathbf{x})$ pointing perpendicularly out from the surface of the g_2 shell at the same point; the two gradients usually point in different directions. Figure 5.13 illustrates for a point $\mathbf{x}$ lying in the curve C that represents the feasible set, where both $g_1 = 0$ and $g_2 = 0$.

In three-dimensional space, there are many vectors that are perpendicular to a one-dimensional curve like C. Together, they form a normal plane that looks like a collar that the curve passes through perpendicularly. An example can be seen in figure 5.14. Every vector in the normal plane is perpendicular to the curve at point $\mathbf{x}$. To define a plane, only two (non-parallel) vectors are required. Any two non-parallel vectors that are perpendicular to the curve C at point $\mathbf{x}$ suffice to define the plane perpendicular to C.

Since C is the intersection of the two surfaces $g_1(\mathbf{x}) = 0$ and $g_2(\mathbf{x}) = 0$, we know that the gradients $\nabla g_1(\mathbf{x})$ and $\nabla g_2(\mathbf{x})$, which are perpendicular to

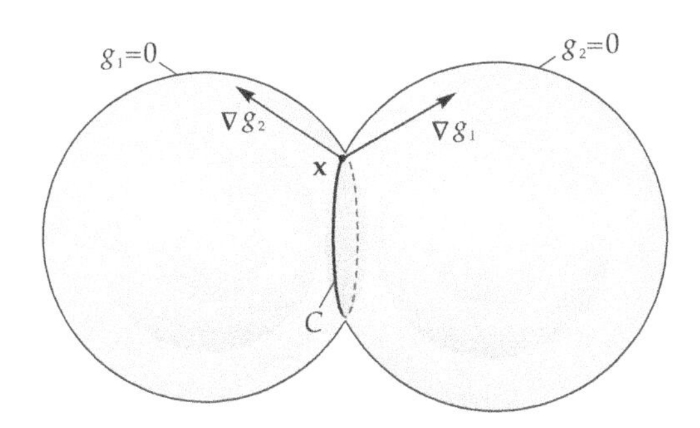

FIGURE 5.13
The gradient $\nabla g_1(\mathbf{x})$ is perpendicular to the surface $g_1 = 0$ at point $\mathbf{x}$, and $\nabla g_2(\mathbf{x})$ is perpendicular to the surface $g_2 = 0$ at point $\mathbf{x}$. Since point $\mathbf{x}$ lies both on the surface $g_1 = 0$ and on the surface $g_2 = 0$, it lies in the curve C that is the intersection of the two surfaces.

their respective constraint surfaces, are both perpendicular to C. Hence they can serve as the two vectors that define the plane perpendicular to C at $\mathbf{x}$. In short, we can define the plane perpendicular to C at $\mathbf{x}$ as the set of vectors of form:

$$\alpha_1 \nabla g_1(\mathbf{x}) + \alpha_2 \nabla g_2(\mathbf{x})$$

in which α_1 and α_2 are allowed to take on any real value.

In the case of an arbitrary number of constraints, in an arbitrary-dimensional space, we have multiple surfaces $g_i(\mathbf{x}) = 0$, each with a normal vector $\nabla g_i(\mathbf{x})$. The feasible set is the intersection C of the constraint surfaces. The surface perpendicular to C at point $\mathbf{x}$ consists of all vectors of form:

$$\sum_i \alpha_i \nabla g_i(\mathbf{x}).$$

The objective function $f(\mathbf{x})$ has the same domain as the constraint functions. In the example we have been using, it is $\mathbf{R}^3$. The "level curves" of f are two-dimensional surfaces in $\mathbf{R}^3$, and the gradient $\nabla f(\mathbf{x})$ is the vector normal to the level surface at $\mathbf{x}$. It points away from the minimizer of f.

In constrained optimization, we are constrained to stay within the feasible set. We can view the problem as one of moving along the feasible-set curve C, looking for a point that minimizes the value of f. That point will not in general be the global minimizer, but will be the "closest" to it that we can reach without moving off the feasible curve.

Consider any point $\mathbf{x}$ on the feasible-set curve C (figure 5.14). There is a normal plane to C at $\mathbf{x}$, representing the set of vectors perpendicular to the

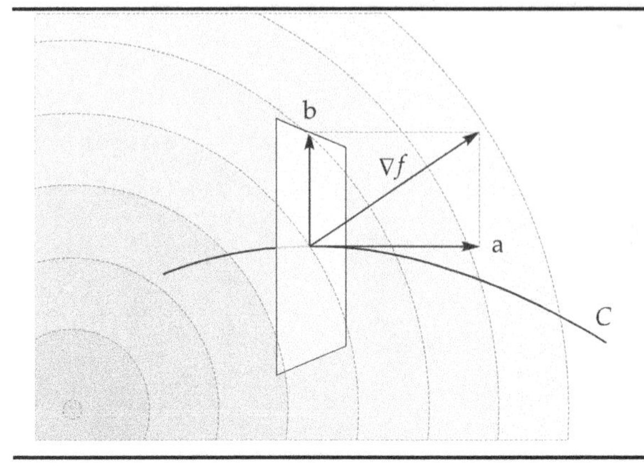

FIGURE 5.14
A point on the feasible-set curve C. The nested shells are the level curves of f. The gradient ∇f consists of a component $\mathbf{a}$ in the direction of the tangent to C and a component $\mathbf{b}$ in a direction lying in the plane that is normal to C.

curve. In most cases, the level surface of f intersects C at an angle, as shown in the figure. Then we can view the gradient of f as the sum of two vectors. One of them is perpendicular to C, and one of them points along C in the direction in which f is increasing. To minimize f, we should move in the opposite direction, in the direction along the curve in which f is decreasing.

We have reached a minimum point when there is no direction of further decrease (figure 5.15). At that point, the gradient of f lies entirely in the plane that is perpendicular to the feasible-set curve. That is, at the constrained minimum $\mathbf{x}$, we have:

$$\nabla f(\mathbf{x}) = \sum_i \alpha_i \nabla g_i(\mathbf{x}). \tag{5.12}$$

If this equation can be solved for $\mathbf{x}$, we have solved the constrained optimization problem.

The constrained minimization of f under equality constraints $g_i(\mathbf{x}) = 0$ is equivalent to an unconstrained minimization problem. The **Lagrangian** of the constrained minimization is the function:

$$L(\mathbf{x}) = f(\mathbf{x}) - \sum_i \alpha_i g_i(\mathbf{x}). \tag{5.13}$$

To do an *unconstrained* minimization of the Lagrangian, we set its gradient to 0 and solve for $\mathbf{x}$:

$$\nabla L(\mathbf{x}) = 0.$$

For any component x_k, we have

$$\frac{\partial}{\partial x_k}\left[f(\mathbf{x}) - \sum_i \alpha_i g_i(\mathbf{x})\right] = \frac{\partial}{\partial x_k} f(\mathbf{x}) - \sum_i \alpha_i \frac{\partial}{\partial x_k} g_i(\mathbf{x}).$$

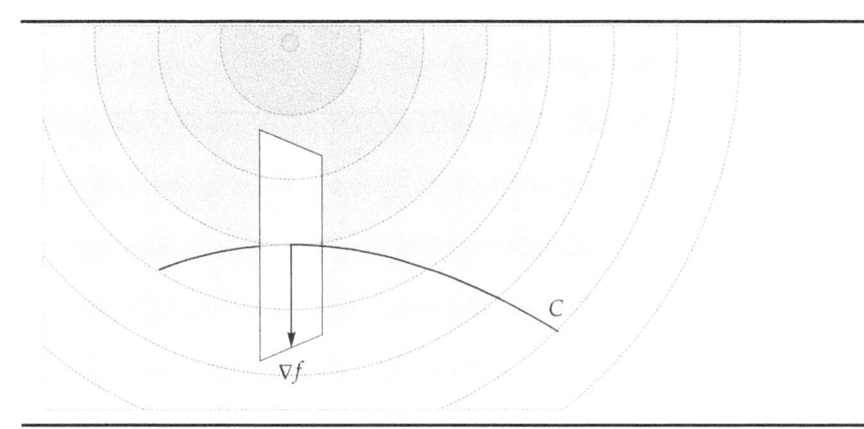

FIGURE 5.15
At the constrained minimum, the gradient ∇f lies completely in the normal plane.

So the gradient, being the vector of partial derivatives, can be expressed as

$$\nabla L(\mathbf{x}) = \nabla f(\mathbf{x}) - \sum_i \alpha_i \nabla g_i(\mathbf{x}).$$

Setting the gradient to zero leads us to the same equation (5.12) that we solved for the constrained minimization of f. That is, unconstrained minimization of the Lagrangian is equivalent to (has the same solution as) constrained minimization of the original objective function.

5.3.3 Inequality constraints

Recall that an inequality constraint $g(x) \leq 0$ defines a convex set whose boundary is the level surface $g(x) = 0$. The set of feasible points is the intersection of the convex sets defined by the constraints. For example, the shaded area in figure 5.16 shows the feasible points in a case with two constraints $g_1 \leq 0$ and $g_2 \leq 0$. The goal is to minimize a function f within the feasible set.

Now consider an example in which f has a three-dimensional domain, and there are two inequality constraints $g_1 \leq 0$ and $g_2 \leq 0$. Each constraint describes a convex solid in the domain. The feasible set is the intersection of the two convex solids, a lentil-shaped solid (figure 5.17). We wish to minimize the value of f, while remaining within the feasible set.

For this example, there are three cases that can arise.

Case 1. The global minimizer of f lies within the feasible set, as in figure 5.17a. Then we can simply ignore the constraints. Solving the unconstrained

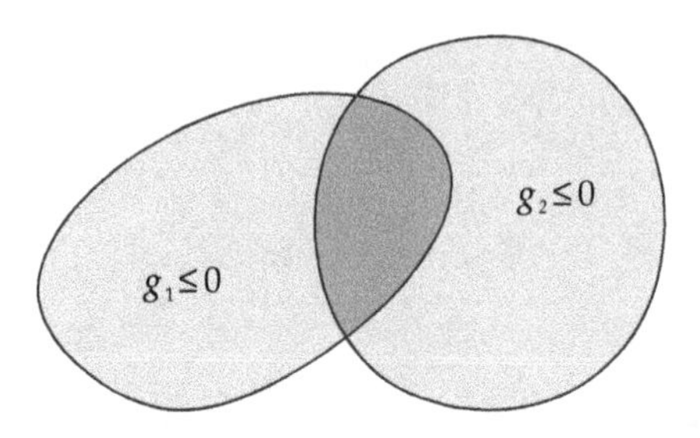

FIGURE 5.16
The intersection of two inequality constraints. The feasible set is the shaded region.

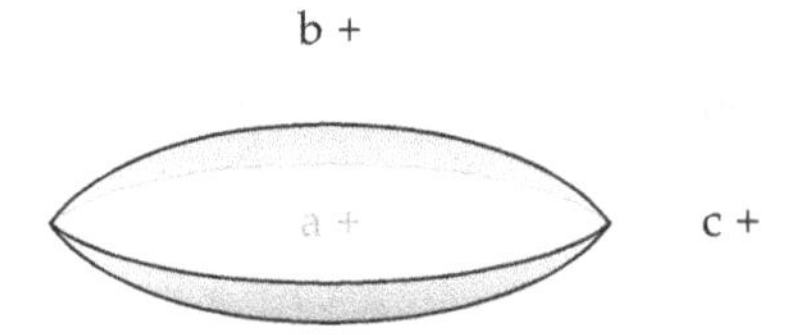

FIGURE 5.17
Three cases: (a) minimum of f within feasible set, (b) minimum of f near surface $g_1 = 0$, (c) minimum of f near curve $g_1 = g_2 = 0$.

problem finds the global minimizer, which is also the solution to the constrained problem.

Case 2. The global minimizer lies outside the feasible set, but nearer to one of the two surfaces of the "lentil" than to its edge. Suppose the minimizer lies near the g_1 surface (figure 5.17b). Let us pick any point within the feasible set, and follow the gradient of f. The gradient points away from the global minimizer of f, so we move "upstream," against the direction of the gradient. If we start at a point in the interior of the feasible set, following the gradient takes us to the surface nearest to the global minimizer. We are not allowed to move out of the feasible set, so we move along the surface, as long as there is a component of ∇f that points along the surface. When we reach a point where ∇f is entirely perpendicular to the surface, we have found the point that minimizes f within the feasible set.

In this case, the g_1 constraint matters, but we can ignore the g_2 constraint. The g_1 constraint is said to be **binding**, and the g_2 constraint is **slack**. The binding constraint behaves like an equality constraint: we follow the constraint surface $g_1(\mathbf{x}) = 0$ until we find a point satisfying $\nabla f = \alpha_1 g_1(\mathbf{x})$. Notice that this is just (5.12) where we limit our attention to binding constraints.

Alternatively, we can describe the solution to the problem with inequality constraints as being the solution to (5.12) where we set $\alpha_i = 0$ for each slack constraint. This description is also valid for Case 1. In Case 1, where the global minimizer lies in the interior of the lentil, all of the constraints are slack, and the solution is found by solving $\nabla f = 0$, which is to say, by solving (5.12) where $\alpha_i = 0$ for all i.

Case 3. The final case is illustrated in figure 5.17c. Here the global minimizer is near the edge of the lentil. Following the gradient of f leads to the edge, and we follow the gradient along the edge until we find the point that satisfies (5.12). Note that the edge is the curve that satisfies both $g_1(\mathbf{x}) = 0$ and $g_2(\mathbf{x}) = 0$. In this case, the problem is equivalent to a problem in which all the inequalities are replaced with equalities. None of the constraints are slack.

We conclude that a solution to the constrained problem is a point that satisfies (5.12), where $\alpha_i = 0$ for all slack constraints. Notice that binding constraints are ones that satisfy $g(\mathbf{x}) = 0$ at the solution point $\mathbf{x}$. Hence we can express the requirement that $\alpha_i = 0$ for slack constraints by requiring that either $g_i(\mathbf{x}) = 0$ (the constraint is binding) or $\alpha_i = 0$. Even more compactly, this can be expressed as a set of equations known as the **Kuhn-Tucker conditions**:

$$\alpha_i g_i(\mathbf{x}) = 0 \qquad \text{for all } i. \tag{5.14}$$

In short, a solution to the constrained problem is a point that satisfies both (5.12) and (5.14).

When the constraints are inequality constraints, we actually know a little more about the coefficients α_i. Let us say that a point $\mathbf{x}$ is a boundary point for constraint g_i if $g_i(\mathbf{x}) = 0$. Every point on the boundary of the feasible set is a boundary point for at least one constraint. If $\mathbf{x}$ is a boundary point for g_i, then the gradient of g_i points away from the interior of the feasible set at $\mathbf{x}$.

Now suppose that there exist binding constraints and $\mathbf{x}$ is the constrained minimum. Since binding constraints exist, the minimum of f lies outside of the feasible set, hence the gradient of f at $\mathbf{x}$ points toward the interior of the feasible set. The binding constraints are the ones that have $\mathbf{x}$ on their boundary, and their gradients point away from the interior, that is, in the opposite direction of ∇f. It follows that $\alpha_i < 0$ in (5.12) for every binding constraint, and we know that $\alpha_i = 0$ for every slack constraint; hence

$$\alpha_i \leq 0 \qquad \forall i. \tag{5.15}$$

Let us sum up. The problem of interest is

$$\text{minimize } f(\mathbf{x}) \qquad \text{subject to } g_i(\mathbf{x}) \leq 0 \quad \forall i. \tag{5.16}$$

It is conventional to reverse the signs of the Lagrange multipliers, and write the Lagrangian as

$$L(\mathbf{x}) = f(\mathbf{x}) + \sum_i a_i g_i(\mathbf{x}) \tag{5.17}$$

where $a_i = -\alpha_i$. With the reversal of signs, (5.15) becomes

$$a_i \geq 0 \qquad \forall i. \tag{5.18}$$

The solution for (5.16) is the same as the solution for

$$\text{minimize } L(\mathbf{x}) \tag{5.19}$$

with no constraints. Namely, the solution is a point $\mathbf{x}*$ that satisfies

$$\nabla f(\mathbf{x}*) + \sum_i a_i \nabla g_i(\mathbf{x}*) = 0. \tag{5.20}$$

We also know that the following is true of the solution point $\mathbf{x}*$:

$$a_i g_i(\mathbf{x}*) = 0 \qquad \forall i. \tag{5.21}$$

A solution satisfying (5.20), (5.21), and (5.18) exists provided that the feasible set is nonempty.

5.3.4 The Wolfe dual

Let us define $\mathbf{a}$ as the vector $(a_1, \ldots, a_k)$, and think of the Lagrangian as a function of both $\mathbf{x}$ and $\mathbf{a}$:

$$L(\mathbf{x}, \mathbf{a}) = f(\mathbf{x}) + \sum_i a_i g_i(\mathbf{x}).$$

The solution for the constrained problem (5.16) is the same as the solution for the unconstrained problem (5.19). Let $\mathbf{x}*$ be the solution to (5.19) and let $\mathbf{a}*$ be the corresponding Lagrange multipliers. (Properly speaking, they are negatives of Lagrange multipliers, but we will use the term a little loosely.) The solution points $\mathbf{x}*$ and $\mathbf{a}*$ satisfy

$$L(\mathbf{x}*, \mathbf{a}*) = \min_{\mathbf{x},\mathbf{a}} L(\mathbf{x}, \mathbf{a}).$$

We have just seen that we can find $\mathbf{x}*$ by setting the gradient of the Lagrangian to zero, that is, by solving (5.20) for $\mathbf{x}*$. But it is possible that solving (5.20) fails to eliminate the Lagrange multipliers a_i. That is, solving (5.20) may give us an expression for $\mathbf{x}*$ as a function of $\mathbf{a}$. The basic problem is that the Lagrangian is a function of both $\mathbf{x}$ and $\mathbf{a}$, but solving (5.20) only minimizes for $\mathbf{x}$. Solving (5.20) actually yields a point $\mathbf{x}'$ such that, for a fixed value of $\mathbf{a}$:

$$L(\mathbf{x}', \mathbf{a}) = \min_{\mathbf{x}} L(\mathbf{x}, \mathbf{a}).$$

Substituting the solution $\mathbf{x}'$ back into the Lagrangian yields the value of the Lagrangian at $\mathbf{x}'$, as a function of $\mathbf{a}$:

$$Q(\mathbf{a}) \equiv L(\mathbf{x}', \mathbf{a}) = \min_{\mathbf{x}} L(\mathbf{x}, \mathbf{a}).$$

The question now is how to eliminate the dependence on $\mathbf{a}$.

We know from (5.18) that all components of $\mathbf{a}*$ are nonnegative, so let us limit attention to points satisfying

$$a_i \geq 0 \qquad \forall i.$$

Then $a_i \geq 0$ and $g_i(\mathbf{x}) \leq 0$ for all i, and it follows that

$$Q(\mathbf{a}) = \min_{\mathbf{x}} \left[f(\mathbf{x}) + \sum_i a_i g_i(\mathbf{x}) \right] \leq L(\mathbf{x}*, \mathbf{a}).$$

Clearly, we have equality at the point $\mathbf{a}*$:

$$Q(\mathbf{a}*) = L(\mathbf{x}*, \mathbf{a}*).$$

Hence, we can eliminate the remaining dependency on $\mathbf{a}$ by *maximizing* $Q(\mathbf{a})$ to find $\mathbf{a}*$. That is, we solve the problem:

$$\text{maximize } Q(\mathbf{a}) \qquad \text{subject to } a_i \geq 0 \quad \forall i. \tag{5.22}$$

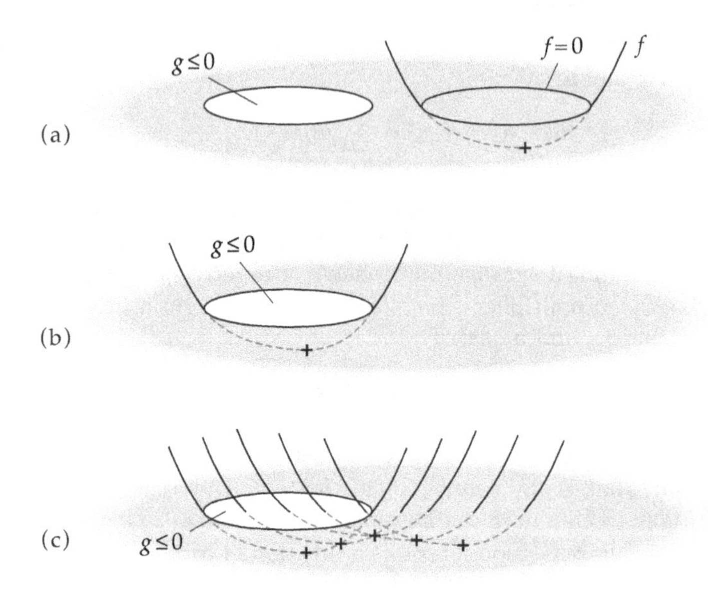

FIGURE 5.18
Mixing f and a single constraint g. (a) When $\|\mathbf{a}\| = 0$, the Lagrangian equals f. (b) When $\|\mathbf{a}\|$ is very large, the Lagrangian approximately equals g. (c) As $\|\mathbf{a}\|$ increases from 0 to infinity, the minimum of the Lagrangian rises and then falls, reaching its maximum at the boundary of the feasible set.

This problem is called the **Wolfe dual** of the constrained problem (5.16).

Geometrically, this is what we are doing. Let us start out with $\mathbf{a} = \mathbf{0}$. Then the Lagrangian is equal to $f(\mathbf{x})$, and minimizing the Lagrangian finds the minimum of f. In the cases of interest, this lies outside of the feasible set, as illustrated in figure 5.18a. For simplicity, we only show the case of a single constraint g. Also note that figure 5.18, unlike the other figures in this section, shows the graph space, using the z axis for the values of f and g. The domain is the plane $z = 0$.

Increasing the magnitude of $\mathbf{a}$ changes the shape of the Lagrangian by mixing in more and more of the g shape. When $\|\mathbf{a}\|$ is very large, $g(\mathbf{x})$ overwhelms $f(\mathbf{x})$ in the Lagrangian, and the minimizer of the Lagrangian falls inside the feasible set (figure 5.18b). By "the minimizer of the Lagrangian," we mean the point $\mathbf{x}'$ such that

$$L(\mathbf{x}', \mathbf{a}) = \min_{\mathbf{x}} L(\mathbf{x}, \mathbf{a}) = Q(\mathbf{a})$$

for the particular choice of $\mathbf{a}$.

The discussion leading up to the Wolfe dual (5.22) shows that, as we increase $\|\mathbf{a}\|$, and $\mathbf{x}'$ moves from the exterior of the feasible set to the interior, the value $L(\mathbf{x}', \mathbf{a})$ first increases, and then falls. It reaches its maximum exactly when $\mathbf{x}'$ lies on the boundary of the feasible set, which is when $\mathbf{x}' = \mathbf{x}*$ (figure 5.18c).

6

Boundary-Oriented Methods

In both clustering and semisupervised learning, a broad distinction can be made between approaches that build on the **cluster hypothesis** – the idea that similar examples should have similar labels, or, in spatial terms, that nearby examples should belong to the same class – and the **separation hypothesis** – the idea that boundaries between classes should lie in sparsely populated regions of the instance space. Approaches that build on the cluster hypothesis seek to characterize what clusters look like. Approaches that build on the separation hypothesis are not concerned with what clusters look like, but rather with what boundary regions look like.

A related distinction is that between generative models and discriminative models. A generative model describes how instances of each class are generated. As we will see in chapter 7, one way of implementing the cluster hypothesis is via a mixture model: choose a class with probability $p(y)$, then generate an instance with probability $p(\mathbf{x}|y)$. A probabilistic classifier $p(y|\mathbf{x})$ arises as a posterior probability, derived by Bayes' rule:

$$p(y|\mathbf{x}) = \frac{p(y)p(\mathbf{x}|y)}{\sum_{y'} p(y')p(\mathbf{x}|y')}.$$

By contrast, a discriminative model does not specify how to generate instances, but constructs the classifier $\hat{y} = g(\mathbf{x})$ directly, by specifying how to divide up the space into class regions. One way of doing so is by implementing the separation hypothesis. Of course, any classifier divides up the instance space into class regions; the issue is whether it does so directly, rather than indirectly by defining a model to generate data points.

Vapnik [227] has prominently advocated the use of discriminative models on the methodological grounds that one should not solve a harder problem than necessary. Estimating the law governing the distribution of data points, he argues, is harder than determining the boundaries between classes. Doing it well requires more data, a particular issue when labeled data is scarce.

Let us consider a concrete example: the classification problem illustrated in figure 6.1. There are two classes, positive and negative. The methods we will consider all construct linear boundaries, which is to say, hyperplanes. The hyperplane separates the instance space into two half-spaces, one for each class.

In "well-behaved" cases, as here, the instances belonging to each class form two clusters with a clear separation between them. A linear boundary sepa-

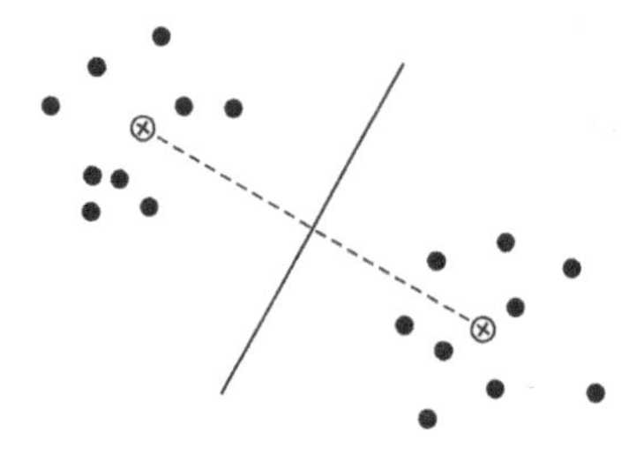

FIGURE 6.1
A linear separator between the positive and negative classes.

rates them cleanly. Less well-behaved cases are possible, in which, no matter how we split the space, some examples will be on the "wrong side" of the separating hyperplane. In the well-behaved case, the points are said to be **linearly separable**. Boundary-oriented methods are often based on intuitions that are clearest in the linearly separable case.

We have already seen that some generative methods (specifically, the Naive Bayes algorithm) construct linear separators. A related generative method is **linear discriminant analysis**. In linear discriminant analysis, it is assumed that the data has a Gaussian (that is, normal) distribution. Restricting attention to the simple case in which there is no correlation among the attributes, and the data is evenly divided between the two classes, the assumption of normality means that the data form two equal-sized circular clusters, as in figure 6.1. In that case, it is optimal (and intuitively natural) to choose as linear separator the bisector of the line segment connecting the centers of gravity of the two clusters, as shown.

However, the linear separator of figure 6.1 requires us to know something about how the data points are distributed. It is optimal only if the assumptions about data distribution are met. If the data distribution is not normal, but heavily skewed, as in figure 6.2, linear discriminant analysis gives less than perfect results. In figure 6.2a, bisecting the line connecting the centers of gravity fails to correctly separate the two classes. But the failure is not due to inseparability: the alternative linear separator in figure 6.2b does correctly classify all the data.

Following Vapnik's methodological precept, discriminative or boundary-oriented methods avoid assumptions about the distribution of the points. They focus on finding linear separators like that in figure 6.2b. That is, they seek linear gaps in the data, regardless of how the data are distributed outside of the boundary region.

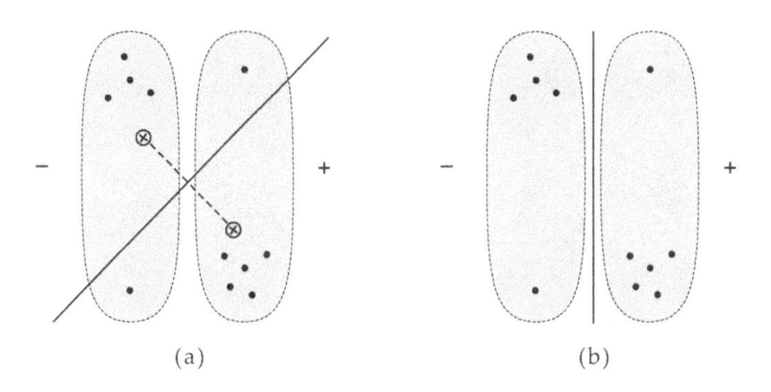

FIGURE 6.2
(a) A linear separator constructed in the same way, but where the data are heavily skewed. (b) A better linear separator for this data.

6.1 The perceptron

6.1.1 The algorithm

The perceptron learning algorithm is a very simple algorithm for directly constructing a linear separator. As discussed in the previous chapter, the separator is represented by its normal vector. The perceptron algorithm begins with a random vector $\mathbf{w}$ (or with the zero vector), and cycles repeatedly through the training data, looking for instances that are misclassified. If we find a misclassified instance $\mathbf{x}$, we update $\mathbf{w}$ in a direction that will get it closer to making the correct prediction for $\mathbf{x}$.

To be more specific, suppose there is a positively labeled example $\mathbf{x}$ that is misclassified, as in figure 6.3. The point $\mathbf{x}$ is on the wrong side of the separator (the negative side). There is an arrow running from the tip of $\mathbf{w}$ to $\mathbf{x}$; this is the vector $\mathbf{x} - \mathbf{w}$. This is the direction in which we would like to move $\mathbf{w}$. For example, the dotted vector and dotted separator are the result if we move halfway along the $\mathbf{x} - \mathbf{w}$ vector; that would suffice to classify $\mathbf{x}$ correctly.

This suggests the update:

$$\begin{aligned}\mathbf{w} &\leftarrow \mathbf{w} + \eta(\mathbf{x} - \mathbf{w}) \\ &= (1 - \eta)\mathbf{w} + \eta\mathbf{x}\end{aligned}$$

where η is a step size. For example, to move halfway along the $\mathbf{x} - \mathbf{w}$ vector, set η to 1/2. If we choose $\eta = 1/2$, we have

$$\mathbf{w} \leftarrow \frac{1}{2}(\mathbf{w} + \mathbf{x}).$$

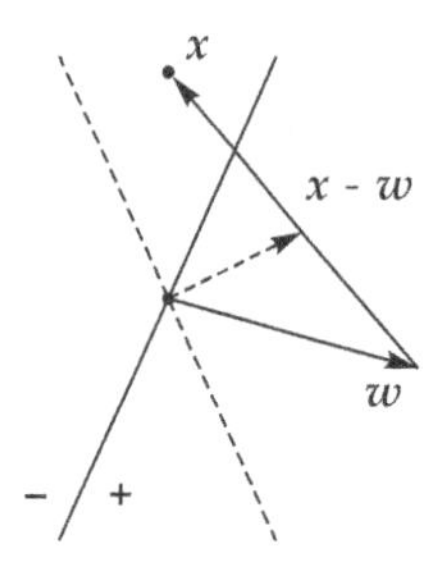

FIGURE 6.3
Adjusting for a misclassified point.

As we observed at the end of section 5.1.1, scaling $\mathbf{w}$ does not affect the location of the separator, only the unit we use to measure distance, so we can scale by two to obtain the simpler form:

$$\mathbf{w} \leftarrow \mathbf{w} + \mathbf{x}.$$

This is the update rule used by the perceptron algorithm. One cycles through the training data, and each time a misclassified example is encountered, one adds it into the weight vector $\mathbf{w}$. The process continues until no examples are misclassified.

However, our discussion so far has only considered positive examples; we must also consider examples with negative labels. There is a convenient trick that allows us to reduce the negative case to the positive case. A misclassified positive example is one for which

$$\mathbf{x} \cdot \mathbf{w} \leq 0.$$

A misclassified negative example is a negative instance for which $\mathbf{x} \cdot \mathbf{w} \geq 0$, which is to say, for which

$$-\mathbf{x} \cdot \mathbf{w} \leq 0.$$

If we replace each negatively labeled instance $\mathbf{x}$ with the negative vector $-\mathbf{x}$, then the criterion is the same for all instances. That is, a negatively labeled instance $\mathbf{x}$ is indistinguishable from a positively labeled instance $-\mathbf{x}$.

Geometrically, taking the negative of a vector reflects it across the origin. Figure 6.4 illustrates. The negatively labeled instances in figure 6.4a are reflected across the origin to obtain figure 6.4b, in which we do not need to distinguish between positive and negative points. We need only find a separator passing through the origin that has all the data points on its positive side.

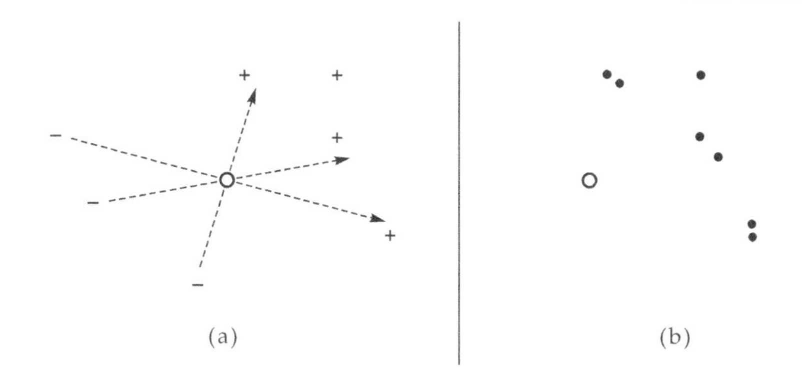

FIGURE 6.4
(a) Reflecting negatively labeled examples across the origin. (b) The result. The goal is now to find a linear separator through the origin that has all instances on one side.

6.1.2 An example

Consider the following data points, characterized by two features F and G; the variable Y represents the label:

	F	G	Y
$\mathbf{x}_1$	-1	2	$-$
$\mathbf{x}_2$	3	0	$-$
$\mathbf{x}_3$	0	4	$+$
$\mathbf{x}_4$	1	5	$+$
$\mathbf{x}_5$	2	2	$+$

We first modify the data points by adding a -1 component:

$$\begin{array}{lll} \mathbf{x}_1 & (-1,2,-1) & - \\ \mathbf{x}_2 & (3,0,-1) & - \\ \mathbf{x}_3 & (0,4,-1) & + \\ \mathbf{x}_4 & (1,5,-1) & + \\ \mathbf{x}_5 & (2,2,-1) & + \end{array}.$$

Then we fold the label into the instance:

$$\begin{array}{ll} \mathbf{x}_1 & (1,-2,1) \\ \mathbf{x}_2 & (-3,0,1) \\ \mathbf{x}_3 & (0,4,-1) \\ \mathbf{x}_4 & (1,5,-1) \\ \mathbf{x}_5 & (2,2,-1) \end{array}.$$

We start the algorithm with a zero weight vector $\mathbf{w} = (0,0,0)$, and begin cycling through the instances $\mathbf{x}_i$. At each point, we take the dot product

$\mathbf{x}_i \cdot \mathbf{w}$. If the result is not positive, we add $\mathbf{x}_i$ into $\mathbf{w}$; otherwise we leave $\mathbf{w}$ unchanged. When we have passed through all the instances, we go back to $\mathbf{x}_1$ and pass through them again. Here is the resulting history of updates:

$\mathbf{w}$	$\mathbf{x}_i$	$\mathbf{x}_i \cdot \mathbf{w}$	$\mathbf{w}$	$\mathbf{x}_i$	$\mathbf{x}_i \cdot \mathbf{w}$
$(0,0,0)$	$(1,-2,1)$	0	$(-2,2,3)$	$(1,5,-1)$	$+5$
$(1,-2,1)$	$(-3,0,1)$	-2	$(-2,2,3)$	$(2,2,-1)$	-3
$(-2,-2,2)$	$(0,4,-1)$	-10	$(0,4,2)$	$(1,-2,1)$	-6
$(-2,2,1)$	$(1,5,-1)$	$+7$	$(1,2,3)$	$(-3,0,1)$	0
$(-2,2,1)$	$(2,2,-1)$	-1	$(-2,2,4)$	$(0,4,-1)$	$+4$
$(0,4,0)$	$(1,-2,1)$	-8	$(-2,2,4)$	$(1,5,-1)$	$+4$
$(1,2,1)$	$(-3,0,1)$	-2	$(-2,2,4)$	$(2,2,-1)$	-4
$(-2,2,2)$	$(0,4,-1)$	$+6$	$(0,4,3)$	$(1,-2,1)$	-5
$(-2,2,2)$	$(1,5,-1)$	$+6$	$(1,2,4)$	$(-3,0,1)$	$+1$
$(-2,2,2)$	$(2,2,-1)$	-2	$(1,2,4)$	$(0,4,-1)$	$+4$
$(0,4,1)$	$(1,-2,1)$	-7	$(1,2,4)$	$(1,5,-1)$	$+7$
$(1,2,2)$	$(-3,0,1)$	-1	$(1,2,4)$	$(2,2,-1)$	$+2$
$(-2,2,3)$	$(0,4,-1)$	$+5$	$(1,2,4)$	$(1,-2,1)$	$+1$

At this point, all five examples are correctly classified, and we are done. In terms of the original data points (attributes F and G), the prediction is

$$f(\mathbf{x}) = \mathbf{x} \cdot (1,2) - 4.$$

We can confirm that it correctly classifies all the original examples:

$\mathbf{x}$	$f(\mathbf{x})$	Y
$(-1,2)$	-1	$-$
$(3,0)$	-1	$-$
$(0,4)$	$+4$	$+$
$(1,5)$	$+7$	$+$
$(2,2)$	$+2$	$+$

6.1.3 Convergence

We wrote above that "the process continues until no examples are misclassified." How do we know that we will ever reach that point?

It is actually possible that no solution exists. It is certainly possible to mix together positive and negative examples so that no linear boundary can separate them. If any linear separator exists that correctly classifies the training data, the data is said to be **linearly separable**.

If the data is linearly separable, it can be shown that the perceptron algorithm will converge to a solution. To see why that is so, let us consider figure 6.5. The data points are the same as in figure 6.4b, but we have added two separators b_1 and b_2, along with their normal vectors w_1 and w_2, respectively. Both b_1 and b_2 are linear boundaries that classify all the data correctly. That

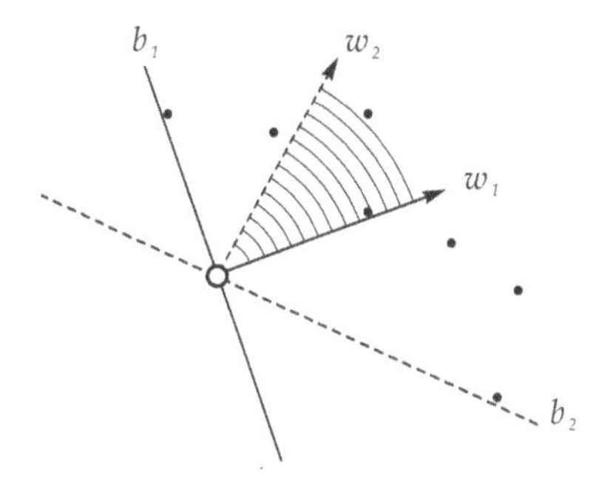

FIGURE 6.5
Two solutions.

is, w_1 and w_2 are solutions to the perceptron learning problem. The separator b_1 is special because it cannot be rotated clockwise any further – it is bumping up against one of the training examples. Similarly, the separator b_2 cannot be rotated counterclockwise any further. As a consequence, all the solution vectors lie in the shaded region between w_1 and w_2.

Now let us consider the "average" solution vector, that points straight down the middle of the solution region, halfway between w_1 and w_2. Call it w_{ave}. Let us consider what happens in one step of the perceptron algorithm (figure 6.6). There is a misclassified instance x, and we update the current normal vector w (not shown) by adding the vector $\mathbf{x}$ to it. The vector $\mathbf{x}$ has a component $\mathbf{a}$ that points in the direction of $\mathbf{w}_{\text{ave}}$, and a component $\mathbf{b}$ that is perpendicular to $\mathbf{w}_{\text{ave}}$. As one cycles through different misclassified instances, the perpendicular component $\mathbf{b}$ will switch between pointing to the right or pointing to the left, causing the current normal vector to swing back and forth, but the component $\mathbf{a}$ will never reverse direction, because all the data lies on the same side of the average separator. As a result, the component $\mathbf{a}$ will accumulate. The current vector $\mathbf{w}$ will gradually increase in length, but its direction will be determined more and more by the accumulation of copies of $\mathbf{a}$, and less and less by the components $\mathbf{b}$, which cancel each other out. That is, the vector $\mathbf{w}$ will make smaller and smaller swings, coming closer and closer to $\mathbf{w}_{\text{ave}}$. Eventually, it will land in the solution region, at which point all the data is correctly classified and no more updates are made.

6.1.4 The perceptron algorithm as gradient descent

It is also possible to view the perceptron algorithm as doing a form of gradient descent. At a given point, the weight vector $\mathbf{w}$ makes an error on instance $\mathbf{x}_i$

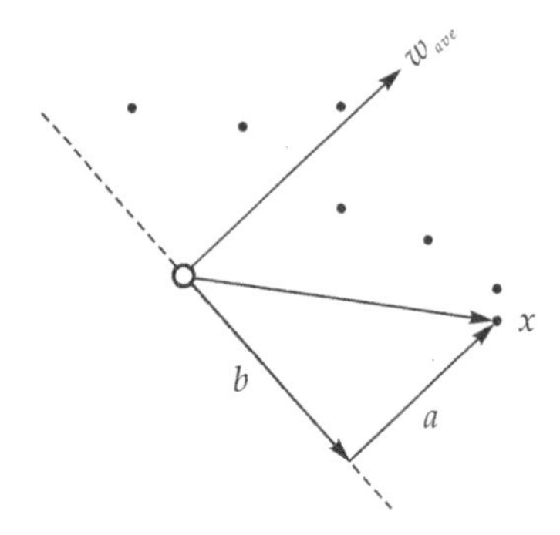

FIGURE 6.6
Each misclassified instance $\mathbf{x}$, viewed as a vector, consists of a component $\mathbf{a}$ in the direction of the average solution vector $\mathbf{w}_{\text{ave}}$ and a component $\mathbf{b}$ perpendicular to it.

just in case $\mathbf{w} \cdot \mathbf{x}_i \leq 0$. That is, for misclassified instances $\mathbf{x}_i$, the amount of error is

$$e_i = -\mathbf{w} \cdot \mathbf{x}_i.$$

We wish to find a value for $\mathbf{w}$ that minimizes the total error, or, more modestly, we wish to find a direction in which to move $\mathbf{w}$ so as to reduce error. That is, we would like to know the gradient of the error function, and move $\mathbf{w}$ in the direction opposite to the gradient.

The error and squared error are monotonically related, and dealing with squared error is more convenient. We can collect the individual errors e_i together into a vector $\mathbf{e} = (e_1, \ldots, e_m)$. We include only instances on which an error is actually made ($e_i \geq 0$). We also collect those instances together as rows of a matrix $\mathbf{X}$. We can write

$$\mathbf{e} = -\hat{\mathbf{y}} = -\mathbf{X}\mathbf{w}.$$

The total squared error is

$$E = \mathbf{e}^{\mathrm{T}}\mathbf{e}.$$

Note that this is the same as squared error in linear regression (section 5.2.4) if we take the target vector $\mathbf{y} = 0$, yielding $\mathbf{e} = \hat{\mathbf{y}}$. Taking the gradient goes the same way:

$$\begin{aligned}
\nabla E &= \mathbf{D}_{\mathbf{w}}\mathbf{e}^{\mathrm{T}}\mathbf{e} \\
&= 2\mathbf{e}^{\mathrm{T}}\mathbf{D}_{\mathbf{w}}\mathbf{e} \\
&= 2\mathbf{e}^{\mathrm{T}}\mathbf{D}_{\mathbf{w}}(-\mathbf{X}\mathbf{w}) \\
&= -2\mathbf{e}^{\mathrm{T}}\mathbf{X}.
\end{aligned}$$

To decrease E, we wish to move $\mathbf{w}$ in the direction $-\nabla E$. We will actually use $-(1/2)\nabla E$, which is a vector pointing in the same direction:

$$\mathbf{w} \leftarrow \mathbf{w} + \eta \mathbf{e}^{\mathrm{T}} \mathbf{X}$$

where η is the "learning rate" or step size.

The product $\mathbf{e}^{\mathrm{T}}\mathbf{X}$ is a row vector times a matrix whose rows are instances. The result is a row vector that is a linear combination of the rows of $\mathbf{X}$, in which each row $\mathbf{x}_i$ is weighted by the corresponding error e_i:

$$\mathbf{e}^{\mathrm{T}}\mathbf{X} = \sum_i e_i \mathbf{x}_i.$$

That is

$$\mathbf{w} \leftarrow \mathbf{w} + \eta \sum_i e_i \mathbf{x}_i. \tag{6.1}$$

We can factor η as $\eta = \alpha\beta$ such that the values βe_i sum to one. That makes it plain that (6.1) moves $\mathbf{w}$ in a direction that is a weighted average of the misclassified instance vectors $\mathbf{x}_i$, with step size α and weights βe_i.

In the perceptron algorithm, instead of moving $\mathbf{w}$ all at once in the direction of the weighted average, we move $\mathbf{w}$ in the direction $\mathbf{x}_i$ for each instance vector in the weighted average, one at a time:

$$\mathbf{w} \leftarrow \mathbf{w} + \eta_i e_i \mathbf{x}_i. \tag{6.2}$$

One can think of η_i as a step size that is different for different instances; the perceptron algorithm chooses $\eta_i e_i = 1$. The stepwise update (6.2) turns out to have the same effect as the simultaneous update (6.1). The stepwise update is known as **stochastic gradient descent**, because in one version of it, we choose an instance $\mathbf{x}_i$ at random according to the weight e_i, and update $\mathbf{w}$ with the randomly chosen instance. The systematic visiting scheme of the perceptron algorithm, cycling through all the instances in turn, has the same effect.

6.2 Game self-teaching

In the discussion of algorithms related to self-training, in section 2.2.5, we mentioned self-teaching in the sense of a program teaching itself to play a game by playing against itself. (There is a different setting that is also called *self-teaching*; it will be discussed in section 9.2.)

The best-known implementation of game self-teaching is Tesauro's backgammon program TD-Gammon [218], though he cites Samuel [200] for the original idea.

In TD-Gammon, the player is based on a function for evaluating a board position. For example, if $\mathbf{w}$ is a weight vector and $\mathbf{x}$ is a vector representing relevant features of the board position, $\mathbf{w}\cdot\mathbf{x}$ can be viewed as a scalar predicted board value that attempts to approximate the probability that the game will end in a win for White. Though a win for White implies a loss for Black, TD-Gammon actually uses two evaluation functions, one for White and one for Black. Also, the TD-Gammon evaluation functions actually use a multilayer perceptron rather than a single perceptron, but the principle is the same. At each step, an exhaustive search is done (every legal move is examined) to find the move that yields a board position that maximizes the probability of winning.

The TD-Gammon learner uses a record of a game to improve its evaluation functions. One can use records of games played by humans (supervised learning), but interestingly, one can also learn very effectively from games that are generated by playing the current evaluation function against itself. Learning proceeds as follows. The only feedback that the learner receives is the win or loss at the end of the game. The target of learning is an evaluation function that returns the probability that the game will end in a win. The exact value is known only at the end of the game, where it is either one (a win) or zero (a loss). The learner propagates that information backward through time, over the course of many training games.

The method of information propagation used by TD-Gammon is known as temporal difference (TD) learning. Let y_t represent the estimated win probability at the t-th time step. If each player plays perfectly, then any difference $|y_{t+1}-y_t|$ is due either to poor estimation of y_t or to the stochasticity in the roll of the dice. Hence learning takes the form of adjusting the evaluation-function weights $\mathbf{w}$ at each time step so as to minimize the difference $|y_{t+1} - y_t|$. Differences due to poor estimation will gradually be eliminated, and differences due to the roll of the dice will be averaged out. TD-Gammon starts out with a random vector $\mathbf{w}$, and improves it by repeatedly playing games and propagating the win/loss information backward through time to make adjustments to $\mathbf{w}$.

At each time t, the gradient $(\nabla y)_t$ with respect to $\mathbf{w}$ is a vector pointing in the direction in the weight space that generates the most rapid increase in y. Note that there is a different gradient at each time step: the best way to change $\mathbf{w}$ depends on the board position. The final time is T, and the value y_T is known. Working backward through time, the weights are adjusted by taking a small step in the direction that moves y_t toward y_{t+1}. If $y_t < y_{t+1}$, then $\mathbf{w}$ is moved in the direction $(\nabla y)_t$, and if $y_t > y_{t+1}$, then $\mathbf{w}$ is moved in the opposite direction. In general, the update is

$$\mathbf{w} \leftarrow \mathbf{w} + \eta(y_{t+1} - y_t)(\nabla y)_t \tag{6.3}$$

where η is the learning rate. Note the similarity to the perceptron update (6.2), viewing $y_t - y_{t+1}$ as the error of the evaluation function on the t-th board position, and viewing $\mathbf{x}_i$ as the perceptron's approximation to the negative of the gradient.

Game self-teaching as thus formulated is a semisupervised learning algorithm. Concatenating together the board positions from many games, we have training instances $(\mathbf{x}_1, \ldots, \mathbf{x}_n)$. The true values of the evaluation function are $(y_1, \ldots, y_n)$, representing the true probability of winning if both players play perfectly in the corresponding board position. The values y_t range over the real interval $[0, 1]$; that is, the problem is a regression problem rather than a classification problem. If all values were known, it would be a supervised learning problem. In fact, only some of the values are known, namely, the ones corresponding to ends of games, making the problem semisupervised.

Game self-teaching has notable similarities to self-training. In lieu of true values, the algorithm takes the predictions of its current evaluation function at face value – or more precisely, it takes the predicted value $\hat{y}_{t+1}$ at face value as the target for time t, on the assumption that the true value is the same at both t and $t+1$. In that respect, it is like self-training without a threshold. However, it also differs from self-training. The fact that game self-teaching learns a real-valued function, the time-sequence nature of the data, and the specific assumptions underlying the update equation (6.3), all distinguish the TD-Gammon version of game self-teaching from self-training.

6.3 Boosting

Boosting is another classifier learning method that constructs a linear separator. It is related to Support Vector Machines (SVMs), and has comparable performance over a wide range of problems, but is rather simpler to explain.

As with perceptrons and, recalling the discussion of section 5.1.4, as with the Naive Bayes algorithm, we take a model to be represented by a weight vector $\mathbf{w}$, and we define the confidence-weighted prediction of the classifier to be

$$f(\mathbf{x}) = \mathbf{x} \cdot \mathbf{w}. \tag{6.4}$$

The sign of $f(\mathbf{x})$ is the predicted class, and the absolute value is interpretable as confidence.

Consider a particular instance $\mathbf{x}$, and suppose that the correct label is y. We assume that y is either $+1$ or -1. If y is positive, then the classifier makes an error just in case $\mathbf{x} \cdot \mathbf{w} \leq 0$. If y is negative, then the classifier makes an error just in case $\mathbf{x} \cdot \mathbf{w} \geq 0$, or equivalently, just in case:

$$(\mathbf{x} \cdot \mathbf{w})y \leq 0.$$

Define $u = (\mathbf{x} \cdot \mathbf{w})y$. With labeled training data, $\mathbf{x}$ and y are given, and u varies as we vary $\mathbf{w}$. Figure 6.7 plots the "error function" as a function of u

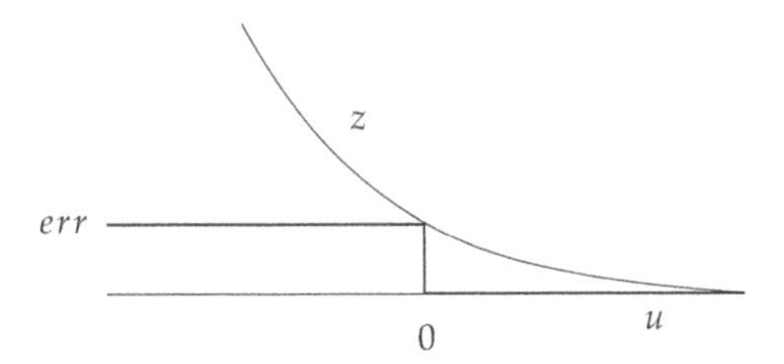

FIGURE 6.7
The error function as a function of $u = (\mathbf{x} \cdot \mathbf{w})y$.

for a single training instance. It is a step function: the number of errors is one if $u \leq 0$ and zero if $u > 0$.

We would like to minimize *err*, but minimizing it directly is difficult because it is a step function. It is much more convenient if we have a continuous function to minimize; its slope tells us which direction to move in order to improve things. The basic idea of boosting is to use an exponential upper bound for *err* as the objective function to minimize. This is also shown in figure 6.7: it is the line marked "z." It is defined as

$$z = e^{-(\mathbf{x} \cdot \mathbf{w})y}.$$

When u is zero, we have $z = e^{-u} = 1$, and the negative sign makes z decrease as u increases. We can think of z as a cost. If the current model classifies the instance correctly, then $z < 1$, and if the current model misclassifies the instance, then $z > 1$. The cost varies exponentially with confidence.

In a complete training set, there are multiple instances $\mathbf{x}_i$, each having its own cost:

$$z_i = e^{-(\mathbf{x}_i \cdot \mathbf{w})y_i}.$$

Since $err(\mathbf{x}_i) \leq z_i$ for each training instance $\mathbf{x}_i$, the total cost:

$$Z = \sum_i z_i = \sum_i e^{-(\mathbf{x}_i \cdot \mathbf{w})y_i} \tag{6.5}$$

is an upper bound for the total number of errors on the training data:

$$\sum_i \text{err}(\mathbf{x}_i) \leq Z.$$

Z is the objective function that boosting seeks to minimize.

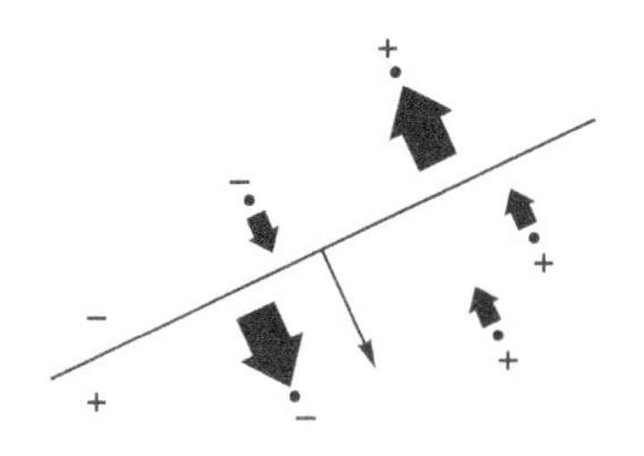

FIGURE 6.8
Costs interpreted as applying pressure to the current separator.

Geometrically, $\mathbf{x}_i \cdot \mathbf{w}$ is the distance from the linear separator to the point $\mathbf{x}_i$. The cost is sensitive only to that distance. It decreases the deeper an instance is in the "right" half-space (the side of the separator that agrees with the instance's label), and increases the deeper an instance is in the "wrong" half-space.

We can think of the cost as exerting a pressure on the separator. In a manner of speaking, correctly classified examples push the separator away in their attempt to move deeper into the right half-space, and misclassified examples pull on the separator, so as to move themselves out of the wrong half-space. The amount of pressure depends on the cost; misclassified examples always exert more pressure than correctly classified examples. Figure 6.8 illustrates. The net effect in this case is to rotate the separator clockwise.

Boosting assumes that we have a collection of "weak predictors" available. A weak predictor h takes an instance as input and makes a prediction $+1$ or -1 concerning the label. The predictor h is "weak" in the sense that its prediction may not be very good. The goal is to find a combination of weak predictors that together do a better job than any of them separately.

Alternatively, we can think of weak predictors simply as features. We represent an instance as a vector of values, one for each weak predictor. Let ω be an instance. We represent it as the vector:

$$\mathbf{x} = (h_1(\omega), \ldots, h_k(\omega)).$$

This is a vector of positive and negative ones.

The weight vector consists of a weight for each weak predictor:

$$\mathbf{w} = (w_1, \ldots, w_k).$$

The basic idea is that good predictors have larger weight and poorer predictors have smaller weight. The combined prediction is the weighted average of the weak predictions $h_i(\omega)$, in which each prediction is weighted by the corresponding weight w_i. That is, the combined prediction is simply $\mathbf{x} \cdot \mathbf{w}$, in concordance with what we have already assumed, in equation (6.4).

For example, suppose we have three instances and four weak predictors, as follows. We also include a column for the labels.

ω	$h_1(\omega)$	$h_2(\omega)$	$h_3(\omega)$	$h_4(\omega)$	y
ω_1	$+1$	$+1$	-1	-1	-1
ω_2	$+1$	-1	-1	$+1$	$+1$
ω_3	-1	$+1$	$+1$	$+1$	-1

Each row corresponds to a value for $\mathbf{x}$. For example, the first row corresponds to

$$\mathbf{x}_1 = (+1, +1, -1, -1).$$

Suppose that the current weight vector is

$$\mathbf{w} = (.5, -.1, -.2, .4).$$

Then

$$\mathbf{x}_1 \cdot \mathbf{w} = (+1)(.5) + (+1)(-.1) + (-1)(-.2) + (-1)(.4) = .2.$$

The cost for the first instance is:

$$z_1 = e^{-(.2)(-1)} = 1.22.$$

Its cost is greater than one, indicating that the instance is misclassified.

Boosting is an iterative algorithm, like the perceptron algorithm. In each iteration, we wish to pick a weak predictor h_j and adjust its weight by adding a quantity δ (which may be either positive or negative). We evaluate a candidate h_j and adjustment δ by considering what the total cost will be if we make the adjustment. The idea is to cycle through all the weak predictors, and for each one, determine the optimal adjustment δ, and the total reduction in cost achievable. We keep the one that maximizes total cost reduction, and fold δ into the weight vector. Then the process is repeated.

The new weight vector $\mathbf{w}'$ will be just like the old one $\mathbf{w}$ except that it has $w_j + \delta$ in place of w_j. For example, suppose that we choose h_2 and set $\delta = +.3$. The new weight vector will be

$$\mathbf{w}' = (.5, -.1 + .3, -.2, .4).$$

The new dot product with instance $\mathbf{x}_1$ will be

$$\begin{aligned}
\mathbf{x}_1 \cdot \mathbf{w}' &= (+1)(.5) + (+1)(-.1 + .3) + (-1)(-.2) + (-1)(.4) \\
&= [(+1)(.5) + (+1)(-.1) + (-1)(-.2) + (-1)(.4)] + (+1)(.3) \\
&= \mathbf{x}_1 \cdot \mathbf{w} + \delta h_2(\mathbf{x}_1).
\end{aligned}$$

Here $\mathbf{w}$ is the current weight vector and $\mathbf{w}'$ is what it will be if we adopt the proposed change. Also, we have written "$h_2(\mathbf{x}_1)$" instead of "$h_2(\omega_1)$." Applied to the vector $\mathbf{x}_1$, the weak predictor h_2 should be understood as simply extracting the second component.

In general, the new cost for the i-th instance will be

$$\begin{aligned} z_i' &= e^{-[\mathbf{x}_i \cdot \mathbf{w} + \delta h(\mathbf{x}_i)]y_i} \\ &= e^{-(\mathbf{x}_i \cdot \mathbf{w})y_i} e^{-\delta h(\mathbf{x}_i)y_i} \\ &= z_i e^{-\delta h(\mathbf{x}_i)y_i}. \end{aligned}$$

The total new cost Z' is

$$Z' = \sum_i z_i e^{-\delta h(\mathbf{x}_i)y_i}. \tag{6.6}$$

There is a way of expressing Z' that will be very useful. Let us define A to be the total cost of instances that the candidate predictor h gets right. The candidate predictor h classifies the i-th instance correctly if $h(\mathbf{x}_i) = +1$ and $y_i = +1$, or if both equal negative one. That is, h classifies $\mathbf{x}_i$ correctly if $h(\mathbf{x}_i)y_i = +1$. Hence we define:

$$A = \sum_i [\![h(\mathbf{x}_i)y_i = +1]\!] z_i.$$

Note that z_i is the current cost, not the "new" cost. We define B similarly to be the total cost of the remaining instances:

$$B = \sum_i [\![h(\mathbf{x}_i)y_i = -1]\!] z_i.$$

Now let us split the sum in (6.6) into a sum over the instances that h gets right and the ones it gets wrong:

$$\begin{aligned} Z' &= \sum_i [\![h(\mathbf{x}_i)y_i = +1]\!] z_i e^{-\delta h(\mathbf{x}_i)y_i} + \sum_i [\![h(\mathbf{x}_i)y_i = -1]\!] z_i e^{-\delta h(\mathbf{x}_i)y_i} \\ &= \sum_i [\![h(\mathbf{x}_i)y_i = +1]\!] z_i e^{-\delta(+1)} + \sum_i [\![h(\mathbf{x}_i)y_i = -1]\!] z_i e^{-\delta(-1)} \\ &= Ae^{-\delta} + Be^{\delta}. \end{aligned} \tag{6.7}$$

Next let us consider how to determine an optimal adjustment for a given candidate predictor h. We want to find the value δ that minimizes Z'. Accordingly, we take the derivative $dZ'/d\delta$ and set it to zero:

$$\begin{aligned} \frac{dZ'}{d\delta} &= A\frac{d}{d\delta}e^{-\delta} + B\frac{d}{d\delta}e^{\delta} \\ &= Be^{\delta} - Ae^{-\delta}. \end{aligned}$$

Setting this to zero, we obtain

$$\begin{aligned} Be^{\delta} &= Ae^{-\delta} \\ e^{2\delta} &= \frac{A}{B} \\ \delta &= \frac{1}{2}\ln\frac{A}{B}. \end{aligned}$$

This gives us the optimal adjustment for a given candidate predictor h.

To determine the new cost if we adopt h, substitute the optimal value for δ into (6.7):

$$\begin{aligned} Z' &= Ae^{-(1/2)\ln(A/B)} + Be^{(1/2)\ln(A/B)} \\ &= A(A/B)^{-(1/2)} + B(A/B)^{(1/2)} \\ &= A\sqrt{B/A} + B\sqrt{A/B} \\ &= 2\sqrt{AB}. \end{aligned} \tag{6.8}$$

In short, the algorithm is this. At each iteration, cycle through all weak predictors h_j. Compute A and B for h_j, and compute the new cost if we adopt h_j as $2\sqrt{AB}$. Keep the h_j that minimizes the new cost. (Actually, since the new cost is a monotone function of AB, it suffices to choose the candidate that minimizes AB.) Compute the optimal adjustment for the winning candidate h_j as $\delta = (1/2)\ln(A/B)$, and adjust the weight vector $\mathbf{w}$ by doing the update:

$$w_j \leftarrow w_j + \delta.$$

This process is repeated until a stopping criterion is met. Often one simply fixes a number of rounds in advance (100, 500, or 1000 rounds being common choices), though other stopping criteria are possible, such as those discussed in section 2.2.2. The resulting final weight vector $\mathbf{w}$ defines the classifier, with its prediction on an arbitrary instance $\mathbf{x}$ being the sign of $\mathbf{x} \cdot \mathbf{w}$.

6.3.1 Abstention

Boosting can readily be generalized to include rules (that is, weak predictors) that abstain. A predictor h abstains from making a prediction on instance $\mathbf{x}$ if $h(\mathbf{x}) = 0$. That is, we allow predictors to range over the values $\{-1, 0, +1\}$ instead of just $\{-1, +1\}$.

Now instead of dividing the training instances into those on which a candidate h makes the correct prediction and those on which it makes the wrong prediction, we must also include a third set: those on which it abstains. We define A and B as before, but we add a third value C:

$$C = \sum_i [\![h(\mathbf{x}_i)y_i = 0]\!] z_i.$$

The expression for the new cost becomes:

$$\begin{aligned} Z' &= \sum_i [\![h(\mathbf{x}_i)y_i = +1]\!] z_i e^{-\delta(+1)} + \sum_i [\![h(\mathbf{x}_i)y_i = -1]\!] z_i e^{-\delta(-1)} \\ &\quad + \sum_i [\![h(\mathbf{x}_i)y_i = 0]\!] z_i e^{-\delta(0)} \\ &= Ae^{-\delta} + Be^{\delta} + C. \end{aligned} \tag{6.9}$$

Since the new term C is insensitive to the choice of δ, the optimal value for δ is unaffected: it is

$$\delta = \frac{1}{2} \log \frac{A}{B}. \tag{6.10}$$

Substituting this value into (6.9) yields

$$Z' = 2\sqrt{AB} + C. \tag{6.11}$$

A round of boosting goes as before, with the exception that one scores a candidate using (6.11) instead of (6.8). The candidate h_j that minimizes the Z' of (6.11) is selected, and its weight is adjusted by the δ of (6.10).

Intuitively, the revised new-cost value (6.11) implies a possible trade-off between the accuracy of the predictor and its coverage. An accurate predictor has a small value for AB, and a broad-coverage predictor has a small value for C. The best predictor is both accurate and broad-coverage.

6.3.2 Semisupervised boosting

A simple way of applying boosting to a mix of labeled and unlabeled data is the following. The learning algorithm receives as input a set of instances $\mathbf{x}_1, \ldots, \mathbf{x}_{m+n}$, the first m of which are labeled $y_1, \ldots, y_m$; the last n instances are unlabeled. To extend the objective function (6.5) to the unlabeled instances, we pretend that the unlabeled instances are labeled correctly, no matter what the classifier predicts. That is, the objective function becomes

$$\sum_{i=1}^{m} \exp\{-\mathbf{w} \cdot \mathbf{x}_i y_i\} + \sum_{i=m+1}^{m+n} \exp\{-|\mathbf{w} \cdot \mathbf{x}_i|\}. \tag{6.12}$$

Geometrically, what is going on is illustrated by the hypothetical example in figure 6.9. The points represent instances, of which some are labeled and some are unlabeled. In figure 6.9, all of the labeled instances are correctly classified, whether we consider hyperplane "A" or hyperplane "B." That is, for either hyperplane, all labeled instances are in the half-space where they belong.

Recall the metaphor that correctly classified points push the separating hyperplane away, in order to move deeper into the correct half-space. The ones that are closest to the hyperplane push hardest, with the effect that the

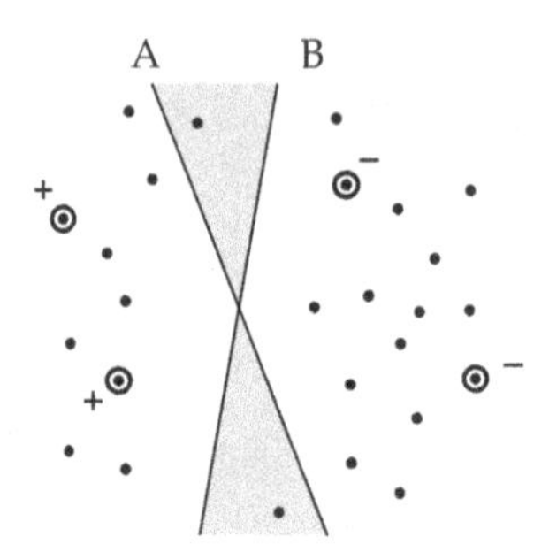

FIGURE 6.9
The "A" hyperplane considers only labeled points; the "B" hyperplane considers all points.

hyperplane settles into the middle of the gap between the positive and negative points, as far away from the nearest points as possible. If the unlabeled points are ignored, the result is the "A" hyperplane.

Now suppose that we assign positive labels to the unlabeled instances in the positive half-space, and negative labels to the unlabeled instances in the negative half-space. We can think of the (originally) unlabeled instances as having "soft" labels. Whenever we move the separating hyperplane, the soft labels automatically change so as to agree with the predictions of the hyperplane. By contrast, the (originally) labeled instances have "hard" labels that do not change.

Taking all the instances into account, "B" is the better hyperplane: among all linear separators that get the hard-labeled instances right, "B" lies in the center of the largest "gap" in the data; it lies as far as possible from the nearest instances, whether labeled or unlabeled. The difference between "A" and "B" is significant because the two hyperplanes make different predictions about the points in the shaded region.

To find the "B" hyperplane, we need to minimize the objective function (6.12). Unfortunately, the absolute value makes it difficult to find a global minimizer. Define

$$Z(\mathbf{w}, \mathbf{y}) = \sum_{i=1}^{m+n} \exp\{-\mathbf{w} \cdot \mathbf{x}_i y_i\}. \tag{6.13}$$

Minimizing (6.12) is equivalent to minimizing the objective function

$$Z(\mathbf{w}) = \min_{\mathbf{y}} Z(\mathbf{w}, \mathbf{y})$$

where $\mathbf{y}$ ranges over labelings that agree with the labeled training data but vary in the labels assigned to unlabeled instances. Naively, one might consider enumerating all such labelings, finding the hyperplane $\mathbf{w}$ that minimizes $Z(\mathbf{w}, \mathbf{y})$ for each labeling $\mathbf{y}$, and choosing the labeling that gives the lowest minimum for Z. Unfortunately, the number of different labelings is 2^n, for n the number of unlabeled instances, hence minimization by enumeration is intractible.

One possibility, suggested by Bennett et al. [15], is reminiscent of self-training. One alternates boosting iterations and relabeling. Before each boosting iteration, use the old weight vector $\mathbf{w}$ to assign labels to the unlabeled instances, producing a complete labeling $\mathbf{y}$. Then do a round of boosting to update $\mathbf{w}$ in the usual way, using objective function (6.13), holding $\mathbf{y}$ constant. Then repeat: use the new weight vector to relabel the unlabeled instances, then use the new labeling to adjust the weight vector. Alternating rounds of relabeling and boosting continue until a stopping criterion is met.

6.3.3 Co-boosting

Another semisupervised variant of boosting is known as co-boosting. Like co-training, co-boosting assumes that there are two "views" of each instance, and that one seeks agreement between predictors trained on the separate views. Co-boosting incorporates agreement explicitly into the objective function. As before, the algorithm receives as input instances $\mathbf{x}_1, \ldots, \mathbf{x}_{m+n}$, the first m of which are labeled $y_1, \ldots, y_m$, and the remaining n of which are unlabeled. We represent the two views by partitioning the weak predictors into view-1 predictors and view-2 predictors. The goal is to construct a pair of confidence-weighted classifiers, represented by weight vectors $\mathbf{w}_1$ and $\mathbf{w}_2$, where $\mathbf{w}_1$ has non-zero components only for view-1 predictors, and $\mathbf{w}_2$ has non-zero components only for view-2 predictors.

For labeled instances, the objective function is (6.5), as in regular boosting, except that terms are required for each of the two classifiers separately. For unlabeled instances, we seek to minimize the disagreement between the two classifiers, which we quantify by a term of the same form as (6.5), but substituting the prediction of the other-view classifier in place of the label. That is, the objective function is:

$$
\begin{aligned}
Z = & \sum_{i=1}^{m} \exp\{-\mathbf{w}_1 \cdot \mathbf{x}_i y_i\} + \sum_{i=1}^{m} \exp\{-\mathbf{w}_2 \cdot \mathbf{x}_i y_i\} \\
& + \sum_{i=m+1}^{m+n} \exp\{-\mathbf{w}_1 \cdot \mathbf{x}_i \hat{y}_i^{(2)}\} + \sum_{i=m+1}^{m+n} \exp\{-\mathbf{w}_2 \cdot \mathbf{x}_i \hat{y}_i^{(1)}\}
\end{aligned} \tag{6.14}
$$

where $\hat{y}_i^{(1)}$ represents the sign of $\mathbf{w}_1 \cdot \mathbf{x}_i$ and $\hat{y}_i^{(2)}$ represents the sign of $\mathbf{w}_2 \cdot \mathbf{x}_i$.

To minimize Z, the procedure is very similar to the regular boosting algorithm. One alternates between the two views. One of the classifiers is the *current* classifier, and the other is *fixed*. Suppose that classifier 1 is fixed and classifier 2 is current. The first term on the right-hand side of (6.14) involves only classifier 1, so it can be ignored. In principle, we should optimize over all three remaining terms. For simplicity, though, we also take the predictions $\hat{y}_i^{(2)}$ to be fixed, and ignore the third term as well as the first. This makes the approach heuristic, but tractable. We are left with

$$\sum_{i=1}^{m} \exp\{-\mathbf{w}_2 \cdot \mathbf{x}_i y_i\} + \sum_{i=m+1}^{m+n} \exp\{-\mathbf{w}_2 \cdot \mathbf{x}_i \hat{y}_i^{(1)}\}$$

as objective function. If we define y_i to be equal to $\hat{y}_i^{(1)}$ for each of the unlabeled instances $m+1 \leq i \leq m+n$, then this can be written simply as

$$\sum_{i=1}^{m+n} \exp\{-\mathbf{w}_2 \cdot \mathbf{x}_i y_i\}$$

which is exactly the objective function for a given round of regular boosting. It is optimized in the same way, by considering each view-2 predictor h_k in turn, computing δ according to (6.10), and choosing the attribute that minimizes the new value of Z according to (6.11). Weight vector $\mathbf{w}_2$ is updated with the adjustment $w_k \leftarrow w_k + \delta$ for the chosen predictor h_k to complete the iteration. The process continues in this fashion, interchanging the roles of view 1 and view 2 after each iteration, until a stopping criterion is met.

6.4 Support Vector Machines (SVMs)

6.4.1 The margin

We noted earlier that, among instances that are correctly classified by the hyperplane, boosting assigns the most weight to the ones that are nearest to the hyperplane. As a result, when there are multiple hyperplanes that correctly classify the training data, boosting tends to find one that lies in the middle of the linear gap between the positives and negatives. For boosting, that is only a tendency. A support vector machine (SVM) learner has the explicit objective of finding the hyperplane that maximizes the distance to the nearest data points.

Figure 6.10 illustrates why that is a desirable objective. It shows two alternative hyperplanes for the same data set. Both hyperplanes classify all the training data correctly, but the "B" hyperplane only just succeeds in doing so. It cuts very close to certain data points; nudging those points even a little bit would put them on the wrong side of the hyperplane. By contrast, the "A" hyperplane is located as far as possible from any data point, and is therefore much more robust to small changes in location of the data points.

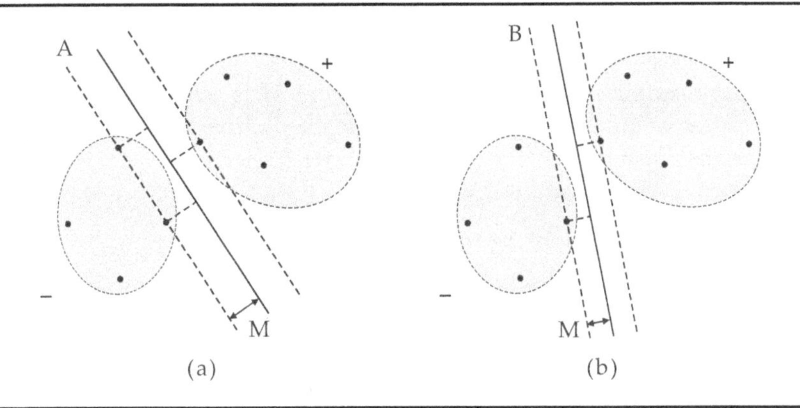

FIGURE 6.10
Two separating hyperplanes for the same data set.

As is now familiar, a hyperplane is defined by a normal vector $\mathbf{w}$ and a distance r from the origin. The signed distance from the hyperplane to a point $\mathbf{x}$ is given by equation (5.1), repeated here for convenience:

$$d = \mathbf{x} \cdot \frac{\mathbf{w}}{\|\mathbf{w}\|} - r.$$

The value d is positive for points on the positive side of the hyperplane, negative for points on the negative side, and zero for points on the hyperplane. Thus, if $y \in \{-1, +1\}$ is the point's label, then $d \cdot y$ is positive for points that are in the correct half-space, and negative for points that are in the wrong half-space. We define the **signed margin** of the point $\mathbf{x}_i$ with label y_i to be

$$m_i = \frac{y_i}{\|\mathbf{w}\|}(\mathbf{x}_i \cdot \mathbf{w} - \rho)$$

where

$$\rho = r\|\mathbf{w}\|.$$

If $\mathbf{x}_i$ is on the correct side of the hyperplane, then m_i represents the distance between $\mathbf{x}_i$ and the hyperplane: the margin of safety preventing $\mathbf{x}_i$ from being misclassified. If $\mathbf{x}_i$ is on the wrong side of the hyperplane, m_i is the negative distance to the hyperplane: the quantity of error that needs to be overcome in order to achieve correct classification for $\mathbf{x}_i$.

Our measure of quality for the hyperplane is the **minimum margin**:

$$M(\mathbf{w}, \rho) = \min_i \frac{y_i}{\|\mathbf{w}\|}(\mathbf{x}_i \cdot \mathbf{w} - \rho). \tag{6.15}$$

This quantity is positive if all instances are correctly classified, and negative otherwise. Its absolute value represents the distance between the hyperplane and the worst point. That is the nearest point to the hyperplane, if all the instances are correctly classified, as in figure 6.10a. The best hyperplane is the one that maximizes M. If a positive value of M is achieved, then all the training instances are correctly classified. Larger positive values of M reflect larger margins of safety.

6.4.2 Maximizing the margin

A support vector machine is a classifier that is constructed by finding the separating hyperplane with the largest possible margin. Given a fixed training set, consisting of labeled points

$$(\mathbf{x}_1, y_1), \ldots, (\mathbf{x}_n, y_n)$$

we wish to find the hyperplane, defined by a pair $(\mathbf{w}, \rho)$, that maximizes M in (6.15).

To maximize a minimum, we rephrase the task as a constrained maximization. We replace the minimum by the constraint that

$$\frac{y_i}{\|\mathbf{w}\|}(\mathbf{x}_i \cdot \mathbf{w} - \rho) \geq U \qquad \forall i \tag{6.16}$$

and we maximize U under that constraint. One can think of U as a claim on a certain amount of buffer space around the separating hyperplane. We want to claim as much space as possible (maximize U), but we also require that the buffer space be devoid of data points. Hence the constraint (6.16), which states that the claimed margin U cannot be larger than the actual margin $M(\mathbf{w}, \rho)$, which is the distance to the nearest data point.

The constraint prevents U from exceeding $M(\mathbf{w}, \rho)$. Concordant with the interpretation of U as the amount of buffer space around the hyperplane, we assume $U > 0$. That will prove convenient shortly, and it is problematic only if $M(\mathbf{w}, \rho) \leq 0$. For the moment, we restrict attention to the linearly separable case, in which case a hyperplane does exist with $M(\mathbf{w}, \rho) > 0$. We will return to the nonseparable case in the next section. We can easily maximize U for a particular choice of $\mathbf{w}$ and ρ by setting it equal to $M(\mathbf{w}, \rho)$. The more interesting question is how to maximize U when $\mathbf{w}$ and ρ are allowed to vary.

A first observation is that we can multiply $\mathbf{w}$ and ρ by any nonzero constant without changing the hyperplane. The hyperplane is determined by the direction of $\mathbf{w}$ and a distance r from the origin. Only the direction of $\mathbf{w}$ matters; changing its length has no effect on the location of the hyperplane. In particular, changing the length has no effect on r, though it obviously does have an effect on ρ, since ρ is defined as $r\|\mathbf{w}\|$. In short, if $\mathbf{w}$ and ρ satisfy the constraint (6.16), then so do $K\mathbf{w}$ and $K\rho$, for any nonzero constant K. One can easily verify that

$$\frac{y_i}{\|K\mathbf{w}\|}(\mathbf{x}_i \cdot K\mathbf{w} - K\rho) = \frac{y_i}{\|\mathbf{w}\|}(\mathbf{x}_i \cdot \mathbf{w} - \rho)$$

for any $K \neq 0$.

Since the length of $\mathbf{w}$ is essentially a free parameter, we can choose to set it equal to $1/U$. That is, if we restrict our attention to pairs $(\mathbf{w}, \rho)$ where $\|\mathbf{w}\| = 1/U$, we will still be able to describe all the same hyperplanes as before. (Here, we rely crucially on the assumption that $U > 0$.) So, restricting attention to pairs $(\mathbf{w}, \rho)$ with $\|\mathbf{w}\| = 1/U$, the constraint (6.16) becomes

$$y_i(\mathbf{x}_i \cdot \mathbf{w} - \rho) \geq 1 \qquad \forall i. \tag{6.17}$$

Another way of arriving at (6.17) is this: Let us interpret U as the unit that we use to measure the margin. The constraint (6.16) requires $m_i \geq 1U$, that is, each point must lie at least one marginal unit from the hyperplane, but otherwise we may choose our unit U as we wish. To maximize the smallest m_i, we should choose U as large as possible. Since the length of $\mathbf{w}$ is a free parameter, we can use it to represent our choice of marginal unit, by defining $U = 1/\|\mathbf{w}\|$. This makes $m_i \geq 1U$ equivalent to (6.17).

We maximize U by minimizing $\|\mathbf{w}\|$. It will actually prove more convenient to minimize $(1/2)\|\mathbf{w}\|^2$. That quantity is monotonically related to $\|\mathbf{w}\|$, so its minimum is the same as the minimum of $\|\mathbf{w}\|$. In short, we have the optimization problem:

$$\begin{aligned} \text{minimize} \quad & \frac{1}{2}\|\mathbf{w}\|^2 \\ \text{subject to} \quad & y_i(\mathbf{x}_i \cdot \mathbf{w} - \rho) \geq 1 \qquad \forall i. \end{aligned} \tag{6.18}$$

We use the methods discussed in section 5.3. By way of reminder, the process is as follows. First we define the Lagrangian $L(\mathbf{w}, \rho, \mathbf{a})$ as a mixture (a linear combination) of the gradient of the objective to be minimized and the gradients of the constraints. The gradients are taken with respect to the independent variables $\mathbf{w}$ and ρ. The vector $\mathbf{a}$ represents the mixing weights for the constraints, that is, Lagrange multipliers. To minimize L with respect to $\mathbf{w}$ and ρ, we set its gradient to zero. That leaves us with a function $Q(\mathbf{a})$ in which only the mixing weights vary. We maximize Q to find the mixture for which the minimizer $\mathbf{x}*$ satisfies the original constraints.

We rewrite the constraint (6.17) in canonical form:

$$1 + \rho y_i - y_i \mathbf{x}_i \cdot \mathbf{w} \leq 0 \qquad \forall i. \tag{6.19}$$

The Lagrangian is

$$L(\mathbf{w}, \rho, \mathbf{a}) = \frac{1}{2}\|\mathbf{w}\|^2 + \sum_i a_i(1 + \rho y_i - y_i \mathbf{x}_i \cdot \mathbf{w}). \tag{6.20}$$

We compute the gradient with respect to $\mathbf{w}$ and ρ by computing each of the partial derivatives. Expanding out $\|\mathbf{w}\|^2$ as $\sum_k w_k^2$, and expanding out $\mathbf{x}_i \cdot \mathbf{w}$ as $\sum_k x_{ik} w_k$, one can verify that

$$\frac{\partial}{\partial w_k}\|\mathbf{w}\|^2 = 2w_k \qquad \frac{\partial}{\partial w_k}\mathbf{x}_i \cdot \mathbf{w} = x_{ik}.$$

Using those two equations, we find that

$$\frac{\partial L}{\partial w_k} = w_k - \sum_i a_i y_i x_{ik}.$$

The partial derivative with respect to ρ is

$$\frac{\partial L}{\partial \rho} = \sum_i a_i y_i.$$

Setting all the partial derivatives to zero, we obtain

$$w_k = \sum_i a_i y_i x_{ik} \qquad \forall k \tag{6.21}$$

$$\sum_i a_i y_i = 0. \tag{6.22}$$

To eliminate the Lagrange multipliers, we maximize the Wolfe dual, which is obtained by substituting (6.21) and (6.22) into the Lagrangian (6.20). First, we observe that

$$\begin{aligned}\|\mathbf{w}\|^2 = \sum_k w_k^2 &= \sum_k \left(\sum_i a_i y_i x_{ik}\right)\left(\sum_j a_j y_j x_{jk}\right) \\ &= \sum_i \sum_j a_i a_j y_i y_j \mathbf{x}_i \cdot \mathbf{x}_j.\end{aligned}$$

Further

$$\begin{aligned}\sum_i a_i y_i \mathbf{x}_i \cdot \mathbf{w} &= \sum_i a_i y_i \sum_k x_{ik} \left(\sum_j a_j y_j x_{jk}\right) \\ &= \sum_i \sum_j a_i a_j y_i y_j \mathbf{x}_i \cdot \mathbf{x}_j.\end{aligned}$$

Hence

$$\begin{aligned}Q(\mathbf{a}) &= \frac{1}{2}\|\mathbf{w}\|^2 + \sum_i a_i + \rho \sum_i a_i y_i - \sum_i a_i y_i \mathbf{x}_i \cdot \mathbf{w} \\ &= \sum_i a_i - \frac{1}{2}\sum_i \sum_j a_i a_j y_i y_j \mathbf{x}_i \cdot \mathbf{x}_j.\end{aligned} \tag{6.23}$$

The dual problem is

$$\text{maximize} \quad Q(\mathbf{a})$$
$$\text{subject to} \left\{ \begin{array}{ll} a_i \geq 0 & \forall i \\ \sum_i a_i y_i = 0 \end{array} \right. .$$

The second constraint is simply a restatement of the condition (6.22). Note that the values $\mathbf{x}_i$ and y_i are all constants; the only unknowns are the a_i. The solution is the optimal value $\mathbf{a}*$ for the Lagrange multipliers. When we have found $\mathbf{a}*$, we can compute $\mathbf{w}*$ using (6.21).

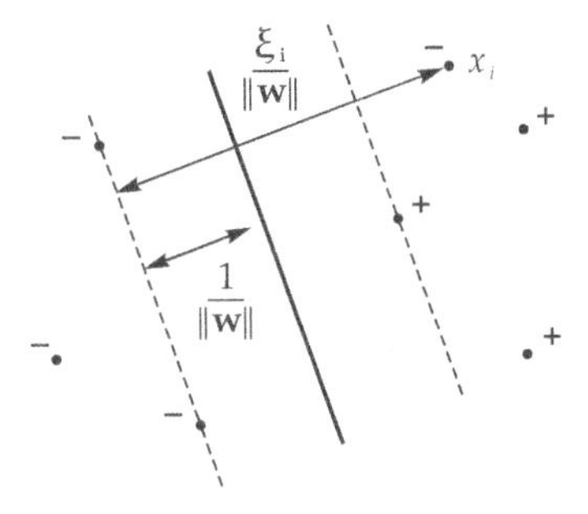

FIGURE 6.11
The nonseparable case.

6.4.3 The nonseparable case

Let us return to the constraint (6.17). Recall that it is equivalent to $m_i \geq 1U$, that is, it states that the signed margin for each point is at least one marginal unit, where we define a marginal unit to be $1/\|\mathbf{w}\|$. If the data is not linearly separable, then the constraint is not satisfiable.

The standard way of dealing with the nonseparable case is to soften the constraint (6.17). The constraint becomes

$$y_i(\mathbf{x}_i \cdot \mathbf{w} - \rho) \geq 1 - \xi_i \qquad \forall i. \tag{6.24}$$

Here the variable ξ_i is a "slack variable" (not to be confused with a slack constraint!) that represents the amount by which data point x_i violates the constraint (figure 6.11). The nominal margin is still $1U$, which is equal to $1/\|w\|$, and we would like to continue to maximize the margin, but we would also like to minimize the total amount of slack that we introduce. Notice that the data point x_i is misclassified just in case $\xi_i > 1$. Hence the total slack $\sum_i \xi_i$ is an upper bound on the number of training errors.

We wish to minimize both $\|\mathbf{w}\|$ and $\sum_i \xi_i$. The usual approach is

$$\text{minimize} \quad \frac{1}{2}\|\mathbf{w}\|^2 + C\sum_i \xi_i \tag{6.25}$$

$$\text{subject to} \left\{ \begin{array}{ll} 1 - \xi_i + \rho y_i - y_i \mathbf{x}_i \cdot \mathbf{w} \leq 0 & \forall i \\ -\xi_i \leq 0 \quad \forall i \end{array} \right. .$$

The value C is a free parameter, fixed by the user, whose value controls the relative weights of the two objectives.

We note in passing that (6.25) can be viewed as a regularized objective function. The idea of regularization was introduced briefly in section 4.2.3:

instead of optimizing only fit to the training data, one optimizes a combination of fit and simplicity. In (6.25), $\sum_i \xi_i$ measures training error; if it is reduced to zero, we have perfect fit to the data. The term $\|\mathbf{w}\|^2$ can be seen as a complexity measure. It depends only on the model, not the data. A "simple" model is one that makes coarse distinctions, and thus requires large boundary gaps in the data distribution – or comes at the cost of large misclassification error if such gaps are not found. As $\|\mathbf{w}\|$ increases, the model can make finer- and finer-grained distinctions.

To minimize (6.25), we form the Lagrangian:

$$L = \frac{1}{2}\|\mathbf{w}\|^2 + \sum_i a_i(1 - \xi_i + \rho y_i - y_i \mathbf{x}_i \cdot \mathbf{w}) - \sum_i b_i \xi_i + C \sum_i \xi_i.$$

There are two constraints, hence two sets of Lagrange multipliers $\mathbf{a}$ and $\mathbf{b}$. Notice that this Lagrangian differs from the previous one (6.20) only by the addition of the term,

$$\sum_i (C - a_i - b_i)\xi_i. \tag{6.26}$$

Hence the derivatives of the Lagrangian with respect to w_i and ρ are unchanged, and setting them to zero yields the equations (6.21) and (6.22), as before. We also take the derivatives of L with respect to ξ_i:

$$\frac{\partial L}{\partial \xi_i} = C - a_i - b_i$$

and setting them to zero yields

$$C = a_i + b_i \qquad \forall i. \tag{6.27}$$

When we substitute (6.27) into the Lagrangian, the new term (6.26) goes away, and $Q(\mathbf{a})$ is the same as before (6.23). However, we must preserve (6.27) as a constraint. The Lagrange multipliers b_i do not appear in (6.23), so they have no effect on maximizing Q. But the b_i are constrained to be nonnegative, so they do have the effect of preventing a_i from exceeding C. That is, the actual effect of (6.27) is that $Q(\mathbf{a})$ must be maximized under the constraint:

$$0 \le a_i \le C \qquad \forall i.$$

In short, the dual problem in the nonseparable case is

$$\text{maximize} \quad Q(\mathbf{a})$$

$$\text{subject to} \begin{cases} 0 \le a_i \le C & \forall i \\ \sum_k a_k y_k = 0 \end{cases}.$$

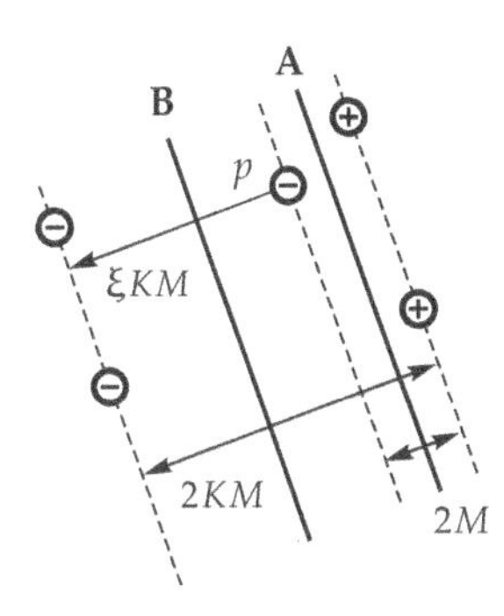

FIGURE 6.12
Using slack in the separable case.

6.4.4 Slack in the separable case

We noted above that the coefficient C in (6.25) represents the trade-off between the two objectives, the two objectives being the squared margin and the total slack. If $C = 1/2$, the two objectives are equally balanced, if $C < 1/2$, more weight is placed on finding a hyperplane with a large margin, and if $C > 1/2$, more weight is placed on avoiding labeling errors.

Could it ever be worthwhile to introduce slack in the separable case, for the sake of obtaining a larger margin? The answer is yes. To keep things simple, let us focus on a single point that requires slack (figure 6.12). The figure shows two different hyperplanes, marked "A" and "B." Assume that the data is separable with margin μ; the "A" hyperplane represents that case. The "B" hyperplane has a larger margin, at the cost of assigning slack ξ to the point marked "p." For simplicity, we ignore all points except p and limit our attention to hyperplanes where the point p is a fixed distance 2μ from the positive margin. We measure the margin in multiples of μ; the margin for a given hyperplane is $M = K\mu$. The "A" hyperplane represents the case $K = 1$, and the "B" hyperplane has some value for K greater than 1.

The slack for point p is ξ, meaning that the distance from p to the negative-side margin is ξ marginal units, which is to say, $\xi K\mu$. Figure 6.12 makes it clear that

$$\xi K\mu = 2K\mu - 2\mu.$$

Hence

$$\xi = 2\left(1 - \frac{1}{K}\right). \tag{6.28}$$

The value of the objective function (6.25) is

$$f(M) = \frac{1}{2}\left(\frac{1}{M^2}\right) + C\xi. \tag{6.29}$$

Since K is proportional to M by $M = K\mu$, we can consider f as a function of K:

$$f(K) = \frac{1}{2\mu^2 K^2} + 2C\left(1 - \frac{1}{K}\right). \tag{6.30}$$

The objective function (6.30) can be thought of as consisting of a "margin term" $1/(\mu^2 K^2)$ weighted by $1/2$, and an "error term" $2(1 - 1/K)$ weighted by C. The margin term represents an expansive force – it improves (that is, decreases) as the margin increases. The error term represents a contractive force – it improves as the margin decreases. The margin term is convex. It starts out large, but falls rapidly and assymptotes to zero. The error term is concave. It is zero when $K = 1$, and it rises rapidly, assymptoting to 2. Their weighted combination, the objective function $f(K)$, is neither concave nor convex. Its first and second derivatives are

$$f'(K) = \frac{2C}{K^2} - \frac{1}{\mu^2 K^3}$$

$$f''(K) = \frac{3}{\mu^2 K^2} - \frac{4C}{K^3}.$$

All the quantities involved are positive, so f is convex $[f''(K) > 0]$ for small values of K and concave for larger values. The flexion point is

$$K = \frac{3}{4}\left(\frac{1}{C\mu^2}\right).$$

Despite the concavity, f is decreasing $[f'(K) < 0]$ only in the region

$$K < \frac{1}{2}\left(\frac{1}{C\mu^2}\right)$$

and it is increasing for larger K. Accordingly, there is a unique minimum, at the point

$$K = \frac{1}{2}\left(\frac{1}{C\mu^2}\right). \tag{6.31}$$

As $K \to \infty$, the objective function never begins decreasing, but assymptotes to $2C$. Figure 6.13 illustrates the particular case where $\mu = 1$ and $C = 1/4$.

Incidentally, if the value $C\mu^2$ is made sufficiently large, then the minimum falls into the region $K < 1$. In that region, however, $\xi < 0$, which is not permitted, so the second term drops out of the objective function, and only the "expansive" margin term remains. Hence, for large values of $C\mu^2$, the optimal hyperplane has $K = 1$; the optimal hyperplane never has $K < 1$.

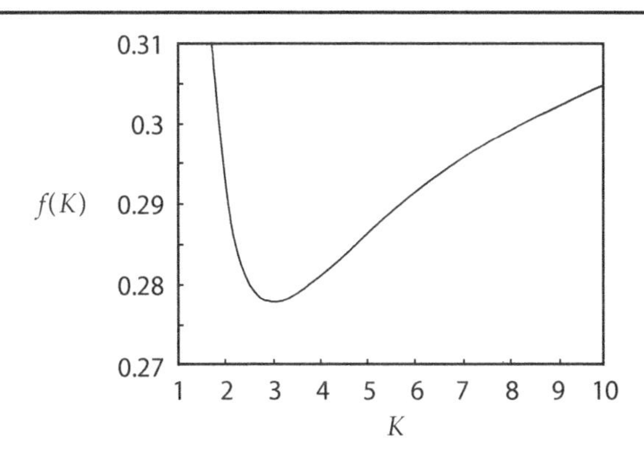

FIGURE 6.13
The objective function for $\mu = 1$, $C = 1/6$.

Let us return to the original question: whether a hyperplane like "B" that misclassifies point "p" might be preferred to "A." The point "p" is misclassified just in case $\xi > 1$. Referring to (6.28), we see that $\xi > 1$ just in case $K > 2$. The optimal value for K is given in (6.31). It exceeds 2 just in case we have set $C < 1/(4M^2)$, which is possible – in fact, figure 6.13 illustrates just such a case.

In short, when we introduce slack variables, the optimal hyperplane may fail to separate the data *even when separation is possible*. This situation is more likely to arise as we decrease C, which is to say, as we assign more weight to the "large margin" objective and less weight to the "no slack" objective.

6.4.5 Multiple slack points

Up to now, we have only considered a single slack variable, for simplicity. Let us consider a more complex case, where there are multiple points with slack. The single slack variable ξ in (6.29) is then replaced with $\sum_i \xi_i$. Suppose there are n^+ positively labeled slack points and n^- negatively labeled slack points. Now let us consider moving the hyperplane a small distance α in the positive direction. We assume α is small enough that all of the slack points remain slack. Let us keep the margin constant, so that the only change in the objective function is determined by the change in $\sum_i \xi_i$.

As a result of moving the hyperplane in the positive direction, the slack must increase for the positively labeled slack points: they are now even deeper in negative territory. The slack increases by the same amount for each, namely, α. The slack for each negatively labeled slack point decreases by α. If there is the same number of positive and negative slack points, then the objective function does not change. But if there are more positives than negatives, the increases outweigh the decreases, and the value of the objective function increases.

In short, if the positive and negative slack points are balanced, the location of the hyperplane is indeterminate (until it moves far enough to change the set of slack points). But if there is any imbalance, then the hyperplane goes as far as it can in the direction the larger group is pushing it, until it bumps up against a new point, or one of the slack points reaches the margin and stops pushing.

Suppose there are more positives than negatives in the dataset. Then a random hyperplane will misclassify more positives than negatives, simply because there are more positives available. This introduces a bias that pushes hyperplanes in the negative direction.

If one wishes to counteract that bias, one can treat the positives and negatives separately:

$$C \sum_i \xi_i = C^+ \sum_{i \in P} \xi_i + C^- \sum_{i \in N} \xi_i$$

where P is the set of positive instances and N is the set of negative instances. Let us introduce "bias corrections" γ^+ and γ^- that allow us to write

$$C \sum_i \xi_i = C \left(\sum_{i \in P} \gamma^+ \xi_i + \sum_{i \in N} \gamma^- \xi_i \right). \tag{6.32}$$

We wish only to correct for bias, not change the net value of the misclassification term.

What we mean by "unbiased" is that moving the hyperplane has no effect on the misclassification term $\sum_i \xi_i$, provided the movement is small enough that the set of slack points remains constant. Moving the hyperplane by a distance δ in the positive direction changes ξ to $\xi + \delta$ for positive slack points, and to $\xi - \delta$ for negative slack points. The net change in the misclassification term is

$$\sum_{i \in P} \gamma^+ \delta - \sum_{i \in N} \gamma^- \delta$$

that is

$$n^+ \gamma^+ \delta - n^- \gamma^- \delta$$

where n^+ is the number of positive slack points and n^- is the number of negative slack points. What we mean by "the unbiased case" is that the net change is zero:

$$n^+ \gamma^+ = n^- \gamma^-.$$

In the case we considered previously, $\gamma^- = \gamma^+ = 1$, implying that the unbiased case has $n^+ = n^-$.

Now suppose that there are N^+ positives in the data set as a whole, and N^- negatives. We would like to make a bias adjustment so that the unbiased case is one in which the number of positive and negative slack points is proportional to the total number of positive and negative points, that is

$$cN^+\gamma^+ = cN^-\gamma^-.$$

That is

$$\frac{\gamma^+}{\gamma^-} = \frac{N^-}{N^+}. \tag{6.33}$$

To fix not just the ratio, but the actual values, recall that we also wish to preserve (6.32). Let us continue to assume that $N^+ \neq N^-$, and that the slack points have the same proportion of positives to negatives as the data set overall. Then (6.32) cannot be true uniformly: moving the hyperplane changes the value on the left-hand side, but not the value on the right-hand side. The natural "neutral" position for the hyperplane is where the average slack is the same on both the negative and positive sides. Let us write $\bar{\xi}$ for the average slack (the same on both sides). Then for the "neutral" hyperplane, (6.32) can be written as

$$(n^+ + n^-)\bar{\xi} = \gamma^+ n^+ \bar{\xi} + \gamma^- n^- \bar{\xi}.$$

Combining this with (6.33) yields

$$\gamma^+ = \frac{N}{2N^+} \qquad \gamma^- = \frac{N}{2N^-}$$

where $N = N^+ + N^-$ is the total number of data points.

6.4.6 Transductive SVMs

Vapnik [227] introduced a setting for learning that he called **transductive inference**. Deductive inference is reasoning from causes to effects and inductive inference is reasoning from effects to causes. Classification in the standard setting uses both: induction to estimate a model from training data, and deduction to predict labels for test data from the model. Transductive reasoning omits the intermediate model and predicts test labels directly, given labeled training data and unlabeled test instances. When contrasted with transductive learning, the standard approach is usually called **inductive learning**; the accompanying deductive step is viewed as the application of what was learned.

A little more concretely, consider training data consisting of instances $\mathbf{x}$ and corresponding labels $\mathbf{y}$, and (unlabeled) test instances $\mathbf{u}$. Inductive learning is the first step in a two-step process. First, the learner λ constructs an estimate of the target function: $\hat{f} = \lambda(\mathbf{x}, \mathbf{y})$ from labeled training data; and, second, the estimated function is applied to unlabeled test data to produce predicted labels $\hat{\mathbf{y}} = \hat{f}(\mathbf{u})$. By contrast, a transductive learner ϕ predicts labels directly: $\hat{\mathbf{y}} = \phi(\mathbf{x}, \mathbf{y}, \mathbf{u})$.

One can easily convert an inductive learner into a transductive learner and vice versa. To convert an inductive learner λ into a transductive learner ϕ, define

$$\phi(\mathbf{x}, \mathbf{y}, \mathbf{u}) \equiv \hat{f}(\mathbf{u}) \quad \text{where} \quad \hat{f} = \lambda(\mathbf{x}, \mathbf{y}).$$

Here we write $\hat{f}(\mathbf{u})$ for the result of applying $\hat{f}$ to each test instance u_i and collecting the results into a vector of predicted labels. To convert a transductive learner ϕ into an inductive learner, define λ such that

$$\lambda(\mathbf{x}, \mathbf{y}) = \hat{f} \quad \text{where} \quad \hat{f}(\mathbf{u}) = \phi(\mathbf{x}, \mathbf{y}, \mathbf{u}).$$

The inductive learner λ simply "memorizes" the training data, and calls ϕ on the training data plus test data, when the test data is made available.

We see that transductive learning is not fundamentally a new kind of learning. The real difference between transductive learning and the standard paradigm is whether one makes use of the unlabeled test data during learning. An inductively trained classifier makes the same prediction on a given test instance u_i no matter what the rest of the test data looks like. A transductive learner may predict different labels for u_i depending on the rest of the test data. The objective function for a (nontrivial) transductive learner does differ from that for an inductive learner, because knowledge of the (unlabeled) test data is included in the objective function for a transductive, but not an inductive, learner.

The semisupervised boosting approach that we considered earlier can be viewed as transductive learning. In figure 6.9, we can interpret the labeled instances as training data and the unlabeled ones as test data. Of the hyperplanes that correctly classify the labeled instances, the "A" hyperplane is the best, in the sense of maximizing the margin for labeled instances, but the "B" hyperplane is the best if one considers all instances. The objective function (6.12) takes both training instances and test instances into account. Specifically, the second term takes test instances into account, seeking to move the hyperplane as far away from them as possible.

Transductive learning as conceived by Vapnik was intended for supervised learning, not semisupervised learning. To be sure, the unlabeled test data $\mathbf{u}$ is not made available to the learner, in the "official" version of the standard setting, but as a practical matter, classifiers are tested on just such a batch of test data, not on individual instances in isolation. Vapnik's insight was to adjust the objective function so as to take advantage of the unlabeled test instances as a batch. Effectively, instead of the optimization (6.25), the objective is

$$\text{minimize} \quad \frac{1}{2}\|\mathbf{w}\|^2 + C\sum_i \xi_i + D\sum_j \zeta_j \tag{6.34}$$

$$\text{subject to} \quad \begin{cases} 1 - \xi_i + \rho y_i - y_i \mathbf{x}_i \cdot \mathbf{w} \leq 0 & \forall i \\ 1 - \zeta_j + \rho \hat{y}_j - \hat{y}_j \mathbf{u}_j \cdot \mathbf{w} \leq 0 & \forall j \\ -\xi_i \leq 0 \quad \forall i \\ -\zeta_j \leq 0 \quad \forall j \end{cases} .$$

Formally, this is simply (6.25) in which we redefine $\mathbf{x}$ to include both training and test instances, and in which we redefine $\mathbf{y}$ to include both training labels and estimated test labels. The formulation (6.34) does, however, explicitly raise the possibility of penalizing errors differently for training and test, by setting $C \neq D$.

Although transductive learning was not originally formulated with semisupervised learning in mind, it applies naturally to the semisupervised setting. In the typical semisupervised case, the learner is provided with labeled training data $(\mathbf{x}, \mathbf{y})$ and additional unlabeled training data $\mathbf{u}$. One can simply apply a transductive learner to $(\mathbf{x}, \mathbf{y}, \mathbf{u})$ as if the unlabeled training data $\mathbf{u}$ were test data.

To be sure, a subtlety arises. The usual semisupervised setting is inductive: in addition to the labeled training data, one has some labeled test data as well to evaluate performance. One can actually distinguish between **inductive semisupervised learning** and **transductive semisupervised learning**. The differences between them are whether the learner is given access to the unlabeled test data in addition to the unlabeled training data, and whether or not predictions will be required for test instances that were not initially given to the learner. Most inductive learners cannot take any advantage of unlabeled test data even if it is provided. Conversely, some transductive learners cannot easily be applied to new test instances. One can always add the new test instances to the unlabeled training data and re-run the learner, but generally at a significant practical cost: the time cost for running a transductive learner on a (large) training set generally significantly exceeds the time cost of applying an inductively trained classifier to a single test instance.

Incidentally, a transductive SVM does produce a classifier, in the form of a separating hyperplane, that can be applied to arbitrary test instances. Hence an SVM that is trained transductively can be applied inductively. Doing so is indistinguishable from using an SVM in an inductive semisupervised setting.

In a transductive semisupervised setting, the (unlabeled) test instances are included in the unlabeled training data that is given to the learner. Since semisupervised learners, by design, can take advantage of unlabeled data, the difference between inductive and transductive semisupervised learning boils down to the type of evaluation done: whether the learner is given access to the unlabeled test instances or not.

6.4.7 Training a transductive SVM

In the previous section, we gave the impression that training an SVM transductively was just like training it inductively, except that the training data is extended to include the test instances. The objective is (6.34), which is just (6.25) applied to the "extended" data set. But there is actually a very important difference between the two: the labels $\mathbf{y}$ are known, but the labels $\hat{\mathbf{y}}$, which appear only in (6.34), are unknown. In other words, *given a particular guess* $\hat{\mathbf{y}}$, we can compute (6.34) by extending the training set to include $(\mathbf{u}, \hat{\mathbf{y}})$ and computing just as we did for (6.25). That is, we have a way of evaluating a particular guess $\hat{\mathbf{y}}$, but not for finding the optimal guess.

The straightforward approach is simply to consider every possible labeling $\hat{\mathbf{y}}$. Since there is a fixed number of unlabeled instances, and only two possible labels for each instance, there is a finite number of possible vectors $\hat{\mathbf{y}}$. However, that number is exponential in the number of unlabeled instances, so the straightforward approach is impractical for all but the smallest unlabeled sets. Bennett and Demiriz [14] took this approach, and were unable to handle datasets with more than about 50 points. But in semisupervised learning, we are frequently interested in *large* unlabeled sets.

For larger unlabeled sets, Vapnik suggests clustering the unlabeled data, and systematically considering all assignments of labels to clusters. All instances within a cluster are assigned the same label. This approach presumes a small number of clusters.

An alternative approach, proposed by Joachims [117], is the following. To begin, train an SVM m_0 using just the labeled data. Then add the unlabeled data. Use m_0 to assign labels to the unlabeled data, and train a new SVM, m_1, on the mixed data (manually labeled and automatically labeled), using a very small value for D in (6.34). By keeping D very small, the learner will seek to maximize the margin even at the cost of misclassifying some examples. This will drive the hyperplane toward regions where the data is sparse and larger margins are obtainable. As a consequence, m_1 generally will not classify all the automatically labeled instances "correctly" (that is, in agreement with m_0). It is to be expected that the progress of the hyperplane into low-density regions was held back by the automatically labeled instances that m_1 misclassified. Accordingly, we relabel those automatically labeled instances where m_1 disagrees with the labels. Instances are relabeled in pairs, to keep the proportion of positives and negatives constant. Then the process repeats, yielding classifiers $m_2, m_3, \ldots$. At some point, one obtains a classifier m_t that agrees with the previous one on all automatically labeled instances. At that point, the process has converged: retraining will simply yield m_t again.

Joachims includes a couple of refinements in the algorithm. First, the user specifies a desired misclassification penalty D'. The learner starts with D very small, and increases it at each iteration until reaching the target D'. Second, the user specifies a target ratio of positives to negatives. In the initial labeling of unlabeled instances, using m_0, the classification threshold is adjusted so that the desired positive-to-negative ratio is obtained. Once established during the initial labeling, the ratio is preserved by the way instances are relabeled in pairs.

6.5 Null-category noise model

There are other approaches that focus on finding low-density regions in which to place boundaries. We will mention just one here: Lawrence & Jordan [134] propose to model lowness of density in boundary regions by means of a probabilistic model of data generation that deletes points that fall near the boundary. We will give only an impression of it here, and refer the reader to the original source and citations there for more detail.

A generative model of the data distribution $p(\mathbf{x}, y)$ is given, where $\mathbf{x}$ is an instance and y is its label, but unlike typical generative models, it is factored into an instance distribution $p(\mathbf{x})$ and a conditional label distribution $p(y|\mathbf{x})$ that has the form of a probabilistic classifier. The probability $p(y|\mathbf{x})$ is based on a distribution $p(\alpha|\mathbf{x})$ of "confidence-weighted labels" α, by which we mean that the sign of α represents the predicted label and the absolute value of α represents the confidence of the prediction. The distribution $p(\alpha|\mathbf{x})$ is Gaussian in form, with mean $\mu(\mathbf{x})$ and variance $\sigma(\mathbf{x})$. This provides us with a probabilistic interpretation of confidence: the probability $p(\alpha|\mathbf{x})$ allows one to measure confidence as $\Pr[\alpha < 0|\mathbf{x}]$ or $\Pr[\alpha > 0|\mathbf{x}]$. As $\mu(\mathbf{x})$ moves toward $+\infty$, the probability $\Pr[\alpha > 0|\mathbf{x}]$ increases to one, and similarly $\Pr[\alpha < 0|\mathbf{x}]$ increases to one as $\mu(\mathbf{x})$ moves toward $-\infty$.

If one thinks of instances $\mathbf{x}$ as points in a horizontal plane, the "instance plane," one can think of $p(\alpha|\mathbf{x})$ as a fuzzy surface, the "label surface," whose altitude at point $\mathbf{x}$ is $\mu(\mathbf{x})$ and whose thickness is $\sigma(\mathbf{x})$. The fuzzy surface is densest at altitude $\mu(\mathbf{x})$ and attenuates rapidly as one moves above and below that altitude. Given a point $\mathbf{x}$, one chooses an altitude α probabilistically, according to the density $p(\alpha|\mathbf{x})$ at that altitude.

The label (or "category") $y \in \{-1, 0, +1\}$ is determined by α. If α is large and positive, then y is $+1$, and if α is large and negative, then y is -1. But the model also countenances a third possibility: the **null category** $y = 0$. Specifically, a parameter w is fixed representing the thickness of a "dead region" just above and below the instance plane. If $\alpha > +w/2$ then $y = +1$, if $\alpha < -w/2$ then $y = -1$, but if α falls in the dead region $[-w/2, +w/2]$, then $y = 0$. Assigning the null category marks a point for deletion.

If $|\mu(\mathbf{x})|$ is much greater than $w/2$, then the probability that α falls into the region $[-w/2, +w/2]$ is very small. Conversely, if $\mu(\mathbf{x}) = 0$ and w is large relative to $\sigma(\mathbf{x})$, then the probability that $-w/2 \leq \alpha \leq +w/2$ is nearly one. That is, at points $\mathbf{x}$ where confidence is high, and the label surface is well above or below the instance plane, the probability of deleting $\mathbf{x}$ is very small, but where confidence is low, and the label surface is close to the instance plane, the probability of deleting $\mathbf{x}$ is very high. As a consequence, where the label surface crosses the instance plane, there is a region in which almost all instances have been deleted.

To complete the data-generation model, if the point is not deleted, a probabilistic decision is made whether or not to "hide" the label. If the label is hidden, the result is an unlabeled sample point, and if the label is not hidden, the result is a labeled sample point. The decision is based solely on the label, so there are only two parameters involved: the probability of hiding a label if it is positive, and the probability of hiding it if it is negative.

Training data given to the learner consists of a mix of labeled and unlabeled sample points generated by the process just described. Deleted points are not included in the sample. The labeled points are used to estimate the parameters of the model $p(y|\mathbf{x})$. Unlabeled points help identify decision boundaries by making it clearer where there are regions that are devoid of instances. Hence, like the other approaches discussed in this chapter, the model is designed to place class boundaries in low-density regions. Unlike the other approaches, however, it is generative rather than discriminative.

7

Clustering

The previous chapters have focussed on semisupervised learning methods that take classification algorithms as their point of departure. In this chapter, we turn to clustering as a basis for semisupervised learning.

7.1 Cluster and label

The simplest idea for using clustering in a semisupervised setting is what we might call **cluster and label**. Apply any clustering algorithm to the data, ignoring the labels if any of the data is labeled. Then assign labels to the clusters based on the labeled data falling within each cluster. The obvious rule is to count each instance labeled y in a given cluster as a vote for label y, and to assign to the cluster the label that receives the most votes. The label predicted for a given unlabeled instance $\mathbf{x}$ is the label assigned to the cluster. The algorithm is summarized in figure 7.1.

There is one point to note. It is assumed that the procedure `cluster` returns a function that is defined everywhere in the instance space. Not all clustering algorithms return such a function; some return only an assignment of clusters

---

procedure `clusterAndLabel` (X, Y, V)
  $X = (\mathbf{x}_1, \ldots, \mathbf{x}_n)$ consists of instances with labels $Y = (y_1, \ldots, y_n)$
  V consists of additional unlabeled instances
  $c(\cdot) \leftarrow$ `cluster`$(X \cup V)$
  for each distinct cluster c_k
    define $\mu(c_k) = \arg\max_y |\{i : c(\mathbf{x}_i) = c_k \wedge y_i = y\}|$
  return function $f(\cdot)$ such that $f(\mathbf{x}) = \mu(c(\mathbf{x}))$

FIGURE 7.1
The cluster-and-label algorithm. It takes a mix of labeled and unlabeled instances as input and returns a classifier.

to the points in the training sample and are undefined at any other point in the instance space. The former kind of clustering algorithm, whose return value is a function defined everywhere in the instance space, is an inductive learner. The latter kind of algorithm, whose return value is defined only for the given sample, is essentially transductive. The reason for the hedge "essentially" is that a transductive learner is defined by two properties: assigning a label to sample points only, and receiving both training and test points in its sample. But a distinction between training and test points is rarely made in clustering, so the lack of generalization beyond the sample is the critical property in this context.

The cluster and label algorithm as stated assumes an inductive clustering algorithm. A simple method for converting a transductive clusterer into an inductive one is to assign each test point to the nearest cluster, where the distance between a point and cluster might be defined in one of several ways, such as those discussed in section 7.3 below. These include the minimum distance to any point in the cluster, the average distance to all points, or the distance to the centroid (defined below).

We turn now to algorithms that can be used as the `cluster` function.

7.2 Clustering concepts

7.2.1 Objective

The idea of clustering is to find "natural groups" within the data. Clustering is easiest and most intuitive when instances are distributed densely in well-defined, approximately circular regions (the clusters), separated by large boundary areas that are devoid of instances. Just as the boundary-oriented semisupervised learning methods discussed in the preceding two chapters contrast with generative methods that describe how the high-density regions are populated (to be discussed in the next chapter), so one commonly distinguishes two approaches to clustering: in agglomerative clustering, the focus is on identifying high-density regions or cluster centers, whereas in divisive clustering, the focus is on identifying good places for boundaries. A third desideratum is to have clusters of approximately equal size, other things being equal; the desire for balance can contribute to the choice of boundary placement.

Defining the quality of a clustering is a somewhat thorny issue. Clustering is put to many purposes, and many different measures of quality have been proposed. Clustering is often used in exploratory data analysis, in order to discover structure that is not known a priori to the researcher. The methods are in many cases purely algorithmic, and do not define an independent measure of quality. Rather they embody "obvious" steps to construct clusters corresponding to geometric intuition – the hierarchical clustering methods that we discuss shortly are good examples.

We are primarily interested in clustering as a form of classification in which the learner is provided no labeled data – what we might call **unsupervised classification**. The objective, and the measure of quality, are the same as in classification: to approximate a target function that assigns classes to instances. The ultimate measure of quality is generalization error.

"Unsupervised classification" may seem a contradiction in terms. If the data distribution is such that clusters are compact and well separated, one can imagine identifying class boundaries correctly using only unlabeled data. But without any labeled data, it is hard to imagine how one could determine which label goes with which cluster.

In some cases, unsupervised classification *is* in principle possible. The difficulty in assigning labels to clusters arises because class labels are usually arbitrary designators – but learning problems do arise in which the classes are not arbitrary, but have intrinsic content with observable distributional reflexes. A prime example is provided by unsupervised grammatical inference, beginning with the problem of inducing parts of speech. One can distinguish function words from content words by several attributes. Function words have higher absolute frequency, and function words have much more uniform distribution. There is intrinsic meaning to the function word/content word distinction that makes it possible to label the clusters once they are identified. Moreover, there is a non-trivial family of problem instances: each language represents a new problem instance, with a different set of function words and content words.

If one defines semisupervised learning broadly enough, one might treat the setting just sketched as semisupervised learning, in which the intrinsic class definitions take the place of seed information. They are comparable to a seed classifier.

Be that as it may, it is in fact more common for class labels to be arbitrary designators. In such a case, unsupervised construction of a classifier is impossible; the best that is achievable is correct identification of the clusters. A clustering algorithm learns a function $c(\mathbf{x})$ assigning cluster labels to instances, and to obtain a classifier $f(\mathbf{x})$, we also require a mapping μ from cluster labels to class labels, allowing us to define:

$$f(\mathbf{x}) = \mu(c(\mathbf{x})).$$

The quality of f is measured as generalization error, in the usual way, and the quality of c is defined to be the quality of the best classifier constructable from c, ranging over choices of μ.

7.2.2 Distance and similarity

The geometric intuitions that we appeal to in clustering depend on having a definition of the distance between instances. A good cluster contains many points that are close to each other, and it is well separated (distant) from other clusters. If instances are represented as real vectors, then Euclidean distance is the obvious measure of distance, but other measures are possible.

A distance measure d is generally taken to satisfy the following conditions, for all instances $\mathbf{x}$ and $\mathbf{y}$:

$$\begin{array}{lll} (i) & d(\mathbf{x},\mathbf{x}) = 0 & \\ (ii) & d(\mathbf{x},\mathbf{y}) \geq 0 & \text{non-negativity} \\ (iii) & d(\mathbf{x},\mathbf{y}) = d(\mathbf{y},\mathbf{x}) & \text{symmetry.} \end{array} \tag{7.1}$$

A distance measure is a *metric* if it also satisfies the triangle inequality:

$$d(\mathbf{x},\mathbf{y}) + d(\mathbf{y},\mathbf{z}) \geq d(\mathbf{x},\mathbf{z}). \tag{7.2}$$

A wide variety of distance measures are available. When the data space is $\mathbf{R}^k$, common choices are Euclidean distance, also known as the L_2 norm (7.3), and Hamming distance, also known as Manhattan distance, city-block distance, or the L_1 norm (7.4):

$$d(\mathbf{x},\mathbf{y}) = \|\mathbf{x} - \mathbf{y}\| = \sqrt{\sum_{i=1}^{k} (x_i - y_i)^2} \tag{7.3}$$

$$d(\mathbf{x},\mathbf{y}) = \sum_{i=1}^{k} |x_i - y_i|. \tag{7.4}$$

Distance – or rather its inverse, proximity – is a geometric representation of the **similarity** between two instances. In some cases, it is most convenient to define similarity directly, and in other cases, it is more convenient to define a measure of distance, and define similarity in terms of distance.

Corresponding to the properties (7.1), a similarity measure s is generally taken to satisfy:

$$\begin{array}{lll} (i) & s(\mathbf{x},\mathbf{x}) = 1 & \\ (ii) & 0 \leq s(\mathbf{x},\mathbf{y}) \leq 1 & \\ (iii) & s(\mathbf{x},\mathbf{y}) = s(\mathbf{y},\mathbf{x}) & \text{symmetry.} \end{array} \tag{7.5}$$

Given the properties just listed, it is obvious that similarity and distance are not opposites in the sense of arithmetic inverse, but rather in the sense that any natural mapping between similarities and distances is monotone decreasing. That leaves open a wide range of possibilities. One common choice for mapping a distance d to a similiarity $s(d)$ is

$$s = \frac{1}{1+d}. \tag{7.6}$$

A more specialized example of a similarity measure derived from a distance metric is the following transform of Euclidean distance that arises naturally in Gaussian random fields:

$$s(\mathbf{x}, \mathbf{y}) = \exp\left(-\frac{\|\mathbf{x} - \mathbf{y}\|^2}{\alpha^2}\right).$$

This is actually a family of similarity metrics, indexed by the parameter α.

The reverse mapping, from similarity to distance, is complicated by the requirements of the triangle inequality (assuming one desires a metric). The trick is to observe (1) that Euclidean distance can be expressed in terms of a dot product:

$$d(\mathbf{x}, \mathbf{y}) = \|\mathbf{x} - \mathbf{y}\| = \sqrt{(\mathbf{x} - \mathbf{y}) \cdot (\mathbf{x} - \mathbf{y})}$$

and (2) that any similarity measure can be expressed as a dot product in an appropriately chosen space.

It is easiest to see why (2) is true in the case of a space containing finitely many points. Suppose we have a space containing three points, $\mathbf{x}_1$, $\mathbf{x}_2$, and $\mathbf{x}_3$. We define a new space in which each dimension corresponds to a pair of original points: in the case of three original points, there are nine dimensions in the new space. Let s_{ij} stand for the similarity $s(\mathbf{x}_i, \mathbf{x}_j)$, and define $\Sigma_i = \sum_{j \neq i} s_{ij}$. The representations for our three points in the new space are as follows. Note that some of the components may be complex.

Dimension

	(1,1)	(1,2)	(1,3)	(2,1)	(2,2)	(2,3)	(3,1)	(3,2)	(3,3)
$\mathbf{x}'_1$	($\sqrt{s_{11} - \Sigma_1}$	$\sqrt{\frac{s_{12}}{2}}$	$\sqrt{\frac{s_{13}}{2}}$	$\sqrt{\frac{s_{21}}{2}}$	0	0	$\sqrt{\frac{s_{31}}{2}}$	0	0)
$\mathbf{x}'_2$	(0	$\sqrt{\frac{s_{12}}{2}}$	0	$\sqrt{\frac{s_{21}}{2}}$	$\sqrt{s_{22} - \Sigma_2}$	$\sqrt{\frac{s_{23}}{2}}$	0	$\sqrt{\frac{s_{32}}{2}}$	0)
$\mathbf{x}'_3$	(0	0	$\sqrt{\frac{s_{13}}{2}}$	0	0	$\sqrt{\frac{s_{23}}{2}}$	$\sqrt{\frac{s_{31}}{2}}$	$\sqrt{\frac{s_{32}}{2}}$	$\sqrt{s_{33} - \Sigma_3}$)

The reader can confirm that the dot product of any two vectors in the new space is equal to the similarity of the corresponding points in the original space. (The construction relies on the symmetry assumption $s_{ij} = s_{ji}$.)

In general, the (i, j)-th component of the new vector $\mathbf{x}'_k$ is equal to

$$\begin{array}{ll} \sqrt{s_{ij}/2} & \text{if } i \neq j \text{ and } k = i \text{ or } k = j \\ \sqrt{s_{kk} - \Sigma_k} & \text{if } i = j = k \\ 0 & \text{otherwise.} \end{array}$$

Then, for $i \neq j$, we have

$$\mathbf{x}'_i \cdot \mathbf{x}'_j = \left(\sqrt{s_{ij}/2}\right)^2 + \left(\sqrt{s_{ji}/2}\right)^2 = s_{ij}$$

and the dot product of a vector with itself is

$$\mathbf{x}'_i \cdot \mathbf{x}'_i = \left(\sqrt{s_{ii} - \Sigma_i}\right)^2 + \sum_{j \neq i} \left[\left(\sqrt{s_{ij}/2}\right)^2 + \left(\sqrt{s_{ji}/2}\right)^2 \right] = s_{ii}.$$

The construction can be generalized to spaces containing infinitely many points, so that any similarity measure is expressible as a dot product in an appropriately chosen space.

As noted above, the Euclidean distance between two points can be expressed in terms of a dot product:

$$\begin{aligned} d(\mathbf{x}', \mathbf{y}') &= \sqrt{(\mathbf{x}' - \mathbf{y}') \cdot (\mathbf{x}' - \mathbf{y}')} \\ &= \sqrt{\mathbf{x}' \cdot \mathbf{x}' - 2\mathbf{x}' \cdot \mathbf{y}' + \mathbf{y}' \cdot \mathbf{y}'}. \end{aligned}$$

We know that Euclidean distance is a metric, and in particular that it satisfies the triangle inequality. In short, we can convert an arbitrary similarity s to a distance metric by defining

$$d(\mathbf{x}, \mathbf{y}) = \sqrt{s(\mathbf{x}, \mathbf{x}) - 2s(\mathbf{x}, \mathbf{y}) + s(\mathbf{y}, \mathbf{y})}. \tag{7.7}$$

If the similarity satisfies $s(\mathbf{x}, \mathbf{x}) = 1$ for all $\mathbf{x}$, (7.7) can be simplified to

$$d(\mathbf{x}, \mathbf{y}) = \sqrt{2(1 - s(\mathbf{x}, \mathbf{y}))}.$$

7.2.3 Graphs

We have introduced similarity as a function that is monotone decreasing with the distance between two instances viewed as points in a space. A similarity measure s is defined over all pairs of points in the space. If we limit attention to a fixed set of n data points $(\mathbf{x}_1, \ldots, \mathbf{x}_n)$, there are two other useful ways of thinking about similarity.

A similarity measure that is defined only at a finite number of points is equivalent to a **similarity matrix**, which is a matrix W with elements $w_{ij} = s(\mathbf{x}_i, \mathbf{x}_j)$. Because of the properties of a similarity measure, a similarity matrix is symmetric and has non-negative entries, with all 1's on the diagonal.

A similarity matrix, in turn, is equivalent to a weighted graph that we call a **similarity graph**. Its vertices are the given data points, and there is an edge $(\mathbf{x}_i, \mathbf{x}_j)$ just in case the similarity between $\mathbf{x}_i$ and $\mathbf{x}_j$ is nonzero. Each edge has a weight $w_{ij} = s(\mathbf{x}_i, \mathbf{x}_j)$.

It is worth emphasizing that the relationship between a similarity measure and a similarity graph is essentially the same as the relationship between the feature space and a finite sample of data points. A similarity measure determines – or is determined by – the geometry of the feature space. The similarity graph is the discrete analogue that results from the restriction to a particular sample.

If similarity values are all either zero or one, the similarity function $s(\mathbf{x}_i, \mathbf{x}_j)$ becomes a relation; we interpret the value one as true ($\mathbf{x}_i$ and $\mathbf{x}_j$ are similar) and the value zero as false (they are not similar). The corresponding graph is unweighted, under the convention that an unweighted graph is the same as a weighted graph whose weights all have value one. The matrix is then called an **adjacency matrix**. It has elements

$$a_{ij} = \begin{cases} 1 \text{ if there is an edge } (\mathbf{x}_i, \mathbf{x}_j) \\ 0 \text{ otherwise} \end{cases}.$$

7.3 Hierarchical clustering

Probably the most widely known, and in that sense the most basic, clustering method is hierarchical clustering. It is an agglomerative method. A "cluster" is simply a set of data points; one begins with a separate singleton cluster for each data point. Then one finds the closest pair of clusters, that is, the pair of clusters (X, Y) that minimizes the distance $d(X, Y)$, and one replaces the clusters X and Y with $X \cup Y$. This step is repeated until a stopping criterion is reached. For example, one might stop when the number of clusters has been reduced to a target number chosen in advance, or when the distance between the nearest pair of clusters exceeds a threshold chosen in advance.

The agglomerative process can be conveniently represented by a **dendogram**, as in figure 7.2. The individual data points are at the leaves along the left edge. The horizontal segments represent clusters, and the vertical "cross-ties" represent the agglomeration of two clusters. The horizontal position of the cross-tie corresponds to the distance between the two clusters. A given stopping criterion determines a vertical cut through the dendogram, and the horizontal lines that the cut intersects are the final clusters.

The method just described is actually a family of methods that differ in the choice of distance measure $d(\mathbf{x}, \mathbf{y})$ between data points, and in the way in which $d(\mathbf{x}, \mathbf{y})$ is extended to a measure of distance $d(X, Y)$ between sets of points (clusters). The choice of distance measure between data points was discussed briefly in the preceding section. The commonest definitions of aggregate distance $d(X, Y)$ are the following:

- *Single linkage.* The distance between clusters X and Y is the minimum distance between any two points $\mathbf{x} \in X$ and $\mathbf{y} \in Y$.

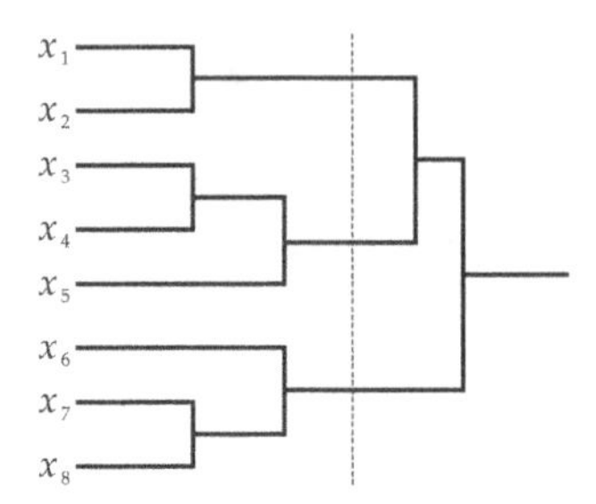

FIGURE 7.2
A dendogram.

- *Complete linkage.* The distance between X and Y is the maximum distance between any $\mathbf{x} \in X$ and $\mathbf{y} \in Y$.
- *Centroid linkage.* The distance between X and Y is the distance between their centroids. Let x_{ij} represent the j-th component of the i-th vector in X. Then the centroid of X is the vector whose j-th component is the average value of x_{ij}, ranging over values of i. The centroid does not generally coincide with any data point in X. The mean is generally used as the definition of "average," but the median is also possible.
- *Average linkage.* The distance between X and Y is the average distance between pairs $\mathbf{x} \in X$ and $\mathbf{y} \in Y$.

It might not be immediately obvious that average linkage gives a different result from centroid linkage. To see that it does, consider the case in which X contains one point $\mathbf{x}$, and Y contains two points $\mathbf{y}_1$ and $\mathbf{y}_2$, equidistant from $\mathbf{x}$, as in figure 7.3. The centroid linkage distance is the length of the dotted line, and the average linkage distance is the length of either solid line. Moreover, if $\mathbf{y}_1$ and $\mathbf{y}_2$ are moved apart from each other without moving the centroid, the average linkage distance increases, but the centroid distance is unchanged.

Hierarchical clustering has the advantage of simplicity. It is easy to grasp and easy to implement, and hence widely used. It has the disadvantage that it is entirely heuristic. Given the aim of producing "compact" clusters, greedily choosing the closest pair of clusters for agglomeration is plausible, but there is no global measure of cluster compactness that the algorithm optimizes.

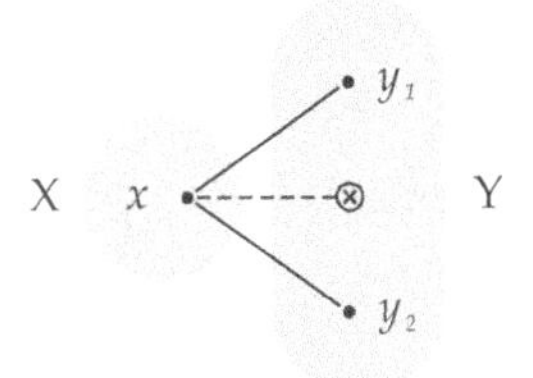

FIGURE 7.3
An example in which centroid linkage and average linkage differ. The centroid linkage distance is the length of the dotted line, and the average linkage distance is the length of either solid line.

7.4 Self-training revisited

7.4.1 k-means clustering

k-means clustering is a popular clustering algorithm that is related to the nearest-neighbor classifier (section 4.1.2). It is agglomerative in the sense that it focuses on identifying centers of density. It also turns out to be related to self-training.

The parameter k represents the number of clusters. Each cluster is represented by a center $\mathbf{m}_j$. As in 1-nearest-neighbor classification, each data point is assigned to the cluster whose center is nearest. That is, the clusters constitute a Voronoi tesselation of the instance space, as in figure 4.4 (page 46), but with the training instances $\mathbf{x}_i$ replaced by the centers $\mathbf{m}_j$. Usually, the number of instances is much greater than the number of centers. The decision boundaries are piecewise linear; they are perpendicular bisectors of the vectors connecting pairs of adjacent centers, as in figure 5.5 (page 71).

The algorithm is given in figure 7.4. Initially, the centers are chosen at random. An iterative algorithm is then used to adjust them. An unlabeled sample of instances $(\mathbf{x}_1, \ldots, \mathbf{x}_n)$ is given as input to the algorithm. Each instance is labeled using the nearest-neighbor rule. If $\mathbf{m}_j$ is the center nearest to $\mathbf{x}_i$, then the j-th label is assigned to $\mathbf{x}_i$. After all instances are labeled, the centers are recomputed. The center $\mathbf{m}_j$ is set to the centroid of the instances

```
procedure kMeansCluster (X)
    X = (x_1, ..., x_n) consists of unlabeled instances
    c = (m_1, ..., m_k) is a random set of centers
    loop until convergence
        L ← label(X, c), using nearest-neighbor rule
        c ← train(L), setting centers to centroids
    end loop
    return c
```

FIGURE 7.4
The k-means clustering algorithm.

labeled j. That is

$$\mathbf{m}_j \leftarrow \frac{1}{|X_j|} \sum_{\mathbf{x} \in X_j} \mathbf{x}$$

where X_j is the set of instances with the j-th label.

Note the resemblance between the k-means clustering algorithm (figure 7.4) and self-training in the "dual" form (figure 2.7, page 26). If we use a labeled seed set (or a seed classifier) in place of the randomly initialized classifier in k-means clustering, we obtain a version of self-training. Self-training and k-means clustering both share the basic principle of alternately labeling instances and training a classifier. The differences are that k-means clustering uses random initialization instead of a seed classifier or labeled seed instances, it does not use a confidence threshold, and it is more specific about the form of the `label` and `train` functions: namely, `label` is nearest-neighbor labeling and `train` sets cluster centers to centroids.

7.4.2 Pseudo relevance feedback

The basic idea of self-training arose independently in the field of information retrieval, where it takes the form of an algorithm called pseudo relevance feedback. We postponed discussing pseudo relevance feedback until now, because it is best understood in terms of the geometric ideas introduced in this chapter. It has commonalities, in particular, with the k-means clustering algorithm. Pseudo relevance feedback involves a binary classification (relevant versus irrelevant), and, as we will see, it resembles k-means clustering in that it represents the classes by their centroids, and defines the decision boundary between classes as a line perpendicular to the vector connecting the centroids.

The most common task in information retrieval is ad hoc document retrieval. Given a query, the problem is essentially one of binary classification of documents. It is assumed that each document is either relevant to the query or not. In one common approach, feature extraction converts each document to a vector $\mathbf{x}$; feature extraction is also applied to the query to produce a vector $\mathbf{q}$. The learning algorithm uses no training data except the query. A document is predicted to be relevant if it is sufficiently similar to the query, where "sufficiently similar" is interpreted as cosine similarity of the vector representations. That is, given query $\mathbf{q}$, a confidence-rated classifier f is defined with

$$f(\mathbf{x}) = \cos(\mathbf{x}, \mathbf{q}) - \theta$$

where θ is a free parameter. The sign of $f(\mathbf{x})$ represents its prediction (positive for relevant and negative for irrelevant), and the absolute value represents confidence.

If we normalize document vectors and query vectors, so that we can assume that $\|\mathbf{x}\| = 1$ and $\|\mathbf{q}\| = 1$, then

$$f(\mathbf{x}) = \mathbf{x} \cdot \mathbf{q} - \theta.$$

This form should by now be familiar as a confidence-weighted predictor, in which the confidence is geometrically interpretable as the distance of point $\mathbf{x}$ from the hyperplane defined by normal vector $\mathbf{q}$ and threshold θ: recall, for example, equation (5.2) and the discussion there (page 68).

Relevance feedback is a form of active learning in which the confidence values of the initial classifier are used to select a sample for manual labeling. The documents with the highest values $f(\mathbf{x})$ are selected to be manually labeled as relevant or irrelevant. That labeled data is then used to construct an improved query vector, and hence an improved confidence-rated classifier.

Speaking of an "improved" query vector implies a measure of quality for query vectors. The measure of quality is based on the predictions of the confidence-rated classifier f. For a given document $\mathbf{x}_i$ with label $y_i \in \{+1, -1\}$, the quality is positive if $f(\mathbf{x}_i)$ and y_i have the same sign, and negative otherwise, and it is weighted by the confidence of $f(\mathbf{x}_i)$. The result is averaged over all documents. In short:

$$\begin{aligned} Q(\mathbf{q}) &= \frac{1}{n} \sum_i y_i f(\mathbf{x}_i) \\ &= \frac{1}{n} \sum_i y_i \cos(\mathbf{x}_i, \mathbf{q}) - \theta \end{aligned}$$

where n is the total number of documents. The *optimal query* is defined to be the query with maximum quality. The optimal query is well-defined if the labels of all documents are known. It is not guaranteed that the query that maximizes Q over a given sample of documents is the same as the optimal query, nor even that the value of Q averaged over the labeled sample agrees with its value over the whole document set. If the sample were drawn i.i.d. from the document set, then the sample value of Q would be known to converge to the population value. But in the case of relevance feedback, the sample is biased by the initial query, and no guarantees are known.

The optimal query has a simple geometric characterization. We can rewrite Q as follows:

$$\begin{aligned} Q(\mathbf{q}) &= \frac{1}{n}\sum_i y_i \frac{\mathbf{x}_i \cdot \mathbf{q}}{\|\mathbf{x}_i\| \cdot \|\mathbf{q}\|} - \theta \\ &= \frac{\mathbf{q}}{\|\mathbf{q}\|} \cdot \frac{1}{n}\sum_i y_i \frac{\mathbf{x}_i}{\|\mathbf{x}_i\|} - \theta. \end{aligned}$$

Let A be the set of relevant documents, and let B be the irrelevant documents. Define:

$$\mathbf{m} = \frac{1}{n}\sum_i y_i \frac{\mathbf{x}_i}{\|\mathbf{x}_i\|}.$$

Notice that $\mathbf{m}$ is not dependent on the choice of $\mathbf{q}$ (nor on the choice of θ, for that matter). The quality of $\mathbf{q}$ can be written as

$$\begin{aligned} Q(\mathbf{q}) &= \frac{\mathbf{q}}{\|\mathbf{q}\|} \cdot \mathbf{m} - \theta \\ &= |\mathbf{m}| \cos(\mathbf{q}, \mathbf{m}) - \theta. \end{aligned}$$

Clearly, for fixed θ, $Q(\mathbf{q})$ is maximized by choosing $\mathbf{q}$ in the same direction as $\mathbf{m}$; that is, any vector in the same direction as $\mathbf{m}$ maximizes Q, including $\mathbf{m}$ itself.

Now, since $y_i = +1$ for $x_i \in A$ and $y_i = -1$ for $x_i \in B$, we can write $\mathbf{m}$ as

$$\begin{aligned} \mathbf{m} &= \frac{1}{|A|}\sum_{i \in A} \frac{\mathbf{x}_i}{\|\mathbf{x}_i\|} - \frac{1}{|B|}\sum_{i \in B} \frac{\mathbf{x}_i}{\|\mathbf{x}_i\|} \\ &= \bar{\mathbf{a}} - \bar{\mathbf{b}}. \end{aligned}$$

In the last line, $\bar{\mathbf{a}}$ represents the centroid of the relevant documents, after each has been normalized (projected onto the unit sphere), and $\bar{\mathbf{b}}$ represents the centroid of the irrelevant documents.

Figure 7.5a illustrates in two dimensions. The normalized documents lie on the unit circle. The circled "+" and "−" are the centroids of the positive and negative documents, and the solid vector is their difference $\mathbf{m}$; it represents the optimal query. The dotted vector is $\mathbf{m}$ translated to the origin. The dotted line perpendicular to $\mathbf{m}$ is the decision boundary for the classifier based on $\mathbf{m}$. The value $\cos(\mathbf{x}, \mathbf{m})$ decreases montonically and symmetrically as one moves around the unit circle away from its intersection with $\mathbf{m}$.

We note that the classifier corresponding to the optimal query $\mathbf{m}$ may fail to separate the relevant and irrelevant classes for any value of θ, even when the data is separable. Figure 7.5b gives an example. The data distribution is skewed, pulling the centroids toward one end of the half-circle for each class. The vector shown is $\mathbf{m}$, and the dashed line represents the decision boundary

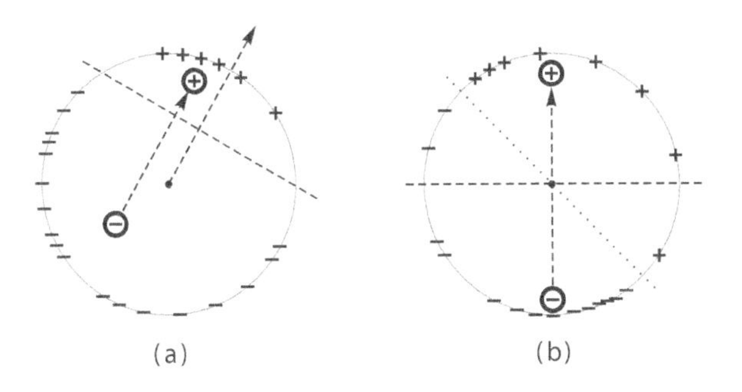

FIGURE 7.5
(a) The geometric interpretation of the optimal query. The optimal query is the vector pointing from the center of gravity of the negatives to the center of gravity of the positives. (b) If the distributions are skewed, the optimal query may fail to separate the classes even when they are linearly separable.

for the corresponding classifier. The decision boundary may be moved up or down by adjusting θ, but it must remain parallel to the dashed line. By contrast, the dotted line represents the decision boundary required to separate the data.

The relevance feedback algorithm is given in figure 7.6. In pseudo relevance feedback, the manual labeling in step 5 is replaced by the assumption that the highest-confidence predictions of f are correct. The algorithm is given in figure 7.7. Pseudo relevance feedback clearly shares critical features of self-training. An initial classifier is given (implicit in $\mathbf{q}_0$). Its high-confidence predictions are taken at face value to label some instances. The classifier is retrained, and the process iterates. The algorithm as given includes throttling (assuming k to be small) and balancing (k is equal for positives and negatives).

7.5 Graph mincut

We turn now to divisive clustering, which, instead of seeking regions with high instance density in which to place cluster centers, seeks low-density regions in which to place boundaries. The most popular divisive methods are graph cut methods.

We shift perspective from the instance space to the similarity graph defined

Input: initial query $\mathbf{q}_0$, unlabeled data $\mathbf{x}_1, \ldots, \mathbf{x}_n$
1. $A, B \leftarrow \emptyset$
2. **For** t from 0 to $T-1$
3. Define $f_t(\mathbf{x}) = \cos(\mathbf{q}_t, \mathbf{x}) - \theta$
4. $S \leftarrow$ the k data points with highest values for f
5. Manually label data points in S; add positives to A, negatives to B
6. $\mathbf{a} \leftarrow$ centroid of A
7. $\mathbf{b} \leftarrow$ centroid of B
8. $\mathbf{q}_{t+1} \leftarrow \mathbf{a} - \mathbf{b}$
9. **End for**
10. **Return** f_T

FIGURE 7.6
The relevance feedback algorithm.

Input: initial query $\mathbf{q}_0$, unlabeled data $\mathbf{x}_1, \ldots, \mathbf{x}_n$
1. $A, B \leftarrow \emptyset$
2. **For** t from 0 to $T-1$
3. Define $f_t(\mathbf{x}) = \cos(\mathbf{q}_t, \mathbf{x}) - \theta$
4. $A \leftarrow A +$ the k data points with highest values for f
5. $B \leftarrow B +$ the k data points with lowest values for f
6. $\mathbf{a} \leftarrow$ centroid of A
7. $\mathbf{b} \leftarrow$ centroid of B
8. $\mathbf{q}_{t+1} \leftarrow \mathbf{a} - \mathbf{b}$
9. **End for**
10. **Return** f_T

FIGURE 7.7
Pseudo relevance feedback.

by the training sample. In the similarity graph, a boundary takes the form of a partitioning, or **cut**, of the graph. The quality of a boundary can be measured in terms of the **cut size**, which is the number of edges that are severed, in an unweighted graph, or the total weight of edges severed, in a weighted graph. The best cut is one that *minimizes* cut size.

The edges of a graph can be viewed in several different ways. (1) As usually defined, the set of edges E contains pairs (v_i, v_j) of vertices. (2) It can be more convenient to identify vertices with their indices, and treat E as a set of pairs (i, j) of node indices. (3) A relation is a set of ordered pairs. Therefore, we may write "$E(i, j)$" as a paraphrase of "$(i, j) \in E$." (4) We may represent the relation $E(i, j)$ as a matrix $\mathbf{E}$ with entry E_{ij} equal to one if (i, j) is an edge, and zero otherwise. This matrix has already been introduced: it is the adjacency matrix of the graph.

An edge connects one vertex to another. One way to identify a class of edges is by providing two vertex sets S and T; we define $E(S, T)$ to be the set of edges directed from a vertex in S to a vertex in T:

$$E(S, T) = \{(i, j) | E(i, j), i \in S, j \in T\}.$$

In an unweighted graph, the size of a set of edges is simply the cardinality of the set:

$$\text{size}(E(S, T)) = |E(S, T)|.$$

In a weighted graph, the appropriate measure of size is the sum of edge weights:

$$\text{size}(E(S, T)) = \sum_{i,j} [\![i \in S, j \in T]\!] w_{ij}.$$

An unweighted graph is the special case in which w_{ij} is equal to one for all edges. We define $w_{ij} = 0$ if there is no edge (i, j), and, contrapositively, we assume that no edge has a weight of zero. We also assume that weights are symmetric, hence that $E(S, T) = E(T, S)$. This is tantamount to assuming an undirected graph.

For a given vertex set S, two edge sets are of particular interest. $E(S, S)$ is the set of edges belonging to the subgraph of S. The **volume** of S is its size:

$$\text{vol}(S) = \text{size}(E(S, S)) = \sum_{i,j} [\![i, j \in S]\!] w_{ij}.$$

The **boundary** of S is the set of edges that begin in S but terminate outside of S:

$$\partial S = E(S, \bar{S})$$

where $\bar{S}$ represents the complement of S within the set of vertices V.

The total weight of edges in $E(S, T)$ is the **cut size** between S and T:

$$\text{cut}(S, T) = \text{size}(E(S, T)).$$

Of particular interest is the cut size between S and its complement, which is the total weight of edges in the boundary of S:

$$\mathrm{cut}(S, \bar{S}) = \mathrm{size}(\partial S). \tag{7.8}$$

If S contains a single node ν, the size of ∂S is the **degree** of ν. We write d_i for the degree of the i-th node:

$$d_i = \sum_j w_{ij}. \tag{7.9}$$

An edge weight often quantifies the similarity between the two nodes connected by the edge, so $\mathrm{cut}(S, T)$ can be thought of as the similarity between vertex sets S and T. At other times, edge weights represent "pipe capacities" or volumes, so the cut can be thought of as the total volume of connection between S and T.

The **mincut** problem is to find a nonempty set $S \subset V$ that minimizes $\mathrm{cut}(S, \bar{S})$. We will discuss methods for finding mincuts, but we must develop some theory first. In particular, we will discuss the Ford-Fulkerson maxflow-mincut algorithm in section 10.6, and we will discuss spectral methods in section 12.3.

7.6 Label propagation

A third broad approach to clustering is to propagate cluster membership information (that is, labels) along the edges of the similarity graph. The idea has arisen in many guises; for example, under the rubric **spreading activation** it can be traced back at least to a 1968 paper by Quillian [184]. Though it has often been proposed with little justification beyond intuitive appeal, it can in fact be placed on a firm mathematical footing, specifically in the form of **random walk** algorithms that we will discuss in chapter 10.

Propagation algorithms can provide a connection between agglomerative and divisive algorithms. It is often natural to formulate them in agglomerative terms, as we do in the next section. But when there is a preference for propagation along high-weight edges and avoidance of low-weight edges, propagation has the effect of placing boundaries along low-weight cuts. That effect is heuristic in the algorithms described here, but under appropriate conditions, a rigorous connection can be established. Details are given in the discussion surrounding the mincut algorithms, in sections 10.6 and 12.3.

7.6.1 Clustering by propagation

The following algorithm is not one that has been proposed in the clustering literature, but serves as an illustration. Choose n vertices at random to serve as cluster seeds, and assign distinct labels to each of them. Choose an unlabeled vertex $\mathbf{x}_i$ that is connected by an edge to a vertex labeled y, and propagate the label y to the vertex $\mathbf{x}_i$. Repeat until no more propagation is possible.

One can imagine many ways of selecting an edge $\mathbf{x}_i$ and label y for propagation. Perhaps the simplest would be the "nearest-neighbor" criterion: find the highest-weight edge that has a labeled node at one terminus and an unlabeled node at the other terminus; let $\mathbf{x}_i$ be the unlabeled node and y be the label of the labeled node. Alternatively, one might use a "voting" criterion. Consider an unlabeled vertex $\mathbf{x}_i$. Define its **neighbors** to be the vertices that it is connected to by an edge. To determine which label should be assigned to $\mathbf{x}_i$, the neighbors vote. In particular, if $\mathbf{x}_j$ is a neighbor labeled y, then $\mathbf{x}_j$ casts a vote in favor of y, with weight w_{ij}. Note that the total vote is $\sum_j w_{ij}$, which is the degree of vertex $\mathbf{x}_i$. The vote in favor of assigning label y to node $\mathbf{x}_i$ is

$$v(\mathbf{x}_i, y) = \sum_j [\![y_j = y]\!] w_{ij}.$$

One chooses the unlabeled node $\mathbf{x}$ and label y that maximize $v(\mathbf{x}, y)$, and one assigns y as the label of $\mathbf{x}$. Ties are broken arbitrarily. After labeling $\mathbf{x}$, the vote assignments $v(\mathbf{x}, y)$ to the remaining unlabeled nodes are recomputed, and a new node is chosen for labeling. The process continues until all nodes have been labeled.

Incidentally, the "nearest-neighbor" criterion can be reduced to the "voting" criterion by defining

$$v(\mathbf{x}_i, y) = \max_j [\![y_j = y]\!] w_{ij}.$$

As just sketched, this is a clustering algorithm, but it is actually more natural as a semisupervised learning algorithm. Instead of choosing cluster seeds at random, we use the labeled data as seed.

7.6.2 Self-training as propagation

It is possible, and informative, to formulate self-training in terms of label propagation in a graph; Yarowsky [239] in fact described self-training informally as label propagation on a graph. The result is quite similar to the algorithm just sketched.

Characterizing self-training as label propagation is simplest when the base learner is a variation of a decision list that we might call a **voting classifier**. Recall that a decision list is a set of rules of the form $F \Rightarrow y$ where F is a single feature and y is a label. We assume that the list contains at most one rule conditioned on a given feature F.

An instance $\mathbf{x}$ is treated as a set of features, and a rule $F \Rightarrow y$ matches $\mathbf{x}$ just in case $\mathbf{x}$ possesses the feature F. The prediction of a standard decision list on instance $\mathbf{x}$ is the prediction of the highest-scoring rule that matches $\mathbf{x}$.

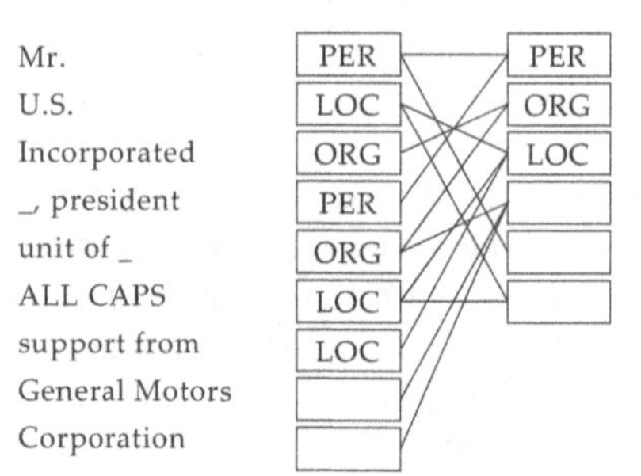

FIGURE 7.8
Bipartite graph.

A voting classifier differs in that rules lack scores; rather each label y receives a vote equal to the total count of rules $F \Rightarrow y$ in the list that match $\mathbf{x}$. The predicted label for $\mathbf{x}$ is the label with the largest vote, or undefined in the case of a tie.

A voting classifier can be represented by introducing nodes corresponding to features, some of which are labeled. The feature node F is labeled y just in case $F \Rightarrow y$ belongs to the rule list. If no rule conditioned on F is present in the list, then the node for feature F is unlabeled. The feature nodes are included together with instance nodes in a bipartite graph, such as that in figure 7.8. There is an edge between instance $\mathbf{x}$ and feature F just in case $\mathbf{x}$ possesses the feature F. The graph is unweighted. The voting classifier's prediction for instance $\mathbf{x}$ can be stated easily in terms of the graph: the label predicted for $\mathbf{x}$ is the one that appears most frequently among the neighbors of $\mathbf{x}$.

A simple version of self-training that constructs a voting classifier is represented by the following label propagation algorithm. Initially, a subset of the nodes is labeled. Let us assume that the initially labeled nodes are all instance nodes, corresponding to a seed consisting of labeled data. (If the seed were a classifier, the initially labeled nodes would be feature nodes. Obviously, a mixture is also possible.)

We begin by propagating labels from instance nodes to feature nodes. The label predicted for a given feature node F is computed in the same way as the label for an instance node: each labeled neighbor of F votes for its own label, and the label with the most votes wins. There is no winner in the case of a tie. The measure of confidence is the percent of the vote that the winner receives. A confidence threshold may be imposed, either in the form

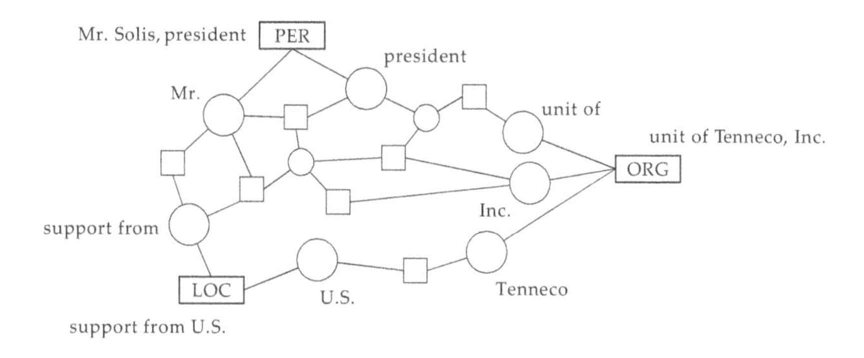

FIGURE 7.9
Propagation into the interior.

of an absolute threshold, or in the form of a quota on the number of feature nodes that receive labels. Persistence may be enforced, either in the sense that the label of a feature node never changes once assigned, or in the sense that a feature node once labeled never becomes unlabeled, though the label may change. It should be clear that label propagation from instance nodes to feature nodes corresponds to the training step in self-training. Its product is a partial labeling of the feature nodes, which, as we have already noted, is equivalent to a voting classifier.

In the second step, labels are propagated from feature nodes back to instance nodes. This corresponds to the labeling step of self-training. The label predicted for an instance node is determined by the vote of its neighbors, which corresponds to the prediction of the voting classifier. Confidence thresholds, throttling, persistence, and the other options previously discussed are available.

The algorithm then repeats, propagating labels from instance nodes to feature nodes and back again, until a stopping criterion is met.

It should be plain that the label propagation algorithm just described is equivalent to self-training, with the exception that it constructs voting classifiers rather than decision lists.

Another way of visualizing the same graph is to place the seed-labeled items at the perimeter and the unlabeled items in the interior, as in figure 7.9. The squares represent instances and the circles represent features. The labeled nodes on the perimeter represent the seed, and the question is how to propagate labels into the interior, effectively partitioning the graph into components, one for each label. This way of viewing the problem is pursued in much greater detail in chapter 10, on graph methods.

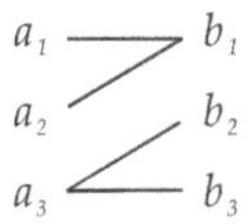

FIGURE 7.10
Bipartite graph of half-instances.

7.6.3 Co-training as propagation

Impressionistically, co-training involves label propagation in a bipartite graph, reminiscent of the bipartite feature-instance graph (figure 7.8) that arose in connection with self-training, but involving the two views. Each instance $\mathbf{x}_i = (\mathbf{a}_i, \mathbf{b}_i)$ consists of two half-instances $\mathbf{a}_i$ and $\mathbf{b}_i$, each being the representation of the instance under one view. That is, $\text{view}_1(\mathbf{x}_i) = \mathbf{a}_i$ and $\text{view}_2(\mathbf{x}_i) = \mathbf{b}_i$. For example, consider the following data set, containing four instances:

$$\mathbf{x}_1 = (a_1, b_1) \qquad \mathbf{x}_2 = (a_2, b_1) \qquad \mathbf{x}_3 = (a_3, b_3) \qquad \mathbf{x}_4 = (a_3, b_2).$$

It can be depicted as the bipartite graph in figure 7.10. The edges in the graph represent individual instances. Labels are assigned to instances (edges), meaning that both half-instances must receive the same label. The graph representation embodies an implicit assumption that a given half-instance is labeled the same way wherever it occurs. The consequence is that labeling any instance in a connected component of the graph effectively labels the entire connected component.

This way of looking at it is actually not quite right. If instances are at all complex, then half-instances will not recur with any appreciable frequency. A half-instance $\mathbf{a}_i$ under the first view is a set of features, as is the half-instance $\mathbf{b}_i$ under the second view. Their disjoint union makes up the features of the instance $\mathbf{x}_i$. Rather than taking half-instances as nodes, it is more reasonable to take individual features as nodes. A pair of views is essentially a partitioning of the universe of features into two sets; a view is one of two disjoint classes of features. If we interpret the nodes in figure 7.10 as features, rather than half-instances, then the edges represent co-occurence. Two features co-occur if there is some instance that possesses them both.

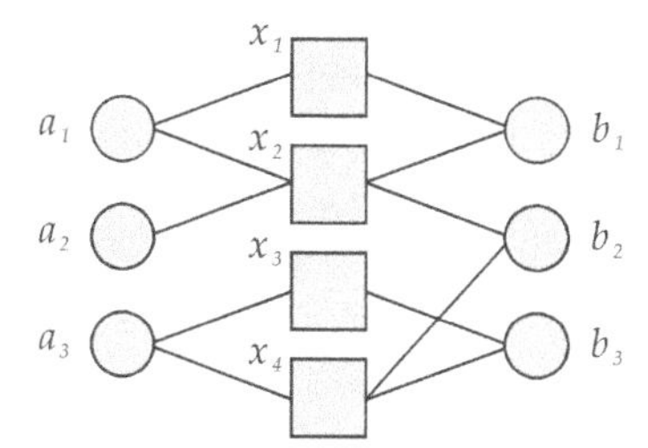

FIGURE 7.11
A tripartite graph representation for co-training. The circles on the left are view-one features, those on the right are view-two features, and the square nodes in the center are instances.

Under that interpretation, however, it is rather too rigid to assume that, if any instance possessing feature F has label y, then *every* instance possessing F has label y. It is reasonable to expect certain features to be highly correlated with a given label – such a correlation makes the rule $F \Rightarrow y$ reliable – but it is too rigid to expect the correlation to be perfect. Label propagation along edges should not be automatic, leading entire connected components to have the same label, but should be somewhat "softer," like the voting method discussed in the previous section for self-training. In fact, for better conformance to the co-training algorithm, we should introduce instances as explicit nodes, as we did with self-training.

These considerations suggest representing instances as follows:

$$\begin{aligned}
\mathbf{x}_1 &= (a_1; b_1) \\
\mathbf{x}_2 &= (a_1, a_2; b_1, b_2) \\
\mathbf{x}_3 &= (a_3; b_3) \\
\mathbf{x}_4 &= (a_3; b_2, b_3)
\end{aligned}$$

in which the a_i represent features belonging to view one, and the b_i represent features belonging to view two; the semicolon separates the features of the two views. For example, the instance $\mathbf{x}_2$ consists of the two half-instances (a_1, a_2) and (b_1, b_2).

The corresponding graph representation is given in figure 7.11. The graph is now tripartite: the square nodes in the center represent instances, the circles on the left represent features of view one, and the circles on the right represent features of view two. In the training step, labels are propagated from the instances outward to the feature nodes of both views, and in the labeling step, labels are propagated back from the features to the instances. The algorithm is actually identical to that described for self-training in the previous section (7.6.2). The distinction between view-one features and view-two features plays no role in the algorithm itself, though it will play a role in our understanding of what the algorithm is trying to accomplish.

We will return to the discussion of co-training in the context of agreement-based methods (chapter 9).

7.7 Bibliographic notes

Many volumes have been written on clustering. I have found Massart & Kaufman [145] useful. The discussion of distance measures largely follows du Toit et al. [80]. The discussion of relevance feedback follows Rocchio [196]; see also [203, 116]. Citations for pseudo relevance feedback include [82, 5, 32, 125].

8

Generative Models

We have previously seen some examples of generative models: the Naive Bayes classifier, for example, and the null category noise model. Generative models characterize the joint distribution $p(\mathbf{x}, y)$ of instances and labels, usually by separately characterizing the **prior** distribution of classes $p(y)$ and the class-specific instance distribution or **likelihood** $p(\mathbf{x}|y)$. This is the source of the name: a generative model represents a hypothesis about how instances are generated for each class. For classification, the generative probability can be reversed using Bayes' rule:

$$p(y|\mathbf{x}) = \frac{p(y)p(\mathbf{x}|y)}{p(\mathbf{x})}.$$

For a given instance $\mathbf{x}$, the denominator is constant, so we can maximize $p(y|\mathbf{x})$ by maximizing the numerator on the right-hand side, the product of prior and likelihood.

Generative models, particularly **mixture models**, are also applicable to clustering. Mixture models have in fact become almost synonymous with clustering in current research. They are the first topic we address in this chapter. In addition to being a prominent approach to clustering, they are also the basis for the earliest examples of semisupervised learning methods, namely, McLachlan's method and Expectation-Maximization (EM).

8.1 Gaussian mixtures

8.1.1 Definition and geometric interpretation

If one has real-valued attributes, a simple model of data generation is a mixture of Gaussians. A separate multi-dimensional Gaussian, or normal, distribution $p(\mathbf{x}|y)$ is assumed for each class, and data is generated by first choosing a class according to $p(y)$ and then generating a data point by sampling from $p(\mathbf{x}|y)$.

A multi-dimensional Gaussian distribution is characterized by a center point or **mean** $\mathbf{m}$, and a **covariance matrix** $\mathbf{C}$ giving the amount of spread in each direction. The distribution has its maximum density at the center, and the

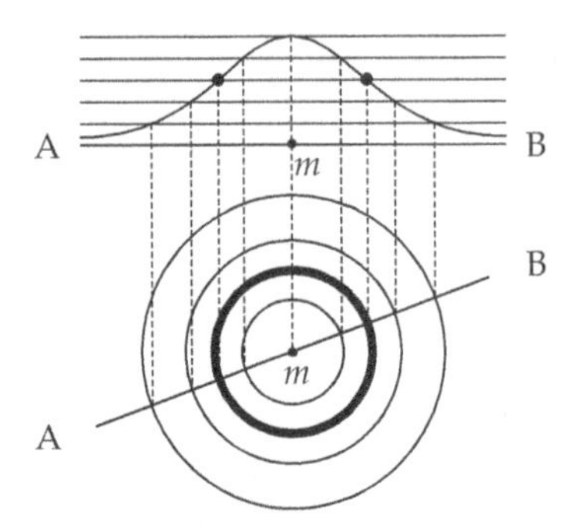

FIGURE 8.1
A two-dimensional Gaussian density.

probability of generating an instance decreases as one moves away from the center. The contours of equal probability form ellipses. If one looks at the probability distribution along a principal axis of the ellipse, its shape is that of a (one-dimensional) normal distribution (figure 8.1).

The normal curve is S-shaped as one moves from the center outward. The distance from the center to the point of recurve is the standard deviation, or the square root of the variance. It can be thought of as the radius of the distribution. In figure 8.1, the mean is the point labeled $\mathbf{m}$, and the contour at one standard deviation (one sigma) from the center is marked with a heavy line.

In a univariate Gaussian, there is one "radius" or "spread" parameter σ, but in a general multivariate Gaussian, the radius depends on the principal-axis direction one measures it in. The single number σ^2 (the variance) is replaced by a matrix (the covariance matrix).

One can understand an arbitrary multi-dimensional Gaussian as the result of scaling, rotating, and translating a standard Gaussian. We start with points $\mathbf{z}$ that have a circular Gaussian distribution with a standard deviation of one, centered at the origin. To scale the distribution, we premultiply by a diagonal matrix

$$S = \begin{bmatrix} \sigma_1 & 0 & \dots & 0 \\ 0 & \sigma_2 & \dots & 0 \\ \vdots & \vdots & \ddots & \vdots \\ 0 & 0 & \dots & \sigma_n \end{bmatrix}$$

whose entries are the standard deviations along each principal axis of the ellipse. We then premultiply by a rotation matrix $\mathbf{R}$, and finally we add $\mathbf{m}$

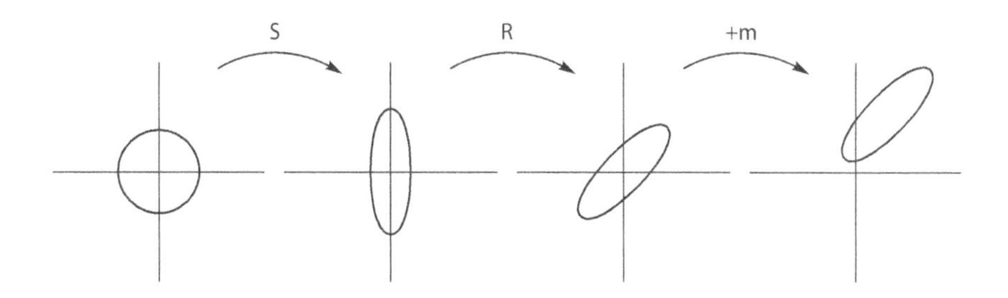

FIGURE 8.2
An arbitrary multivariate Gaussian is related to a standard Gaussian via scaling $\mathbf{S}$, rotation $\mathbf{R}$, and translation.

to re-center the result. These steps are illustrated in figure 8.2. In short, we define

$$\mathbf{x} = \mathbf{m} + \mathbf{RSz}.$$

The inverse transformation takes the actual data points $\mathbf{x}$ and maps them to a coordinate space in which their distribution is a standard normal distribution:

$$\mathbf{z} = \mathbf{S}^{-1}\mathbf{R}^{-1}(\mathbf{x} - \mathbf{m}).$$

This transformation is known as **whitening**. (Why is it called "whitening"? As we will see in chapter 12, the spectrum of a matrix corresponds to its radius along each of its principal axes. White noise has equal power throughout its spectrum, meaning that it has the same radius along all principal axes: it is circular.)

The multivariate standard normal $p(\mathbf{z})$ is defined as

$$p(\mathbf{z}) = \frac{1}{Z} \exp\left\{-\tfrac{1}{2}\|\mathbf{z}\|^2\right\} \tag{8.1}$$

where Z is a normalizing constant. The probability density falls off exponentially with distance from the origin, and the contours of equal density are circles centered on the origin.

A mixture of Gaussians is a distribution in which each class y is associated with a different Gaussian. We will limit attention to mixtures in which the Gaussians have the same covariance matrix but different centers. In such a case, it is convenient to apply the scaling and rotation to the entire space, mapping the original data to what we might call the "whitened space." We use primes for vectors in the whitened space:

$$\mathbf{x}' = \mathbf{S}^{-1}\mathbf{R}^{-1}\mathbf{x}$$

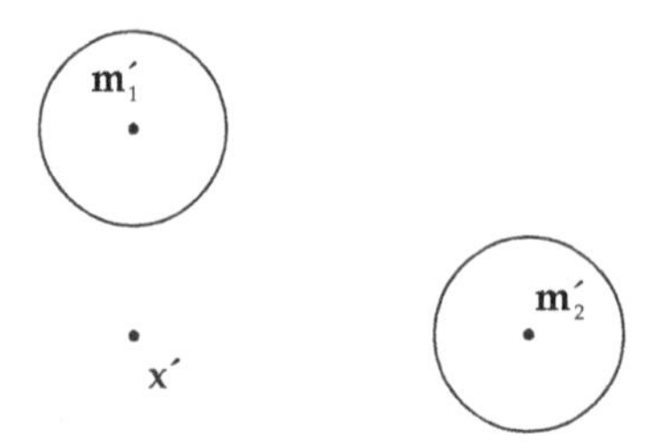

FIGURE 8.3
Two Gaussians.

$$\mathbf{m}'_i = \mathbf{S}^{-1}\mathbf{R}^{-1}\mathbf{m}_i.$$

In the whitened space, (8.1) becomes

$$p(\mathbf{x}'|y_i) = \frac{1}{Z}\exp\left\{-\tfrac{1}{2}\|\mathbf{x}' - \mathbf{m}'_i\|^2\right\}. \tag{8.2}$$

Figure 8.3 illustrates the case with two Gaussians, showing only the centers $\mathbf{m}'_1$ and $\mathbf{m}'_2$ and the one-sigma contours. The contours are the same size and shape for both, meaning that both Gaussians have the same covariance matrix, and the contour shape is circular, meaning that the covariance matrix is a diagonal matrix with the same constant value for every diagonal entry. We can take that value to be one. This illustrates the situation after whitening, whatever the covariance matrix was for the original data.

An instance $(\mathbf{x}', y)$ is generated by choosing $y = 1$ or $y = 2$ at random, according to a prior distribution $p(y)$, and then choosing $\mathbf{x}'$ from the Gaussian distribution centered at $\mathbf{m}'_y$. Knowing only the point $\mathbf{x}'$, as in figure 8.3, there are two possibilities. It might have been generated from the distribution centered at $\mathbf{m}'_1$ or the distribution centered at $\mathbf{m}'_2$. Intuitively, it seems clear that class 1 (centered at $\mathbf{m}'_1$) is the more likely source in this case, but the probability is nonzero for either distribution. Depending on the prior $p(y_2)$, it might even turn out that class 2 is the more likely source. But in the simplest case, which we henceforth assume, the classes have equal prior probabilities, and our original intuition is valid.

8.1.2 The linear discriminant decision boundary

Linear discriminant analysis addresses the classification problem in the case we are considering, namely, a mixture of Gaussians with a common covariance matrix and equal priors. It can be shown that the optimal decision boundary between two adjacent centers consists of the set of points equidistant from the centers in the whitened space. This is the same linear boundary as with a nearest-neighbor algorithm, as in figure 5.5 (page 71).

That the decision boundary is equidistant from the two centers is not hard to see in the one-dimensional case. The univariate Gaussian distribution is

$$p(x|y_j) = \frac{1}{Z} \exp\left\{ -\tfrac{1}{2} \left(\frac{x - m_j}{\sigma} \right)^2 \right\}.$$

The log is easier to work with:

$$\log p(x|y_j) = -\log Z - \tfrac{1}{2} \left(\frac{x - m_j}{\sigma} \right)^2.$$

To determine whether y_1 or y_2 is the more likely source, it suffices to consider the ratio of the probabilities: if $p(x|y_1)/p(x|y_2) > 1$, then y_1 is more likely, and if it is less than one, y_2 is more likely. The decision boundary lies where the likelihood ratio is equal to one, which is to say, where the log likelihood ratio ℓ is equal to zero:

$$\begin{aligned} \ell &= \log \frac{p(x|y_1)}{p(x|y_2)} \\ &= \frac{1}{2} \left(\frac{x - m_2}{\sigma} \right)^2 - \frac{1}{2} \left(\frac{x - m_1}{\sigma} \right)^2 \\ &= \frac{1}{2\sigma^2} \left[2x(m_1 - m_2) + m_2^2 - m_1^2 \right] \\ &= \frac{m_1 - m_2}{\sigma^2} \left(x - \frac{m_1 + m_2}{2} \right). \end{aligned} \tag{8.3}$$

The value $a = (m_1 + m_2)/2$ is the midpoint between the two centers. If m_1 lies to the left of m_2, then $m_1 - m_2$ is negative, and the log likelihood ratio (8.3) is proportional to the "signed distance" $a - x$. It is positive when $x < a$, that is, when x and m_1 both lie to the left of a. It is negative when x is on the same side as m_2. Hence a positive value indicates m_1 and negative indicates m_2, with the decision boundary lying at a, the point that is equidistant between m_1 and m_2.

A similar characterization holds in the multidimensional case as well. Using the probability distribution (8.2), the log likelihood ratio is

$$\begin{aligned} \ell &= \tfrac{1}{2} \|\mathbf{x}' - \mathbf{m}_2'\|^2 - \tfrac{1}{2} \|\mathbf{x}' - \mathbf{m}_1'\|^2 \\ &= \tfrac{1}{2} (\mathbf{x}' - \mathbf{m}_2') \cdot (\mathbf{x}' - \mathbf{m}_2') - \tfrac{1}{2} (\mathbf{x}' - \mathbf{m}_1') \cdot (\mathbf{x}' - \mathbf{m}_1') \\ &= (\mathbf{m}_1' - \mathbf{m}_2') \cdot \mathbf{x}' - \frac{\|\mathbf{m}_1'\|^2 - \|\mathbf{m}_2'\|^2}{2} \\ &= (\mathbf{m}_1' - \mathbf{m}_2') \cdot \mathbf{x}' - (\mathbf{m}_1' - \mathbf{m}_2') \cdot \frac{\mathbf{m}_1' + \mathbf{m}_2'}{2}. \end{aligned} \tag{8.4}$$

We see, by the form of (8.4), that ℓ corresponds to the distance from $\mathbf{x}'$ to a separating hyperplane whose normal vector is $\mathbf{m}_1' - \mathbf{m}_2'$. The point that makes $\ell = 0$ is $\mathbf{x}' = (\mathbf{m}_1' + \mathbf{m}_2')/2$, the midpoint of the line segment $\mathbf{m}_1' - \mathbf{m}_2'$.

That is, the decision boundary is the separating hyperplane that is halfway between $\mathbf{m}_1'$ and $\mathbf{m}_2'$, perpendicular to the line segment connecting them, and the log likelihood ratio ℓ is proportional to the signed distance from $\mathbf{x}'$ to the decision boundary. The sign of ℓ indicates the nearest center (positive for $\mathbf{m}_1'$ and negative for $\mathbf{m}_2'$), and $|\ell|$ is the distance away from the boundary. As in the nearest-neighbor case, the sign can be interpreted as a prediction, and the absolute value can be interpreted as a confidence.

A final point requires clarification. In the previous section, we mentioned the covariance matrix $\mathbf{C}$ and implied that it is connected to the amount of spread in each principal axis direction, that is, to the matrix $\mathbf{S}$. But to this point, we have not actually drawn a formal connection between $\mathbf{C}$ and $\mathbf{S}$. Writing out (8.2) in terms of the original $\mathbf{x}$ and $\mathbf{m}_i$, we have

$$\begin{aligned} p(\mathbf{x}|y_i) &= \frac{1}{Z} \exp\left\{-\tfrac{1}{2}[\mathbf{S}^{-1}\mathbf{R}^{-1}(\mathbf{x}-\mathbf{m}_i)]^{\mathrm{T}}[\mathbf{S}^{-1}\mathbf{R}^{-1}(\mathbf{x}-\mathbf{m}_i)]\right\} \\ &= \frac{1}{Z} \exp\left\{-\tfrac{1}{2}(\mathbf{x}-\mathbf{m}_i)^{\mathrm{T}}\mathbf{R}\mathbf{S}^{-2}\mathbf{R}^{\mathrm{T}}(\mathbf{x}-\mathbf{m}_i)\right\}. \end{aligned}$$

Here we have taken advantage of two facts: since $\mathbf{S}^{-1}$ is diagonal, transposing it has no effect, and since $\mathbf{R}$ is a rotation matrix, its inverse is equal to its transpose.

Though we will not prove it, it turns out that the covariance matrix $\mathbf{C}$ can be expressed as

$$\mathbf{C} = \mathbf{R}\mathbf{S}^2\mathbf{R}^{\mathrm{T}}. \tag{8.5}$$

The rotation $\mathbf{R}$ is the same as we have seen all along: it determines the directions of the principal axes along which the scaling $\mathbf{S}^2$ is applied. Since $\mathbf{S}$ is a diagonal matrix whose elements are the standard deviations in each axis direction, the scaling implicit in $\mathbf{C}$, namely $\mathbf{S}^2$, is a diagonal matrix whose elements are the *variances* in each principal axis direction.

The decomposition (8.5) is significant. Any symmetric matrix (the covariance matrix is symmetric) can be decomposed into the form $\mathbf{Q}\mathbf{\Lambda}\mathbf{Q}^{\mathrm{T}}$ where $\mathbf{Q}$ represents a rigid rotation and $\mathbf{\Lambda}$ represents a scaling. Equivalently, $\mathbf{Q}$ is an orthonormal matrix, meaning that its columns are perpendicular unit vectors, and $\mathbf{\Lambda}$ is a diagonal matrix; for this reason, the decomposition $\mathbf{Q}\mathbf{\Lambda}\mathbf{Q}^{\mathrm{T}}$ is called **diagonalization**. Further, the columns of $\mathbf{Q}$ are the **eigenvectors** of the decomposed matrix, and the elements along the diagonal of $\mathbf{\Lambda}$ are the corresponding **eigenvalues**. In the case of $\mathbf{C}$, we have $\mathbf{Q} = \mathbf{R}$ and $\mathbf{\Lambda} = \mathbf{S}^2$. We will examine diagonalization in more detail in chapter 11.

Since $\mathbf{R}^{\mathrm{T}}\mathbf{R} = \mathbf{R}\mathbf{R}^{\mathrm{T}} = \mathbf{I}$ and $\mathbf{S}^2\mathbf{S}^{-2} = \mathbf{I}$, we have

$$(\mathbf{R}\mathbf{S}^2\mathbf{R}^{\mathrm{T}})(\mathbf{R}\mathbf{S}^{-2}\mathbf{R}^{\mathrm{T}}) = \mathbf{I}$$

implying that:

$$\mathbf{C}^{-1} = \mathbf{R}\mathbf{S}^{-2}\mathbf{R}^{\mathrm{T}}.$$

This yields the multivariate Gaussian distribution in the form one most commonly encounters:

$$p(\mathbf{x}|y_i) = \frac{1}{Z} \exp\left\{-\tfrac{1}{2}(\mathbf{x} - \mathbf{m}_i)^{\mathrm{T}}\mathbf{C}^{-1}(\mathbf{x} - \mathbf{m}_i)\right\}.$$

And the log likelihood ratio (8.4) becomes

$$\ell = (\mathbf{m}_1 - \mathbf{m}_2)^{\mathrm{T}}\mathbf{C}^{-1}\left(\mathbf{x} - \frac{\mathbf{m}_1 + \mathbf{m}_2}{2}\right). \tag{8.6}$$

8.1.3 Decision-directed approximation

From this point, we will assume a whitened space, but drop the primes. Let us represent the locations of the centers as $M = (\mathbf{m}_1, \ldots, \mathbf{m}_k)$. Each possibility for M represents a different **model**. We have seen that, given a point $\mathbf{x}$ and a choice between two labels, the label that maximizes $p(\mathbf{x}|y, M)$ is represented by the sign of the log likelihood ratio, and it is the label that corresponds to the nearer of the two centers. Since this is true for every pairwise choice, it follows that the label that maximizes $p(\mathbf{x}|y, M)$ overall is the label represented by the nearest center. By Bayes' rule:

$$p(y|\mathbf{x}, M) = \frac{p(y|M)p(\mathbf{x}|y, M)}{p(\mathbf{x}|M)}.$$

Since $p(y|M)$ is uniform across choices of y, by assumption, and $p(\mathbf{x}|M)$ is constant with respect to y, it follows that maximizing the likelihood $p(\mathbf{x}|y, M)$ is the same as maximizing the posterior $p(y|\mathbf{x}, M)$. This provides us with a theoretical justification for the nearest-neighbor classifier: under the assumptions given, it maximizes the probability $p(y|\mathbf{x}, M)$. If X is an i.i.d. sample of instances, and Y is an assignment of labels to each, it follows that applying the nearest-neighbor classifier to each instance independently yields the labeling Y that maximizes $p(Y|X, M)$ for the given values of X and M.

In unsupervised learning, we are given X but not M. A Bayesian approach would ask for the labeling that maximizes $p(Y|X)$, integrating over all possible values of M. Since X is fixed, we can maximize $p(Y|X)$ by maximizing

$$p(X, Y) = \int_M p(X, Y|M)dp(M). \tag{8.7}$$

If there are k centers, each being an m-dimensional vector, then a model M can be represented as a point in a km-dimensional space. For a given choice of Y, and remembering that X is fixed, $p(X, Y|M)$ is a function defined for each point M in the model space. The prior density $p(M)$ can be thought of as a distortion of the model space, stretching regions of high prior probability and shrinking regions of low prior probability, so that the area of a region in

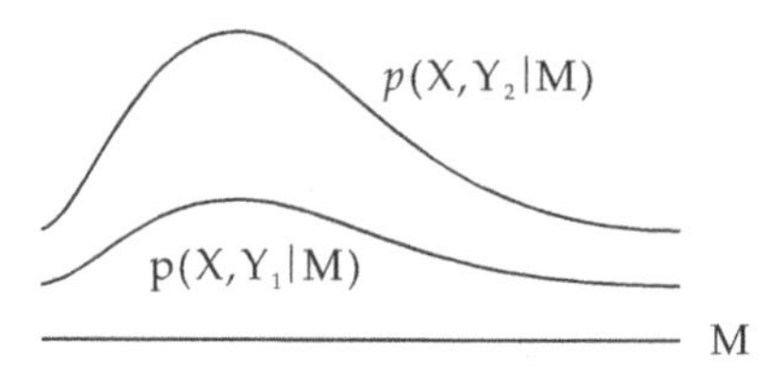

FIGURE 8.4
Cross sections through two surfaces $p(X,Y|M)$. The area under the surface is $p(X,Y)$.

the distorted space corresponds to its prior probability. Then one can think of $p(X,Y|M)$ as a surface over the distorted space. The value of interest, $p(X,Y)$, is the area under that surface. Figure 8.4 shows a cross section through two different hypothetical surfaces, for two different choices of Y.

When we are ignorant of the prior density $p(M)$, we cannot compute (8.7). An expedient in that case is to assume that the surfaces $p(X,Y|M)$ for different values of Y have similar shape but differ in height (and hence, area under the graph) and in the location of their highest point. Then the value $p(X,Y|M)$ at the highest point can be used as a proxy for the area under the graph, and instead of maximizing (8.7), we maximize $p(X,Y|M)$ for both Y and M.

When Y is known, choosing M so as to maximize $p(X,Y|M)$ is the **maximum likelihood principle**. In the case of a mixture of Gaussians, it can be shown that the value M that maximizes the likelihood $p(X,Y|M)$, for fixed X and Y, is obtained by taking the vector means of the training instances that share a given label. That is, for each label j,

$$\hat{\mathbf{m}}_j = \frac{1}{|X_j|} \sum_{\mathbf{x} \in X_j} \mathbf{x}$$

where X_j is the set of training instances that are labeled with the j-th label. This provides justification for choosing centroids as cluster centers, as in k-means clustering – again, under the assumption of a mixture of Gaussians with uniform priors, and remembering that we are working in the whitened space.

When Y is unknown, a plausible heuristic is to start out with some convenient value for M, maximize Y while holding M constant, then hold the

FIGURE 8.5
Alternating maximization of M and Y may not find global maximum.

new Y constant and maximize M. By repeated alternation, one hopes to find the global maximum. This is a heuristic, because it can fail to find the global maximum even if one exists. Think of a space with values of M as one dimension and values of Y as another. The alternating method does searches in axis-parallel directions, as in figure 8.5. More realistically, the searches are not done along lines (one-dimensional subspaces), but within very high-dimensional subspaces.

The k-means algorithm is an application of this alternating maximization heuristic to the case of a mixture of circular Gaussians with uniform priors. We begin with M fixed but arbitrary and choose Y to maximize $p(X, Y|M)$. Given the assumption that $p(Y|M)$ is the same for all Y, this is the same as maximizing $p(X|Y, M)$, which is what the nearest-neighbor algorithm does. Then we hold the new Y fixed and choose M to maximize $p(X, Y|M)$, which is what the centroid rule does. Alternating nearest-neighbors with centroid estimation until convergence yields the k-means algorithm.

The alternating maximization heuristic can be applied more generally. Let `train` be any supervised learning algorithm that chooses M to maximize likelihood $p(X, Y|M)$ for fixed X, Y, and let `label` be a prediction rule that chooses Y so as to maximize $p(Y|X, M)$ for fixed X and M. We continue to assume uniform $p(Y|M)$, so that the Y that maximizes $p(Y|X, M)$ is the same Y that maximizes $p(X, Y|M)$. Using these relaxed definitions of `train` and `label` in the k-means algorithm (figure 7.4, page 139) is known as **decision-directed approximation**.

If we replace the random initialization in decision-directed approximation with either labeled seed instances or a seed classifier, then we obtain a version of self-training without the confidence threshold.

8.1.4 McLachlan's algorithm

One of the earliest forms of self-training was proposed by McLachlan [149] as a way of extending linear discriminant analysis to the case of mixed labeled and unlabeled data. In form, the algorithm is identical to decision-directed approximation, but the assumption of Gaussian mixtures and the particular choice for `train` and `label` make it an exact, rather than heuristic, algorithm.

The algorithm receives a mixture of labeled and unlabeled instances as input. It begins by using the labeled instances to estimate the means and covariance matrix one needs for the log likelihood (8.6), yielding a confidence-weighted classifier $f_0(\mathbf{x})$ whose value is ℓ in (8.6). The classifier is applied to the unlabeled data, and its predictions are taken at face value. A new classifier f_1 is trained on both the labeled data and unlabeled data, using the predictions of f_0 as labels for the unlabeled data. Then the process repeats.

In a little more detail, let P_0 be the set of positive labeled instances (in the labeled seed) and N_0 be the negative labeled instances. Define P_t to be the union of P_0 and the unlabeled instances $\mathbf{x}$ for which $f_t(\mathbf{x}) \geq 0$, and define N_t to be the complement. The classifier f_t is defined as

$$f_t(\mathbf{x}) = (\mathbf{p}_t - \mathbf{n}_t)^{\mathrm{T}} \mathbf{C}_t^{-1} \left(\mathbf{x} - \frac{\mathbf{p}_t + \mathbf{n}_t}{2} \right)$$

where $\mathbf{p}_t$ is the centroid of P_t, $\mathbf{n}_t$ is the centroid of N_t, and the covariance matrix $\mathbf{C}_t$ is computed by pooling both P_t and N_t. The class that f_t predicts for $\mathbf{x}$ is the sign of $f_t(\mathbf{x})$.

McLachlan proves that the algorithm converges, and reduces risk below what is achieved using the labeled data alone. The algorithm is like self-training in that the predictions of the classifier are taken at face value, and the resulting labeled data is used to train a new classifier. It should be noted, however, that there is no thresholding and no throttling.

It is intuitively clear that the two-Gaussian assumption creates conditions under which unlabeled data helps. Consider a sample of instances like that in figure 8.6. Most of the instances are unlabeled, but a few are labeled "+" or "−". The centroids of the labeled data are marked by the circled plus "⊕" and circled minus "⊖". One can tell by the distribution of data points (both labeled and unlabeled) that there is significant error in the means due to sampling error; the small size of the labeled sample distorts the covariance estimates as well. It seems clear that one can improve the estimates by any reasonable method of assigning unlabeled data points to the two classes. Only for a minority of instances in the vicinity of the decision boundary will there be uncertainty about the correct classification.

The fact that McLachlan uses a hard assignment of labels makes the algorithm similar to self-training. An obvious alternative is to make a soft assignment. The idea is to split each unlabeled instance into two fractional instances, one labeled positive and one labeled negative. The relative size of the two fractional instances is determined by one's certainty or uncertainty

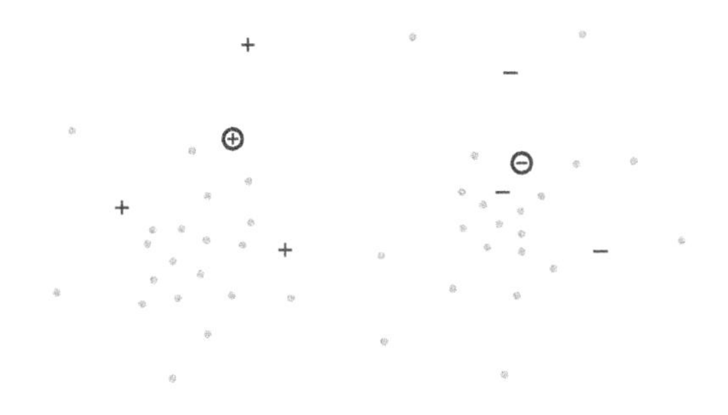

FIGURE 8.6
Mixture of labeled and unlabeled data.

about the true label. Instances near the decision boundary are split nearly in half, while instances further away are assigned almost entirely to the class in whose region they lie. This idea leads to the EM algorithm, which we discuss next.

8.2 The EM algorithm

Self-training is arguably the simplest semisupervised learning algorithm, but it was not the first that was known in computational linguistics. Statistical methods were introduced to computational linguistics from speech recognition, and the Expectation-Maximization (EM) algorithm was a mainstay of speech recognition, in the guise of the Baum-Welch algorithm for unsupervised training of Hidden Markov Models (HMMs).

The EM algorithm is a very general tool, and its use for semisupervised learning is only one application, and not necessarily the best known. It is more commonly used for dealing with missing attributes, or for doing maximum likelihood calculations when direct calculation proves intractable. We will restrict our attention to its use for semisupervised learning.

8.2.1 Maximizing likelihood

To explain the EM algorithm, we begin by considering maximum-likelihood estimation. We have already encountered likelihood, and gotten some idea of how common it is as an objective function. Likelihood is simply the probability of the training data $(\mathbf{X}, \mathbf{y})$ according to a model θ, viewed as a function of θ:

$$L(\theta) = p(\mathbf{X}, \mathbf{y}; \theta).$$

Formally, $\mathbf{X}$ is a matrix whose rows are individual instances $\mathbf{x}_i$ and $\mathbf{y}$ is a column vector whose elements y_i are the instance labels.

Since the instances in the training sample are drawn i.i.d. from the instance distribution, the likelihood can be written:

$$L(\theta) = \prod_i p(\mathbf{x}_i, y_i; \theta).$$

Taking the log does not change the identity of the minimizer, and it will make the maximization easier:

$$\ell(\theta) = \log L(\theta) = \sum_i \log p(\mathbf{x}_i, y_i; \theta).$$

We can convert from a summation over the indices of the training sample to a summation over the instance space by the trick of replacing $p(\mathbf{x}_i, y_i)$ with $\sum_{\mathbf{x},y} [\![\mathbf{x} = \mathbf{x}_i, y = y_i]\!] p(\mathbf{x}, y)$:

$$\ell(\theta) = \sum_i \sum_{\mathbf{x},y} [\![\mathbf{x} = \mathbf{x}_i, y = y_i]\!] \log p(\mathbf{x}, y; \theta) \tag{8.8}$$

$$= \sum_{\mathbf{x},y} c(\mathbf{x}, y) \log p(\mathbf{x}, y; \theta) \tag{8.9}$$

where $c(\mathbf{x}, y)$ represents the **count** of the labeled instance $(\mathbf{x}, y)$ in the data:

$$c(\mathbf{x}, y) = \sum_i [\![\mathbf{x} = \mathbf{x}_i, y = y_i]\!].$$

8.2.2 Relative frequency estimation

We will illustrate the usefulness of the log likelihood (8.9) by deriving the maximum-likelihood solution for a family of models that includes a wide variety of commonly occurring cases. These are cases in which the probability $p(\mathbf{x}, y)$ is a product of parameters θ_{jk} representing probabilities of some aspect or part of $(\mathbf{x}, y)$. For example, in the Naive Bayes model, the quantities θ_{jk} correspond to conditional probabilities of assigning a particular value (k) to a particular attribute (j), given the label y. Or in a Markov model, the quantities θ_{jk} represent transition and emission probabilities, that is, the conditional probability of a transition to state k given that the current state is j, or the conditional probability of emitting the k-th symbol, given that the current state is j.

In general, the subscript j ranges over different kinds of stochastic choices that are faced while generating $(\mathbf{x}, y)$, and k ranges over different outcomes of the stochastic choice. The stochastic choice may be conditionalized on parts of the structure that have already been generated. That is, the stochastic choice is identified with a fixed distribution over outcomes, and the subscript j distinguishes not only the kind of choice but also any aspect of the context that affects the distribution of outcomes. Let us write E_{jk} for the event in which the result of the stochastic choice j is outcome k. The parameter θ_{jk} represents the conditional probability of E_{jk}, given that the stochastic choice being made is stochastic choice j.

A given event may in principle occur more than once, or not at all, in a given labeled instance. We define $f_{jk}(\mathbf{x}, y)$ to be the number of times that the event E_{jk} occurs in $(\mathbf{x}, y)$. Then the probability of the complete labeled instance can be written in the form:

$$p(\mathbf{x}, y; \theta) = \prod_{jk} \theta_{jk}^{f_{jk}(\mathbf{x},y)}. \tag{8.10}$$

This is known as **exponential form**. The model $\theta = (\ldots, \theta_{jk}, \ldots)$ consists of a choice of value for each parameter θ_{jk}.

For models that can be expressed in exponential form, the log likelihood (8.9) becomes

$$\begin{aligned} \ell(\theta) &= \sum_{\mathbf{x},y} c(\mathbf{x}, y) \log \prod_{jk} \theta_{jk}^{f_{jk}(\mathbf{x},y)} \\ &= \sum_{\mathbf{x},y} c(\mathbf{x}, y) \sum_{jk} f_{jk}(\mathbf{x}, y) \log \theta_{jk}. \end{aligned} \tag{8.11}$$

In the second line, "$c(\mathbf{x}, y)$" represents the count of the labeled instance $(\mathbf{x}, y)$, that is, $n\tilde{p}(\mathbf{x}, y)$. Define the total count of event E_{jk} to be

$$c_{jk} = \sum_{\mathbf{x},y} c(\mathbf{x}, y) f_{jk}(\mathbf{x}, y).$$

Then (8.11) can be written simply as

$$\ell(\theta) = \sum_{jk} c_{jk} \log \theta_{jk}.$$

We wish to maximize ℓ, but we must do so under the constraint that the sum of probabilities of outcomes for any given stochastic choice is one. That is

$$\sum_{k} \theta_{jk} = 1 \qquad \forall j.$$

We can place the constraints in the form $\mu_t = 0$ by defining

$$\mu_t = \sum_{k} \theta_{tk} - 1.$$

To do a constrained maximization, we set the gradient of the likelihood equal to a linear combination of the gradients of the constraints:

$$\frac{\partial}{\partial \theta_{jk}} \ell = \sum_t \lambda_t \frac{\partial}{\partial \theta_{jk}} \mu_t. \tag{8.12}$$

To solve this, we first find expressions for each of the partial derivatives:

$$\frac{\partial}{\partial \theta_{jk}} \ell = \frac{c_{jk}}{\theta_{jk}}$$

$$\frac{\partial}{\partial \theta_{jk}} \mu_t = [\![j = t]\!].$$

Substitute these expressions into (8.12) and solve, to obtain

$$\frac{c_{jk}}{\theta_{jk}} = \sum_t \lambda_t [\![j = t]\!]$$

$$\theta_{jk} = \frac{c_{jk}}{\lambda_j}. \tag{8.13}$$

Now we use the constraint that $\sum_k \theta_{jk} = 1$:

$$\sum_k \frac{c_{jk}}{\lambda_j} = 1$$

$$\lambda_j = \sum_k c_{jk}.$$

Substituting this expression for λ_j into (8.13) yields

$$\theta_{jk} = \frac{c_{jk}}{\sum_{k'} c_{jk'}}. \tag{8.14}$$

Very simply, for a particular stochastic choice j, the maximum likelihood estimate of the probability of the k-th outcome is its relative frequency. That is, of the times when the j-th stochastic choice occurred in the training data, what proportion resulted in the k-th outcome?

8.2.3 Divergence

The expression we derived above for log likelihood (8.9) can be re-expressed as a comparison between two probability distributions. One is obvious: it is the distribution $p(\mathbf{x}, y; \theta)$. The other is implicit in the counts $c(\mathbf{x}, y)$. Namely, if we normalize counts, we obtain a probability distribution called the **empirical distribution**, and conventionally written $\tilde{p}$. If there are n training instances, then $\tilde{p}(\mathbf{x}, y) = c(\mathbf{x}, y)/n$, and equation (8.9) becomes

$$\ell(\theta) = n \sum_{\mathbf{x},y} \tilde{p}(\mathbf{x}, y) \log p(\mathbf{x}, y; \theta).$$

In this form, ℓ is closely related to an information-theoretic quantity known as **cross entropy**. In the absence of a standard notation, we will write the cross entropy of p and q as $H(p\|q)$, defined

$$H(p\|q) = -\sum_{\omega} p(\omega) \log q(\omega).$$

The log likelihood (8.9) can be written as

$$\ell(\theta) = -nH(\tilde{p}\|p_\theta) \tag{8.15}$$

where p_θ represents the probability distribution $p(\cdot;\theta)$.

To explain the significance of cross entropy, a quick review of the basic concepts of information theory is useful. The **point entropy** $\eta(\omega)$ of an item ω is a measure of the unexpectedness of ω, and is inversely related to its probability, namely

$$\eta(\omega) = \log \frac{1}{p(\omega)}.$$

Entropy is the average value of point entropy, measured using p:

$$H(p) = -\sum_{\omega} p(\omega) \log p(\omega).$$

It quantifies our uncertainty about which item we get if we sample from p; it is maximized by the uniform distribution, and it is minimized, with value zero, by a point distribution in which a single item has probability one.

The cross entropy $H(p\|q)$ can be thought of as an estimate of $H(p)$ where we sample from p but compute point entropies using q. For example, p may be the unknown distribution underlying a physical process, and q is a hypothesis about it. We can physically sample from p, and we can use the guess q to compute point entropies.

The cross entropy can be expressed as a sum:

$$\begin{aligned} H(p\|q) &= -\sum_{\omega} p(\omega) \log q(\omega) \\ &= -\sum_{\omega} p(\omega) \log p(\omega) + \sum_{\omega} p(\omega) \log \frac{p(\omega)}{q(\omega)} \\ &= H(p) + D(p\|q). \end{aligned}$$

The quantity $D(p\|q)$ is the **divergence** of p from q. It can be shown that divergence is nonnegative, and is equal to zero just in case the distributions p and q are identical. Here is a quick proof. The graph of the natural log function $y = \ln x$ touches the line $y = x - 1$ at the point $x = 1$, $y = 0$ (figure 8.7), and lies below it everywhere else. That is

$$\ln x \leq x - 1 \qquad \text{with equality iff } x = 1.$$

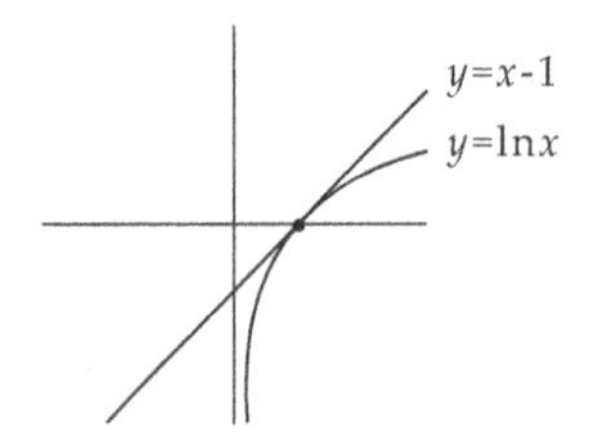

FIGURE 8.7
The graph of $\ln x$ lies everywhere below the graph of $x - 1$, except at the point $x = 1$.

Set x equal to the ratio $q(\omega)/p(\omega)$ for an arbitrary item ω:

$$\ln \frac{q(\omega)}{p(\omega)} \leq \frac{q(\omega)}{p(\omega)} \qquad \text{with equality iff } p(\omega) = q(\omega).$$

Since this is true for arbitrary ω, it is true if we average over choices of ω, and in particular if we take the expectation with respect to p:

$$\sum_{\omega} p(\omega) \ln \frac{q(\omega)}{p(\omega)} \leq \sum_{\omega} p(\omega) \frac{q(\omega)}{p(\omega)} - 1 \tag{8.16}$$

with equality if and only if $p(\omega) = q(\omega)$ everywhere, that is, if and only if $p = q$. Now observe that the left-hand side of (8.16) is the negative of divergence, and the right-hand side simplifies to zero. That is, $D(p\|q) \geq 0$, with equality if and only if $p = q$.

This property makes divergence somewhat like a measure of distance between two distributions, though it is not strictly speaking a distance metric – in particular, it is not symmetric, and it does not satisfy the triangle inequality. But the fact that divergence is nonnegative means that

$$H(p\|q) \geq H(p).$$

In other words, if q is a hypothesis about p, and we sample from p but use q to compute point entropies, then the estimated entropy $H(p\|q)$ is an overestimate unless q is identical to p. If p is fixed, and we vary q, we can find a best approximation to p by minimizing $H(p\|q)$. Indeed, if p is fixed, then minimizing cross entropy is the same as minimizing divergence, and minimizing divergence can be thought of as finding the hypothesis q that is most similar to the unknown distribution p.

Returning to (8.15), we have

$$\ell(\theta) = -n[H(\tilde{p}) + D(\tilde{p}\|p_\theta)].$$

Since $H(\tilde{p})$ is constant across choices of θ, it follows that maximizing log likelihood is the same as minimizing divergence:

$$\arg\max_\theta \ell(\theta) = \arg\min_\theta D(\tilde{p}\|p_\theta). \tag{8.17}$$

As an objective function, divergence from the empirical distribution is equivalent to likelihood – both sort hypotheses in the same way, and the model that minimizes divergence is the same as the one that maximizes likelihood. Or to say it another way, adopting likelihood as an objective function means that, among the distributions p_θ in our hypothesis space, we seek the one that is nearest (in the sense of divergence) to the sample distribution.

8.2.4 The EM algorithm

With this background, we turn to the EM algorithm. As usually formulated, EM estimates a density $p(\mathbf{x}, y)$, rather than learning a classifier. But a classifier is implicit in the density: for a given instance $\mathbf{x}$, choose the label that maximizes $p(\mathbf{x}, y)$. This is equivalent to maximizing

$$p(y|\mathbf{x}) = \frac{p(\mathbf{x}, y)}{p(\mathbf{x})}$$

since the denominator is constant across choices of label.

If we had complete, labeled training data $(\mathbf{X}, \mathbf{y})$, and an appropriate model class, we could use maximum likelihood estimation as described in the previous section. When we have only partial information about the training data, EM comes into play. There are several settings in which some of the training data might be missing. For example, in the classic missing data scenario, some attribute values are unknown for some instances, that is, part of $\mathbf{X}$ is missing. Alternatively, if the labels $\mathbf{y}$ are missing, we have an unsupervised setting, in which case EM is essentially a clustering algorithm. Or again, some of the labels (part of $\mathbf{y}$) may be absent and the rest present, in which case we have a semisupervised learning problem; this last case is the one of particular interest to us.

Let us write $\mathbf{v}$ for the part of $(\mathbf{X}, \mathbf{y})$ that is provided ("visible"), and $\mathbf{u}$ for everything that is missing ("unseen"). In the case of particular interest, $\mathbf{v}$ represents all the instances, along with the labels for the labeled instances, and $\mathbf{u}$ represents the labels for the unlabeled instances. The log likelihood of a model θ becomes

$$\ell(\theta) = \log p(\mathbf{v}; \theta) = \log \sum_{\mathbf{u}} p(\mathbf{u}, \mathbf{v}; \theta). \tag{8.18}$$

In this case, taking the log does not help, because what is inside the log is a sum, not a product.

The EM algorithm provides a way to deal with this difficulty. It is an iterative algorithm. Instead of maximizing log likelihood (8.18) directly, it starts with a guess $\theta^{(0)}$, and constructs a sequence $\theta^{(0)}, \theta^{(1)}, \ldots, \theta^{(t)}, \ldots$ of models with increasing values of likelihood.

The initial model $\theta^{(0)}$ can come from anywhere. We could choose one at random. In the semisupervised setting, we could obtain an initial model by ignoring the unlabeled instances and training just on the labeled instances. That would give us the model that maximizes likelihood on the labeled instances, which is not necessarily the same as the model that maximizes $p(\mathbf{v}; \theta)$ (recall that $\mathbf{v}$ includes not only the labeled instances but also additional instances without labels), but is a natural place to start.

At each step of the algorithm, we use $\theta^{(t)}$ to do a soft labeling of the unlabeld instances, and then we maximize log likelihood of the resulting pseudo-labeled data. That is, we compute a new model $\theta^{(t+1)}$ by maximizing expected log likelihood, averaging across all possibilities for $\mathbf{u}$ in accordance with the probabilities assigned by the *old* model $\theta^{(t)}$:

$$\theta^{(t+1)} = \arg\max_{\theta} \sum_{\mathbf{u}} p(\mathbf{u}|\mathbf{v}; \theta^{(t)}) \log p(\mathbf{u}, \mathbf{v}; \theta). \tag{8.19}$$

The expression that is being maximized is called the EM **auxiliary function**. It is usually written as $Q(\theta; \theta^{(t)})$, but to emphasize its role as a pseudo log likelihood, I prefer to write:

$$Q_t(\theta) = \sum_{\mathbf{u}} p(\mathbf{u}|\mathbf{v}; \theta^{(t)}) \log p(\mathbf{u}, \mathbf{v}; \theta). \tag{8.20}$$

The value $Q_t(\theta)$ in (8.20) is *not* equal to $\ell(\theta)$ in (8.18). But repeatedly applying (8.19) does yield a sequence of models with increasing likelihood.

Before examining how we know that the update (8.19) increases likelihood, let us first consider why maximizing $Q_t(\theta)$ in (8.20) is any easier than maximizing $\ell(\theta)$ in (8.18). For the sake of concreteness, let us assume the semisupervised setting, in which some labels are known and some are unknown. Let us write L for the indices of the labeled instances and U for the indices of the unlabeled instances, and let $\mathbf{u}$ range over labelings of the unlabeled instances; $\mathbf{v}$ includes all instances as well as the known labels. Then (8.20) takes the form:

$$\begin{aligned} Q_t(\theta) &= \sum_{\mathbf{u}} p(\mathbf{u}|\mathbf{v}; \theta^{(t)}) \log \left[\prod_{i \in L} p(\mathbf{x}_i, y_i; \theta) \prod_{j \in U} p(\mathbf{x}_j, u_j; \theta) \right] \\ &= \sum_{i \in L} \log p(\mathbf{x}_i, y_i; \theta) + \sum_{j \in U} \sum_{\mathbf{u}} p(\mathbf{u}|\mathbf{v}; \theta^{(t)}) \log p(\mathbf{x}_j, u_j; \theta). \end{aligned}$$

We again use the trick of writing $p(\mathbf{x}_i, y_i)$ as $\sum_{\mathbf{x},y} [\![\mathbf{x}_i = \mathbf{x}, y_i = y]\!] p(\mathbf{x}, y)$ in order to switch to a summation over the entire space of instances and labels:

$$Q_t(\theta) = \sum_{i \in L} \sum_{\mathbf{x},y} [\![\mathbf{x} = \mathbf{x}_i, y = y_i]\!] \log p(\mathbf{x}, y; \theta) + \sum_{j \in U} \sum_{\mathbf{u}} p(\mathbf{u}|\mathbf{v}; \theta^{(t)}) \sum_{\mathbf{x},y} [\![\mathbf{x} = \mathbf{x}_j, y = u_j]\!] \log p(\mathbf{x}, y; \theta). \quad (8.21)$$

To simplify this expression, we define the count $c_L(\mathbf{x}, y)$ of the instance $(\mathbf{x}, y)$ in the labeled data, and the **expected count** $c_U(\mathbf{x}, y)$ in the unlabeled data, averaging over possible labelings:

$$c_L(\mathbf{x}, y) = \sum_{i \in L} [\![\mathbf{x}_i = \mathbf{x}, y_i = y]\!] \quad (8.22)$$

$$c_U(\mathbf{x}, y) = \sum_{j \in U} [\![\mathbf{x}_j = \mathbf{x}]\!] \sum_{\mathbf{u}} [\![u_j = y]\!] p(\mathbf{u}|\mathbf{v}; \theta^{(t)}). \quad (8.23)$$

Let us consider the summation in (8.23) over possible labelings $\mathbf{u}$ for the unlabeled data. A labeling $\mathbf{u}$ is a vector each of whose elements u_j is a particular choice of label for one unlabeled instance. We are limiting attention to those labelings for which $u_j = y$. Since probabilities of label assignments are independent for different instances, we can split every labeling $\mathbf{u}$ into its assignment at the j-th position, u_j, and its assignments at all other positions; let us write $\mathbf{u}'$ for the rest of $\mathbf{u}$ excluding u_j. Then the sum over $\mathbf{u}$ in (8.23) becomes

$$p(y|\mathbf{x}_j; \theta^{(t)}) \sum_{\mathbf{u}'} p(\mathbf{u}'|\mathbf{v}; \theta^{(t)}).$$

The sum over $\mathbf{u}'$ is unconstrained – we are summing the probabilities of all possibilities – so its value is one. In short:

$$c_U(\mathbf{x}, y) = \sum_{j \in U} [\![\mathbf{x}_j = \mathbf{x}]\!] p(y|\mathbf{x}; \theta^{(t)}). \quad (8.24)$$

Note that the right-hand side is the count of $\mathbf{x}$ in the unlabeled data, times the probability of y given $\mathbf{x}$ according to the old model.

With definitions (8.22) and (8.24) in hand, we can simplify (8.21) to

$$Q_t(\theta) = \sum_{\mathbf{x},y} c_L(\mathbf{x}, y) \log p(\mathbf{x}, y; \theta) + \sum_{\mathbf{x},y} c_U(\mathbf{x}, y) \log p(\mathbf{x}, y; \theta).$$

Defining $c(\mathbf{x}, y) = c_L(\mathbf{x}, y) + c_U(\mathbf{x}, y)$, this becomes simply

$$Q_t(\theta) = \sum_{\mathbf{x},y} c(\mathbf{x}, y) \log p(\mathbf{x}, y; \theta). \quad (8.25)$$

Now we observe that, apart from the difference in how $c(\mathbf{x}, y)$ is defined, (8.25) is identical to (8.9), and can be maximized in exactly the same way.

For example, if $p(\mathbf{x}, y; \theta)$ can be put into exponential form (8.10), then $Q_t(\theta)$ is maximized by relative frequency estimation, as discussed in section 8.2.2.

Now we turn to the question of how we know that the update (8.19) increases likelihood. In the previous section, we showed that maximizing likelihood is the same as minimizing divergence from the empirical distribution (8.17), and minimizing divergence is the same as minimizing the cross entropy $H(\tilde{p}\|p)$. Now consider the EM auxiliary function (8.20). It almost has the form of a cross entropy in which the old-model probability $p(\mathbf{u}|\mathbf{v};\theta^{(t)})$ plays the role of the empirical distribution. The connection that we saw between the old model and the pseudo counts of (8.25) lends strength to the idea of looking at cross entropy (or divergence) from the old distribution.

Specifically, we look at the divergence of the new distribution $p(\mathbf{u}|\mathbf{v};\theta)$ from the old distribution $p(\mathbf{u}|\mathbf{v};\theta^{(t)})$. We know that the divergence is non-negative, and it is strictly positive unless the two distributions are identical:

$$\begin{aligned} 0 &\leq \sum_{\mathbf{u}} p(\mathbf{u}|\mathbf{v};\theta^{(t)}) \log \frac{p(\mathbf{u}|\mathbf{v};\theta^{(t)})}{p(\mathbf{u}|\mathbf{v};\theta)} \\ &= \sum_{\mathbf{u}} p(\mathbf{u}|\mathbf{v};\theta^{(t)}) \log \left[\frac{p(\mathbf{u},\mathbf{v};\theta^{(t)})}{p(\mathbf{v};\theta^{(t)})} \cdot \frac{p(\mathbf{v};\theta)}{p(\mathbf{u},\mathbf{v};\theta)} \right] \\ &= Q_t(\theta^{(t)}) - \ell(\theta^{(t)}) + \ell(\theta) - Q_t(\theta). \end{aligned}$$

That is

$$\ell(\theta) - \ell(\theta^{(t)}) \geq Q_t(\theta) - Q_t(\theta^{(t)}). \tag{8.26}$$

In each step of the EM algorithm, we compute the value

$$\theta* = \arg\max_{\theta} Q_t(\theta)$$

and we set $\theta^{(t+1)} \leftarrow \theta*$. Substituting $\theta*$ for θ in (8.26), we have

$$\Delta\ell \geq Q_t(\theta*) - Q_t(\theta^{(t)}). \tag{8.27}$$

We know that $\Delta\ell$ cannot be negative, because then we would have $Q_t(\theta*) < Q_t(\theta^{(t)})$, but in fact $\theta*$ maximizes $Q_t(\theta)$. Further, if the EM update chooses a model $\theta^{(t+1)}$ that differs from $\theta^{(t)}$, then we know that $\Delta\ell > 0$. Namely, since $\theta*$ maximizes $Q_t(\theta*)$, if $\theta*$ differs from $\theta^{(t)}$, then $Q(\theta*) - Q(\theta^{(t)})$ must be strictly positive, in which case $\Delta\ell$ must also be strictly positive.

Unfortunately, the contrapositive is not true. The inequality (8.27) does not exclude the existence of a model θ' with $\ell(\theta') > \ell(\theta^{(t)})$, but $Q(\theta') < Q(\theta^{(t)})$. That is, when the EM algorithm converges, with $\theta^{(t+1)} = \theta^{(t)}$, we cannot assert that we have found the model that maximizes $\ell(\theta)$. We can only assert that each step of the EM algorithm, *until* convergence, increases likelihood.

We close by noting similarities between the EM algorithm and self-training. One way of interpreting (8.24) is as follows. The model $\theta^{(t)}$ represents a "soft" classifier in the sense that, instead of predicting a fixed label $f(\mathbf{x})$

when presented with instance $\mathbf{x}$, it chooses a label at random, with probability $p(y|\mathbf{x};\theta^{(t)})$. Suppose that we generate a very large corpus by multiplying the training data N times. In the large corpus, there are N copies of each labeled instance $(\mathbf{x}_i, y_i)$ for $i \in L$, and N copies of each unlabeled instance $\mathbf{x}_j$ for $j \in U$. We label the unlabeled instances by taking the predictions of $\theta^{(t)}$ at face value. Those predictions are probabilistic, however, so the result is $Np(y|\mathbf{x}_j;\theta^{(t)})$ copies of $(\mathbf{x}_j, y)$, for each label y. The result is that the count of $(\mathbf{x}, y)$ in the generated corpus, for any $(\mathbf{x}, y)$, is $Nc_L(\mathbf{x}, y) + Nc_U(\mathbf{x}, y) = Nc(\mathbf{x}, y)$. The resulting empirical distribution is

$$\tilde{p}(\mathbf{x}, y) = \frac{Nc(\mathbf{x}, y)}{Nn} = \frac{c(\mathbf{x}, y)}{n}$$

where n is the actual number of training instances. Maximum likelihood estimation is sensitive only to relative frequencies, not absolute frequencies: for example, the parameter estimates (8.14) are unchanged if we multiply the counts in numerator and denominator by N. Or to say it more directly, multiplying the counts by N in (8.25) would have no effect on the maximization: the model that maximizes $NQ_t(\theta)$ is the same as the one that maximizes $Q_t(\theta)$.

In short, we can think of EM as a particular case of self-training in which the `train` step involves maximum-likelihood estimation and produces a probabilistic classifier. Instead of limiting attention to the high-confidence predictions, EM's `label` step conceptually generates a large sample from the predictions of the probabilistic classifier, for all unlabeled instances. Those predicted labels are taken at face value to train the next version of the classifier, and the cycle repeats.

9

Agreement Constraints

Viewed in one way, semisupervised learning of a classifier should be impossible. The usual objective in classifier construction is minimization of training error, that is, minimization of the number of instances for which the classifier's prediction $f(\mathbf{x}_i)$ differs from the actual label y_i. Unless we say something more, adding unlabeled training data is no help: since the labels are not given, we are free to label the examples to agree with the classifier predictions, no matter which classifier we choose. That is, the unlabeled data does not help us choose among classifiers; it is equally happy with them all.

What this argument points out is that something more must be assumed, beyond the objective of reducing training error. We have already seen at least one example of what that "something more" might be. In chapter 6 we saw that, among separating hyperplanes that are equally good at classifying the labeled training data, there may be differences in the size of margin achievable when one also takes unlabeled data into account.

This chapter looks at a different "something more" that might be said. If one assumes two independent **views** of the data, one can look for pairs of classifiers, one for each view, that agree in their predictions on the unlabeled data. Requiring agreement imposes a non-trivial constraint on classifier selection. If the two views of the data are independent, then it should be difficult to find pairs of classifiers that predict the labeled data well *and* agree in their predictions on unlabeled data. If the two views are uncorrelated except for the structure imposed by the true classification, then significant agreement cannot be by chance, but reveals the structure of the true classification.

9.1 Co-training

Co-training was introduced in section 2.3, but the presentation there was brief. The algorithm was given in figure 2.8, repeated here as figure 9.1.

Like self-training, co-training is fundamentally algorithmic. Though predicated on an intuition – in this case, the assumption that each view is independently sufficient to learn the target function, and that the two views are not merely redundant – it is not explicitly designed to optimize an objective function. Nevertheless, the motivating intuition can be sharpened and shown

```
procedure cotrain (L, U)
1  L is labeled data, U is unlabeled data
2  P ← random selection from U
3  loop until stopping criterion is met
4      f1 ← train(view1(L))
5      f2 ← train(view2(L))
6      L ← L + select(label(P, f1)) + select(label(P, f2))
7      Remove the labeled instances from P and replenish P from U
8  end loop
```

FIGURE 9.1
Co-training.

to be valid.

The algorithm can be understood as optimizing accuracy on the labeled data and agreement on the unlabeled data. Co-training constructs two classifiers, f_1 and f_2, each a function of only one view. The algorithm is iterative. In each round of training, the labeled data L consists of the labeled seed data as well as the high-confidence predictions of each of the two classifiers. Training f_1 on L constructs a classifier that is consistent with the seed, with the predictions of (previous versions of) f_2, and incidentally with the predictions of previous versions of f_1. Self-training is used to gradually increase the firing rate while keeping conditional accuracy high, which in this case implies consistency with the other-view classifier.

9.1.1 The conditional independence assumption

Let us consider how to sharpen the two co-training assumptions, namely, (1) that each view is independently sufficient to learn the target function, and (2) that they are not merely redundant. As for the assumption (1), something weaker will actually suffice. We assume that pairs (f_1, f_2) exist in the hypothesis space that are arbitrarily close to (have arbitrarily small error with respect to) the target function. As for assumption (2), Blum & Mitchell's original paper on co-training [21] formalizes "not redundant" in a way that is actually much stronger. We consider each instance to consist of two half-instances $\mathbf{x} = (\mathbf{a}, \mathbf{b})$, one half-instance for each view of the data. Blum & Mitchell's version of assumption (2) is patterned after the Naive Bayes assumption. It states that labeled instances are generated by choosing a label y at random according to a fixed distribution $p(y)$, then choosing each half-instance $\mathbf{a}$ and $\mathbf{b}$ independently, from fixed distributions $p(\mathbf{a}|y)$ for view one and $p(\mathbf{b}|y)$ for view two. It follows that the half-instances $\mathbf{a}$ and $\mathbf{b}$ are

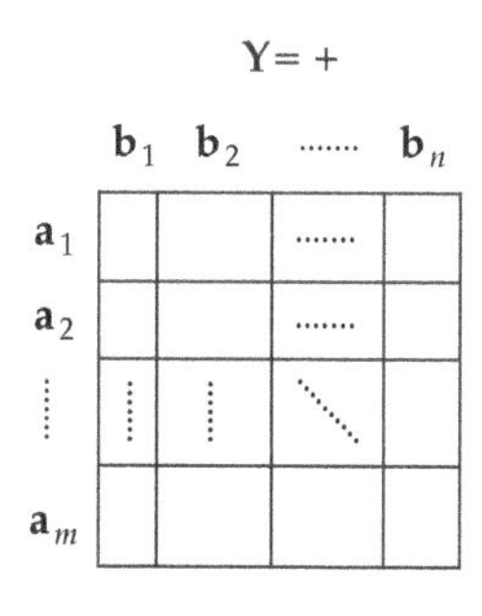

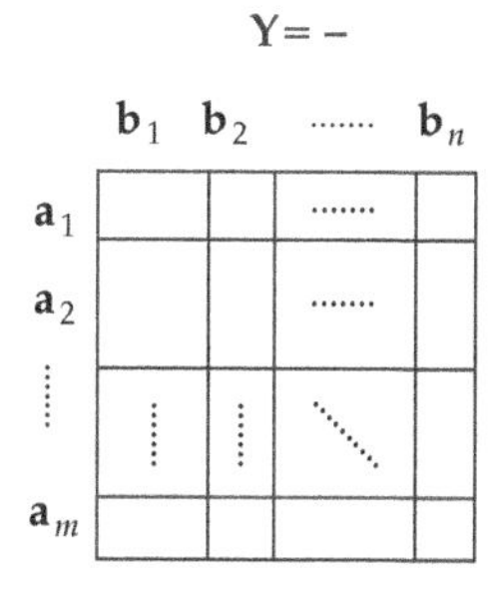

FIGURE 9.2
Conditional independence of half-instances. Area represents probability. Areas are distributed identically in each column, yielding unbroken horizontal lines.

conditionally independent given y.

A graphical depiction may be useful. Let Y be a random variable representing the label; it takes on values "+" and "−." Let A be a random variable representing view one, taking on values $\mathbf{a}_1, \ldots, \mathbf{a}_m$, and let B be a random variable representing view two, taking on values $\mathbf{b}_1, \ldots, \mathbf{b}_n$. The sample space is the set of all labeled instances $(\mathbf{a}, \mathbf{b}, y)$ distributed according to a fixed (but unknown) distribution D. The conditional independence assumption is

$$\Pr[A = \mathbf{a} | B = \mathbf{b}, Y = y] = \Pr[A = \mathbf{a} | Y = y]. \tag{9.1}$$

Let us fix $Y = y$ and consider the probability $p(\mathbf{a}|\mathbf{b}, y)$ as we vary the value of $\mathbf{b}$. By (9.1), $p(\mathbf{a}|\mathbf{b}, y) = p(\mathbf{a}|y)$; that is, for fixed y, the probability assigned to $\mathbf{a}$ is the same regardless of the choice of $\mathbf{b}$. The situation can be depicted graphically as in figure 9.2. The cells contain instances, classified by their values for Y, A, and B. The area of a given cell corresponds to the total probability of instances that fall in that cell. Within the columns, each representing a fixed choice of $\mathbf{b}$, probability is distributed the same way among the $\mathbf{a}_i$, with the result that both the vertical lines and horizontal lines are continuous. The probability distribution among the $\mathbf{a}_i$ differs in the left square, where $Y = +$, and the right square, where $Y = -$, but it is the same within a square.

By contrast, figure 9.3 illustrates lack of conditional independence. The probability distribution over values A differs depending on the value of B, with the consequence that horizontal lines are discontinuous.

Now let us consider any classifier f_1 that depends only on view one. The function f_1 can be entirely characterized by its decision boundary, that is, the set of instances (out of the entire population of instances) for which $f_1(\mathbf{x}) = +$,

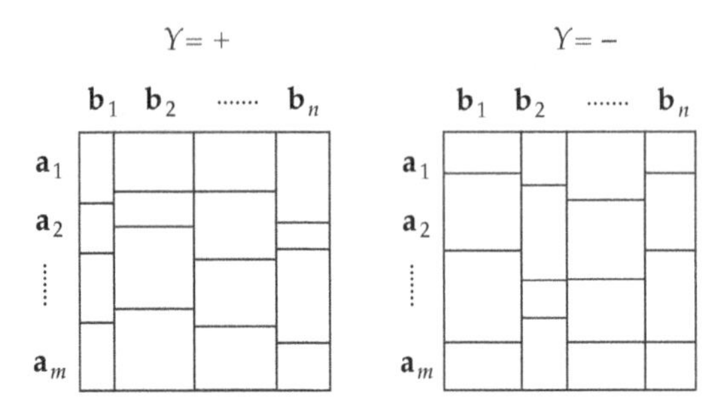

FIGURE 9.3
Conditional dependence of half-instances. Areas are distributed differently in different columns, yielding broken horizontal lines.

and the complement set where $f_1(\mathbf{x}) = -$. Since $f_1(\mathbf{x})$ is sensitive only to the view-one half-instance, the decision boundary is determined by the set of half-instances $\{\mathbf{a}_i\}$ that make the value of f_1 positive. Without loss of generality, let us assume that the first k half-instances $\mathbf{a}_1, \ldots, \mathbf{a}_k$ make f_1 positive. In figure 9.2, the region where f_1 is positive consists of the first k rows of either square. By similar reasoning, letting f_2 be an arbitrary classification function that depends only on view two, the region where f_2 is positive consists of the first ℓ columns of either square. This is shown in figure 9.4. What the figure should make clear is that conditional independence of the half-instances implies conditional independence of the two classifiers: the heavy lines (representing the decision boundaries of the two classifiers) are unbroken (implying conditional independence). The argument applies to any two classifiers, provided that one is sensitive only to view one, and the other is sensitive only to view two.

9.1.2 The power of conditional independence

The conditional independence of arbitrary view-one and view-two classifiers is extremely powerful. It makes labeled data almost unnecessary. In fact, it allows us to cluster the data perfectly. We only need a single labeled data point to tell us which cluster is the positive class and which is the negative class.

To explain why, we need to introduce a couple of definitions. Let f_1 and f_2 be fixed but arbitrary classifiers, with f_1 sensitive only to view one, and f_2 sensitive only to view two. Let us define the **disagreement regions** of f_1 and f_2 to consist of the instances $\mathbf{x}$ such that $f_1(\mathbf{x}) \neq f_2(\mathbf{x})$. The disagreement regions of f_1 and f_2 are shaded in figure 9.5a. The **disagreement probability**

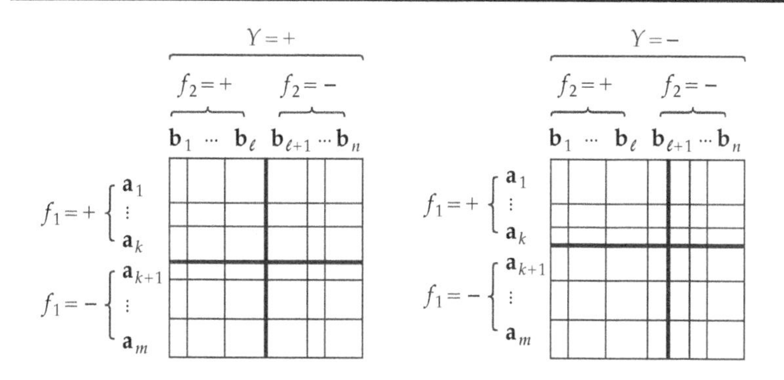

FIGURE 9.4
Conditional independence of half-instances implies conditional independence of single-view classifiers f_1 and f_2.

is the area of the shaded regions.

Let us also define the **(conditional) minority value** of f_1 given that $Y = y$ to be the value u (either positive or negative) with the smaller probability:

$$\Pr[f_1 = u|Y = y] \leq \Pr[f_1 = -u|Y = y]$$

The minority value depends on y. The **minority-value regions** are the regions where f_1 takes on its minority value; they are shaded in figure 9.5b.

As should be clear from the figure, the probability of disagreement is an upper bound on the probability of the minority values. The two probabilities differ in that the disagreement probability includes the areas marked "M," whereas the minority-values probability substitutes the areas marked "m." By definition of minority values, the areas marked "m" must be smaller, or at least no larger, than the areas marked "M."

Finally, let us define the **error regions** of f_1 to be the regions in which the value of f_1 disagrees with the correct label. In figure 9.5b, it happens that the error regions are identical to the minority-values regions. In this case, the disagreement rate between f_1 and f_2 is an upper bound for the error rate of f_1. That is significant, because the disagreement rate can be measured on unlabeled data, but the error rate cannot. If a pair of classifiers (f_1^*, f_2^*) exists that perfectly classify the data, then f_1^* and f_2^* necessarily have a disagreement rate of zero. Conversely, if we find a pair of classifiers with a disagreement rate of zero, their error rate must also be zero, so we have in fact found the target functions f_1^* and f_2^*. Even if the target function is not in the family of classifiers that we consider, by finding a pair (f_1, f_2) with low disagreement rate ϵ, we can guarantee that either of them has an error rate not greater than ϵ.

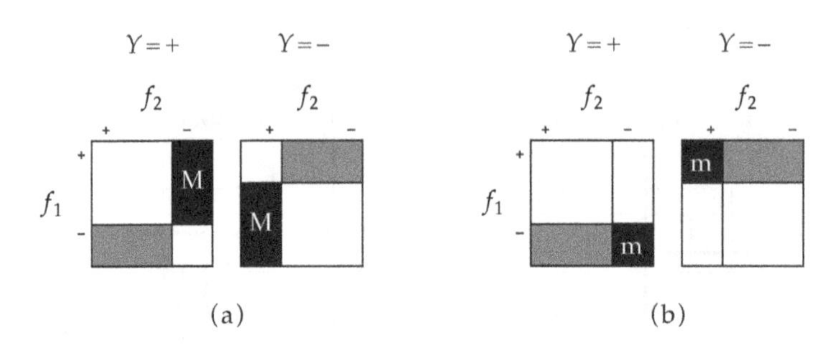

FIGURE 9.5
(a) Shaded and black areas are regions of disagreement; (b) shaded and black areas are regions of minority values. The difference is the replacement of the areas marked "M" with the areas marked "m."

The fly in the ointment is that figure 9.5b illustrates only one possibility for the relationship between minority-value regions and error regions. There are actually four possibilities, as illustrated in figure 9.6. Case (a) is the same as figure 9.5b; the minority-value regions are the same as the error regions, and disagreement upper bounds error. In case (b), the minority-value regions are the opposite of the error regions. In this case, f_1 makes the wrong predictions most of the time, which means that reversing its predictions yields a good classifier. In particular, disagreement upper bounds the error of $-f_1$. If the disagreement rate is zero, then either f_1 or $-f_1$ is the target function, and a single labeled instance will tell us which it is. If the disagreeement rate is nonzero but small, then either f_1 or $-f_1$ has a low error rate, and a small labeled sample will suffice to determine which, with high probability.

In the (c) and (d) cases, the minority values are the same regardless of the true label. In these cases, disagreement does not tell us anything about error, so we need to be able to detect and avoid these cases. Fortunately, it is possible to do so. Let p be the smaller of the two class probabilities: $p = \min_y \Pr[Y = y]$, and let $\hat{p}$ be the smaller of the class probabilities as predicted by f_1, that is, $\hat{p} = \min_y \Pr[f_1 = y]$. Note that $\hat{p}$ is not sensitive to the true label Y – unlike the minority probabilities, $\hat{p}$ can be determined using unlabeled data only. In the (c) and (d) cases, $\hat{p}$ is equal to the minority-values probability. In the (a) and (b) cases, $\hat{p}$ is greater than the minority-values probability.

Now, we would like to find pairs (f_1, f_2) whose disagreement rate ϵ is small. If the target function is in the hypothesis space, we can push ϵ to zero, and even if not, if arbitrarily good approximations are available, we can push ϵ arbitrarily close to zero. We would certainly like to be able to find pairs

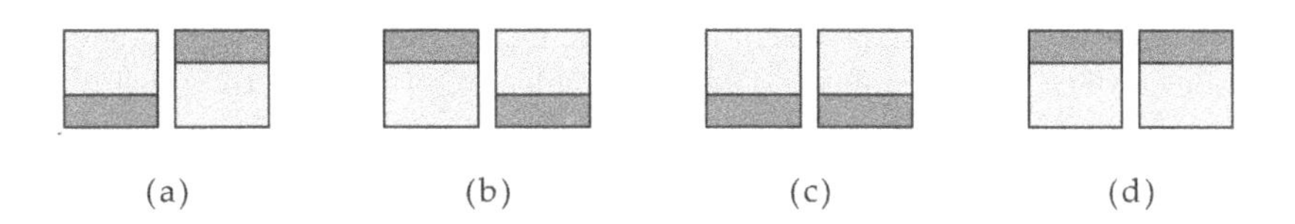

FIGURE 9.6
(a) Minority values equal error values, (b) Minority values of $-f_1$ equal error values, (c–d) Minority values are the same regardless of Y.

whose disagreement rate is significantly less than p. If $\hat{p}$ is close to p, then the disagreement rate will also be significantly less than $\hat{p}$. Let us define a "good pair" to be a pair satisfying $\epsilon < \hat{p}$. Unlabeled data suffices to determine whether a pair is good or not.

We have already seen that the disagreement rate upper bounds the minority-values probability, and that $\hat{p}$ is equal to the minority-values probability in the (c) and (d) cases. So, for the (c) and (d) cases, we have $\epsilon > \hat{p}$, hence no good pairs fall in the (c) and (d) cases. It follows that, for good pairs, disagreement upper bounds true error rate. If we can find good pairs with arbitrarily low disagreement rates, then we can find arbitrarily good approximations to the target function, using unlabeled data only. The only proviso is that we cannot know which value of f_1 represents true positives without at least one labeled instance.

The discussion here is based on a paper by Dasgupta, Littman, and McAllester [67]. We have focussed exclusively on generalization error, that is, expected error on instances drawn according to the population distribution. They give a formal theorem and proof that covers not only generalization error, but also deals with the sampling error inherent in a finite training sample. They include the possibility that a classifier abstains from making a prediction ($f = \perp$), and they show in particular that, under the Blum & Mitchell conditional independence assumption,

$$\Pr[f_1 \neq Y | f_1 \neq \perp] \leq \frac{\epsilon}{\hat{p} - \epsilon}$$

where, as before, ϵ is the rate of disagreement between f_1 and f_2 (where both make predictions) and $\hat{p}$ is the smaller of the two label probabilities according to f_1.

To summarize the discussion, we have concluded that co-training is essentially a version of the "cluster and label" idea introduced in section 7.1. Seeking paired classifiers that agree on unlabeled data effectively clusters the data, and if Blum & Mitchell's conditional independence assumption is true, the resulting clusters, as sets of instances, exactly match the true classes. The labeled data is used only to determine which label belongs to which cluster.

9.2 Agreement-based self-teaching

As we have just seen, the distinctive principle of co-training is agreement between two views that are conditionally independent given the value of target function. The principle is older than Blum & Mitchell's paper. It is already enunciated clearly by de Sa in her dissertation [72] and earlier work [71].

Her approach is based on a supervised learning method called Learning Vector Quantization (LVQ), proposed by Kohonen [129]. The classifier's prediction is determined by a nearest-neighbor rule. Each class y is represented by a point $\mathbf{m}_y$ in instance space, and the predicted label for a new instance $\mathbf{x}$ is the class whose center is nearest:

$$f(\mathbf{x}) = \arg\min_y \|\mathbf{x} - \mathbf{m}_y\|.$$

Recall that $\mathbf{x} - \mathbf{m}_y$ is the vector whose tail is at the point $\mathbf{m}_y$ and whose head is at the point $\mathbf{x}$, and $\|\mathbf{x} - \mathbf{m}_y\|$ is its length.

The algorithm is iterative. It begins with an arbitrary placement of the class centers $\mathbf{m}_j$, and then cycles through the instances until a stopping criterion is met, updating the locations of the centers at each step. Let $\mathbf{x}$ be the current instance, let $\mathbf{m}_y$ be the center for the correct class, and let $\mathbf{m}_z$ be the nearest incorrect class center. Intuitively, z is the "distractor" class. The aim is to move the centers $\mathbf{m}_y$ and $\mathbf{m}_z$ so that $\mathbf{x}$ lies closer to the correct center $\mathbf{m}_y$ and further from the distractor $\mathbf{m}_z$.

The situation is depicted in figure 9.7. The current instance is the solid point $\mathbf{x}$, and its true label is y. The center for class y is the circled cross $\mathbf{m}_y$, and the center for the distractor class z is $\mathbf{m}_z$. Because of the nearest-neighbor criterion, the decision boundary between classes y and z is the perpendicular bisector of the line connecting the centers $\mathbf{m}_y$ and $\mathbf{m}_z$. The connecting line and the decision boundary are shown as solid lines. (Note that linear discriminant analysis yields the same decision boundary, as discussed in section 8.1.2.) In figure 9.7, the current instance $\mathbf{x}$ happens to fall on the wrong side of the decision boundary and our aim is to move the boundary to correct that. If it had fallen on the correct side of the boundary, the aim would have been to move it deeper into "friendly territory."

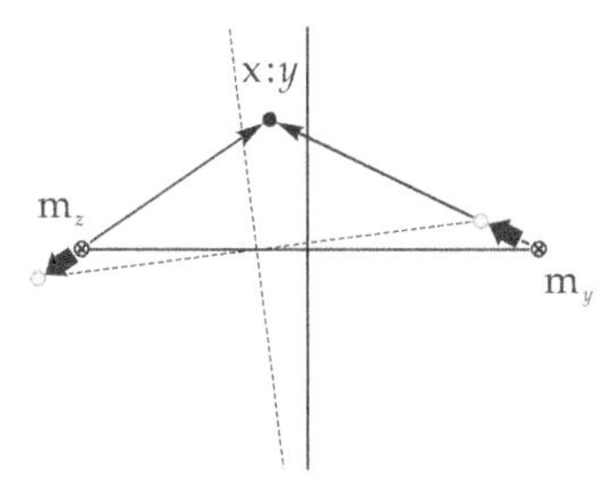

FIGURE 9.7
The LVQ update.

The (thin) vector with its tail at $\mathbf{m}_y$ and its head at $\mathbf{x}$ is the vector $\mathbf{x} - \mathbf{m}_y$, and the comparable vector with tail at $\mathbf{m}_z$ is $\mathbf{x} - \mathbf{m}_z$. We move the center for the correct label a step of size ϵ toward $\mathbf{x}$, and we move $\mathbf{m}_z$ an ϵ-sized step away from $\mathbf{x}$. The steps are shown as thick arrows. The new centers are dotted circles, and the new connecting line and decision boundary are shown as dashed lines. In this particular case, the adjustment has moved $\mathbf{x}$ over to the correct side of the decision boundary.

Algebraically, the update is

$$\begin{aligned} \mathbf{m}_y &\leftarrow \mathbf{m}_y + \epsilon \tfrac{\mathbf{x} - \mathbf{m}_y}{\|\mathbf{x} - \mathbf{m}_y\|} \\ \mathbf{m}_z &\leftarrow \mathbf{m}_z - \epsilon \tfrac{\mathbf{x} - \mathbf{m}_z}{\|\mathbf{x} - \mathbf{m}_z\|}. \end{aligned} \tag{9.2}$$

We normalize the vectors to get unit vectors in the desired direction, then we scale the unit vectors to length ϵ.

The situation depicted in figure 9.7 has a couple of properties that we would like to state as requirements. First, we assumed that $\mathbf{m}_y$ and $\mathbf{m}_z$ are the two nearest centers to $\mathbf{x}$. By definition, $\mathbf{m}_z$ must be one of the two nearest centers, but nothing guarantees that $\mathbf{m}_y$ is. We add that as a requirement. If $\mathbf{m}_y$ is not one of the two nearest centers, then no update is done. Second, the figure depicts $\mathbf{x}$ lying near the boundary between y and z. We add that as a requirement as well: an update is done only if $\mathbf{x}$ lies within a margin c of the boundary, where c, like ϵ, is a free parameter of the algorithm. Recall that we can compute the distance between $\mathbf{x}$ and the boundary by projecting $\mathbf{x}$ onto the line connecting the centers: see equation 5.3, page 70.

The center for a given class y is initially placed at the centroid of the instances belonging to class y. Then multiple passes are taken through the data, using the LVQ update just described to improve the positioning of decision boundaries. The step size ϵ and margin c are gradually decreased with each pass.

To apply this algorithm to a mixture of labeled and unlabeled data, de Sa proposes using two modalities (views), training separate classifiers for each modality, and using agreement between the classifiers in lieu of labeled data. As in co-training, each instance is taken to consist of two half-instances, $\mathbf{x}^{(1)}$ and $\mathbf{x}^{(2)}$. The first half-instances $\mathbf{x}^{(1)}$ represent points in the instance space for the first view, and the second half-instances $\mathbf{x}^{(2)}$ represent points in the instance space of the second view. We construct separate classifiers for each space. In this case, the classifiers are nearest-neighbor classifiers, represented by the locations of the class centers: $(\mathbf{m}_1^{(1)}, \ldots, \mathbf{m}_L^{(1)})$ for the first view and $(\mathbf{m}_1^{(2)}, \ldots, \mathbf{m}_L^{(2)})$ for the second view. The prediction of the k-th classifier ($k \in \{1, 2\}$) is

$$f_k(\mathbf{x}) = \arg\min_y \|\mathbf{x}^{(k)} - \mathbf{m}_y^{(k)}\|.$$

The update rule (9.2) is essentially unchanged, except that it applies to each view separately:

$$\begin{aligned} \mathbf{m}_y^{(k)} &\leftarrow \mathbf{m}_y^{(k)} + \epsilon \frac{\mathbf{x}^{(k)} - \mathbf{m}_y^{(k)}}{\|\mathbf{x}^{(k)} - \mathbf{m}_y^{(k)}\|} \\ \mathbf{m}_z^{(k)} &\leftarrow \mathbf{m}_y^{(k)} - \epsilon \frac{\mathbf{x}^{(k)} - \mathbf{m}_y^{(k)}}{\|\mathbf{x}^{(k)} - \mathbf{m}_y^{(k)}\|}. \end{aligned}$$

The critical difference is that, whereas y was previously provided by having labeled data, it is now defined to be the label predicted by the other classifier, computed before the update. (The two classifiers are updated in parallel.) The conditions remain in force that $\mathbf{m}_y^{(k)}$ must be one of the two closest centers to $\mathbf{x}^{(k)}$, and $\mathbf{x}^{(k)}$ must be within margin c of the decision boundary between $\mathbf{m}_y^{(k)}$ and $\mathbf{m}_z^{(k)}$, otherwise there is no update.

With a mix of labeled and unlabeled data, the labeled data is used to initialize the class centers, and is included in the passes through the data. The updates with labeled data, of course, use the given labels, as in the supervised algorithm.

9.3 Random fields

9.3.1 Applied to self-training and co-training

Another way of capturing (soft) agreement constraints is via a random field. A random field is a probability function defined on a labeled graph. The vertices of the graph are ordered arbitrarily $x_1, \ldots, x_n$, and a corresponding assignment of labels $\mathbf{y} = (y_1, \ldots, y_n)$ is called a **configuration**. An energy function $H(\mathbf{y})$ is defined in terms of the labels. For our purposes, "energy" can be thought of as "error" or "cost" – high energy is bad, low energy is good. We are particularly interested in fields in which the labels are drawn from $\{+1, -1\}$, and the energy is defined as

$$H(\mathbf{y}) = -\sum_{(i,j)\in E} y_i y_j \tag{9.3}$$

where E is the set of edges of the graph. In words, if two vertices connected by an edge have different labels, then a cost $(+1)$ is contributed to H, and if two vertices connected by an edge have the same label, then a credit (-1) is contributed. That is, agreement between neighboring vertices is good and disagreement is bad.

Fields of the form (9.3) are related to **Ising models**. The physicist Ernst Ising proposed a model of ferromagnetism in which the graph is a rectangular grid, and the energy is

$$H(\mathbf{y}) = -\frac{1}{kT}\left(J\sum_{(i,j)\in E} y_i y_j - mB\sum_i y_i\right)$$

where T is temperature, k is a constant determining the influence of temperature, B is the intensity of a constant external magnetic field, and J and m are constants determined by physical properties of the magnetic material. Our energy (9.3) is the special case where there is no external field and $J = kT$; we also abstract away from Ising's rectangular grid.

Recall that we can view self-training and co-training as involving bipartite graphs with labeled vertices, where vertices representing features alternate with vertices representing instances. The self-training graph was illustrated in figure 7.8 (page 148), and the co-training graph was illustrated in figure 7.11 (page 151). The co-training graph is bipartite in the sense of involving alternating instance and feature nodes, but tripartite in the sense that features can be divided into view-one features and view-two features.

In the training step, self-training selects rules $F \Rightarrow y$ such that instances possessing feature F are almost always labeled y. Selecting rule $F \Rightarrow y$ is equivalent to labeling the feature vertex F with label y, and the instances that possess feature F are represented by the instance vertices connected to F by an edge. Thus, in terms of the graph, the training step assigns labels to features so as to maximize agreement with neighboring labeled instances. The labeling step, in turn, assigns labels to instances $\mathbf{x}$ according to the prediction of rules $F \Rightarrow y$ for features F possessed by $\mathbf{x}$. That is, the labeling step assigns labels to instance vertices so as to maximize agreement with neighboring labeled features.

Co-training does the same, though it adds the complication that it partitions the features into two views and handles each view separately, a complication that we ignore here. What is important for present purposes is that both self-training and co-training seek to maximize agreement between the labels of neighboring vertices in a graph. That is, both aim to minimize the energy (9.3).

A random field uses the energy function $H(\mathbf{y})$ to define a probability distribution over configurations, namely

$$p(\mathbf{y}) = \frac{1}{Z}e^{-H(\mathbf{y})}$$

where Z is a normalizing constant. The distribution assigns high probability to configurations with low energy (low cost) and low probability to configurations with high energy (high cost). Notice that the negative sign here cancels the negative sign in (9.3): if neighboring vertices agree in label, probability is increased, and if they disagree, probability is decreased.

A field of the sort under discussion is readily adapted to semisupervised learning. Labeled data is represented as instance vertices whose labels are fixed. The goal is to assign labels to the remaining vertices (features and instances) so as to maximize $p(\mathbf{y})$. The situation in which one is given a seed classifier instead of labeled data is also easily represented: the seed classifier consists in an initial assignment of labels to some features. One again seeks to assign labels to the remaining vertices so as to maximize $p(\mathbf{y})$.

9.3.2 Gibbs sampling

A simple and well-known algorithm for finding a configuration that maximizes $p(\mathbf{y})$ is Gibbs sampling. Consider a particular vertex x with neighbors $x_1, \ldots, x_T$. The vertex x may represent either an instance or a feature. We define a local probability distribution over the labels of x as

$$p(y) = \frac{e^{-H(y)}}{e^{-H(y)} + e^{-H(-y)}} \tag{9.4}$$

where

$$H(y) = -\sum_{i=1}^{T} y \cdot y_i$$

where y_i represents the label of neighboring vertex x_i. Let p represent the number of positives among the neighboring labels $y_1, \ldots, y_T$, and let n represent the number of negatives. Then

$$-H(+1) = p - n \qquad -H(-1) = n - p.$$

The Gibbs sampling algorithm begins with a labeled graph. The initial labeling is immaterial. The algorithm proceeds iteratively. At each step, a vertex is chosen uniformly at random. Cycling systematically through the vertices works just as well, but the mathematical justification is more involved, so we assume random choice. Let the chosen vertex be x, and let its neighbors be $x_1, \ldots, x_T$. A new label y is selected for x, by sampling from $p(y)$ as defined in (9.4). It is permissible for the new label to be the same as the old. At each iteration, a single vertex is relabeled in this manner. A *sweep* is completed when every vertex has been relabeled once. Multiple sweeps yield a sequence of configurations $\mathbf{y}_1, \mathbf{y}_2, \ldots, \mathbf{y}_t, \ldots$.

As the number of sweeps increases, the relative frequency of a given label at a given vertex converges to its true probability. That is, if we run the Gibbs sampler for enough iterations, and we accumulate the relative frequencies of labels at each vertex, we can determine the probability that a given vertex has a given label, according to the random field.

We can view Gibbs sampling as a stochastic version of self-training. In the case of a bipartite feature-instance graph, a single step of the Gibbs sampler corresponds either to a labeling step, if the selected vertex is an instance vertex, or a training step, if it is a feature vertex. A labeling step consists in relabeling an instance according to a stochastic vote by the features it possesses. A training step corresponds to the replacement of a rule $F \Rightarrow y$ with a rule $F \Rightarrow y'$, according to a stochastic vote by the instances that possess feature F. After a sufficient number of iterations, the relative probability of labels for a given instance can be determined by relative frequency in the sequence of configurations constructed, and likewise a final classifier can be constructed of rules $F \Rightarrow y$ with probability determined by their relative frequency (among rules conditioned on F) in the sequence of configurations.

9.3.3 Markov chains and random walks

We will not attempt to give a full justification for the Gibbs sampler, but we will at least give the sketch of an account. Doing so will give us occasion to introduce the concepts of Markov chains and random walks, concepts which will arise again when we discuss label propagation in the next chapter.

The Gibbs sampler cycles through the nodes of the graph. When choosing nodes at random, it may take longer to finish a sweep, which is to say, to visit each node once, but eventually all nodes will be selected at least once. When the sampler visits node x_i, it chooses a new label by sampling from $p(y|x_i)$. The possibility that the new label will be the same as the old one is not ruled out. Define the **possible labels** for x_i to be those with $p(y|x_i) > 0$, and define a **possible state** of the system to be any way of assigning a possible label to each node. Clearly, after we have made one pass through the nodes, visiting each node in turn, the state we obtain might be any possible state (though not all possible states will be equally probable).

The action of the Gibbs sampler defines a **Markov chain**: that is, a matrix $\mathbf{P}$ in which the entry p_{ji} represents the probability of a transition from state i to state j. Note that i is now ranging over states of the complete graph, not over individual nodes. The i-th column of $\mathbf{P}$ represents the "next-state" probability distribution for state i. Be warned that many authors use rows to represent next-state distributions, instead of columns, but I prefer following the usual convention of interpreting vectors (here, probability distributions over states) as column vectors. If there are n nodes in the graph, there is a nonzero probability of completing a sweep through a graph in n steps. Since any state j is reachable from any other state i after one sweep through the graph, there is a nonzero probability of reaching any j from any i after n steps. The probability of reaching j from i after n steps is $\mathbf{P}^n_{ji}$, so we have just concluded that the matrix $\mathbf{P}^n$ has all entries strictly positive. A probability matrix with this property, namely, that some power of the matrix has all entries strictly positive, is said to be **ergodic**.

Suppose we have a probability distribution $\mathbf{x}$ over states. We can run a number of Gibbs samplers simultaneously, and think of each one as a particle moving through the state space. Then x_i is the proportion of particles that are in the state i at a particular time. Alternatively, we can think of a single particle moving through the state space. Then x_i represents the probability of the particle being in state i, if we choose a time at random.

The distribution $\mathbf{x}$ is a function of time, so we should actually write $\mathbf{x}_t$ for the distribution at time t. We will write $x_i^{(t)}$ for the i-th element of $\mathbf{x}_t$. The probability of being in state j at time $t+1$ is the weighted average of probabilities of being in state i at time t and making the transition from i to j:

$$x_j^{(t+1)} = \sum_i p_{ji} x_i^{(t)}.$$

That is, the j-th element in $\mathbf{x}_{t+1}$ is the dot product of the j-th row of $\mathbf{P}$ with $\mathbf{x}_t$. (Rows correspond to destination states.) Hence

$$\mathbf{x}_{t+1} = \mathbf{P}\mathbf{x}_t.$$

Since the transition matrix $\mathbf{P}$ is unchanging, it follows that

$$\mathbf{x}_{t+k} = \mathbf{P}^k \mathbf{x}_t$$

and in particular,

$$\mathbf{x}_k = \mathbf{P}^k \mathbf{x}_0. \tag{9.5}$$

Earlier, in the discussion of Gaussian mixtures (section 8.1.2), we introduced the idea of diagonalization. Intuitively, when one multiplies by a matrix, the effect can be decomposed into a rotation and a scaling. Further, the directions in which the scaling is applied are represented by the eigenvectors of the matrix, and the amount of scaling in each eigenvector direction is represented by the corresponding eigenvalue.

In that earlier discussion, the matrix in question was symmetric. Now we are interested in the probability matrix $\mathbf{P}$, which is not usually symmetric. Even so, it can be diagonalized. In the earlier discussion, we introduced eigenvectors as the directions in which the scaling is applied. The more precise definition is that an eigenvector represents a direction in which the effect of the matrix is a scaling but no rotation. That is, $\mathbf{u}$ is an eigenvector of $\mathbf{P}$ just in case

$$\mathbf{P}\mathbf{u} = \lambda\mathbf{u}.$$

The scalar λ is the eigenvalue corresponding to the eigenvector $\mathbf{u}$.

If $\mathbf{P}$ is an $n \times n$ matrix, it may have as many as n distinct eigenvectors. Consider what happens when we collect those eigenvectors together as the columns of a matrix:

$$\mathbf{P}\left[\mathbf{u}_1 \ldots \mathbf{u}_n\right] = \left[\lambda_1 \mathbf{u}_1 \ldots \lambda_n \mathbf{u}_n\right]. \tag{9.6}$$

The right-hand side can be expressed as a matrix product of form:

$$\left[\mathbf{u}_1 \ldots \mathbf{u}_n\right] \begin{bmatrix} \lambda_1 & \ldots & 0 \\ \vdots & \ddots & \vdots \\ 0 & \ldots & \lambda_n \end{bmatrix}.$$

Writing $\mathbf{U}$ for the matrix whose columns are eigenvectors and $\mathbf{\Lambda}$ for the diagonal matrix containing eigenvalues, the equation (9.6) becomes

$$\mathbf{PU} = \mathbf{U\Lambda}$$

$$\mathbf{P} = \mathbf{U\Lambda U}^{-1}. \tag{9.7}$$

This is the more general form of diagonalization, which is valid provided that the inverse $\mathbf{U}^{-1}$ exists.

In our earlier dicussion, we considered diagonalization of symmetric matrices. When a symmetric matrix is diagonalized, it can be shown that the eigenvectors in $\mathbf{U}$ are mutually perpendicular unit vectors, that is, that $\mathbf{U}$ describes a rigid rotation, and hence that $\mathbf{U}^{-1}$ exists, and in fact equals $\mathbf{U}^{\mathrm{T}}$. In the case of current interest, $\mathbf{P}$ is usually not symmetric. It is, however, ergodic, and ergodicity can also be shown to imply the existence of $\mathbf{U}^{-1}$. Ergodicity also implies two important properties of the eigenvalue matrix $\mathbf{\Lambda}$: namely, that the largest eigenvalue in $\mathbf{\Lambda}$ equals 1, and that all the remaining eigenvalues are nonnegative but strictly less than one. Without loss of generality, we can assume that the eigenvalues are sorted in increasing order, so that $\lambda_n = 1$ and $0 \leq \lambda_i < 1$ for $i < n$.

These properties imply immediately that $\mathbf{P}$ has a unique stationary distribution. A **stationary distribution** is a probability distribution $\mathbf{x}$ such that

$$\mathbf{Px} = \mathbf{x}. \tag{9.8}$$

In other words, a stationary distribution is an eigenvector whose eigenvalue is 1. For an ergodic matrix, such an eigenvector exists and is unique. Namely, the stationary distribution of $\mathbf{P}$ is the eigenvector $\mathbf{u}_n$ corresponding to the eigenvalue $\lambda_n = 1$.

The stationary distribution is the unique fixed point for the operation of premultiplying by $\mathbf{P}$. Once we arrive at the stationary distribution, applying $\mathbf{P}$ will never take us elsewhere. It is natural to ask, then, whether applying $\mathbf{P}$

repeatedly to an arbitrary distribution will eventually bring us to the stationary distribution. That is, does the sequence $\mathbf{x}_0, \mathbf{x}_1, \ldots$ defined by (9.5) converge to the stationary distribution of $\mathbf{P}$, or does it fail to converge? Since the stationary distribution is the only fixed point, the sequence cannot converge to something else, and since the vectors $\mathbf{x}_k$ are all probability distributions, hence of bounded size, the sequence cannot diverge, but it could in principle settle into a limit cycle.

To analyze (9.5) further, we need to consider the powers of $\mathbf{P}$. The diagonalization (9.7) makes it easy to compute them. Squaring $\mathbf{P}$ yields

$$\mathbf{P}^2 = \mathbf{U\Lambda U}^{-1}\mathbf{U\Lambda U}^{-1} = \mathbf{U\Lambda}^2\mathbf{U}^{-1}.$$

In general:

$$\mathbf{P}^k = \mathbf{U\Lambda}^k\mathbf{U}^{-1}. \tag{9.9}$$

This simplifies matters greatly, because one can take the power of a diagonal matrix simply by taking the powers of its elements:

$$\mathbf{\Lambda}^k = \begin{bmatrix} \lambda_1^k \ldots 0 \\ \vdots \ \ddots \ \vdots \\ 0 \ \ldots \lambda_n^k \end{bmatrix}.$$

The invertibility of $\mathbf{U}$ implies that its columns are linearly independent. That is, its columns form a basis for n-dimensional space. Hence we can express any n-dimensional vector as a linear combination of the eigenvectors $\mathbf{u}_i$. This is true in particular for $\mathbf{x}_0$:

$$\mathbf{x}_0 = \sum_i a_i \mathbf{u}_i = \mathbf{Ua} \tag{9.10}$$

for some set of weights $\mathbf{a}$.

Combining equation (9.10) with equations (9.5) and (9.9) yields

$$\begin{aligned} \mathbf{x}_k &= \mathbf{U\Lambda}^k\mathbf{U}^{-1}\mathbf{Ua} \\ &= \sum_i \mathbf{u}_i \lambda_i^k a_i. \end{aligned}$$

Not only $\mathbf{x}_0$, but every $\mathbf{x}_k$ can be expressed as a linear combination of the eigenvectors $\mathbf{u}_i$. If $\mathbf{P}$ is ergodic, then $\lambda_n = 1$ and all the other λ_i are less than one, hence $\lambda_n^k = 1$ and $\lambda_i^k \to 0$ for $i < n$. It follows that $\mathbf{x}_\infty = a_n \mathbf{u}_n$. But since $\lambda_n = 1$, we must have $\mathbf{Pu}_n = \mathbf{u}_n$, implying that $\mathbf{x}_\infty = \mathbf{u}_n$, which is to say, that $a_n = 1$. This conclusion follows from the fact that $\lambda_n = 1$. We cannot conclude that the other eigenvectors of $\mathbf{P}$ represent probability distributions, nor can we conclude that the other coefficients a_i equal one, or are even positive. What we *can* conclude is that the sequence $\mathbf{x}_0, \mathbf{x}_1, \ldots$ converges to $\mathbf{u}_n$,

$$\mathbf{x}_\infty = \mathbf{u}_n.$$

We have shown that, whatever the initial distribution, with time the chain converges to its stationary distribution. If we think of a population of particles each moving from state to state, the stationary distribution represents the proportion of particles in each state. Even though the individual particles continue to move, their distribution settles down to the stationary distribution, and the distribution no longer changes once settled. Alternatively, if we think of a single particle moving through the space, then no matter which state the particle starts in, if one chooses a random time that is far enough into the future to allow convergence, the probability distribution over possible locations for the particle is given by the stationary distribution. (If the particle starts out in state i, one can think of the distribution $\mathbf{x}_0$ as having a 1 in the i-th position and 0's elsewhere.)

To connect the discussion of stationary distributions back to the Gibbs sampler, an additional concept will be useful. A distribution $\mathbf{x}$ and a Markov chain $\mathbf{P}$ are in **detailed balance** if, for all states i and j:

$$p_{ji}x_i = p_{ij}x_j.$$

In words, the probability of being in state i and then transitioning to state j is equal to the probability of being in state j and then transitioning to state i. If $\mathbf{x}$ and $\mathbf{P}$ are in detailed balance, then we have

$$\sum_i p_{ji}x_i = \left(\sum_i p_{ij}\right) x_j = x_j.$$

That is, taking the dot product of the j-th row of $\mathbf{P}$ with $\mathbf{x}$ yields x_j. Hence:

$$\mathbf{Px} = \mathbf{x}.$$

In short, detailed balance implies that $\mathbf{x}$ is the stationary distribution for the chain.

We have already noted that the Gibbs sampler defines an ergodic Markov chain $\mathbf{P}$. We will use $\mathbf{y} = (y_1, \ldots, y_n)$ to represent an arbitrary assignment of labels to the nodes of the graph; $\mathbf{y}$ also represents a state visited by the sampler in the course of its random walk. Let us write $G(\mathbf{y}_j|\mathbf{y}_i)$ for p_{ji}, the probability that the sampler goes from state $\mathbf{y}_i$ to state $\mathbf{y}_j$ in one step. Note that $\mathbf{y}_i$ and $\mathbf{y}_j$ are two different vectors assigning labels to nodes in the graph. We would like to show that the transition probability $G(\mathbf{y}_i|\mathbf{y}_j)$ stands in detailed balance with the true probability over labelings, $p(\mathbf{y})$. If so, then $p(\mathbf{y})$ is the stationary distribution of the transition matrix, meaning that the probability of the Gibbs sampler visiting state $\mathbf{y}$ converges to the true probability $p(\mathbf{y})$. What we would like to show is the following:

$$G(\mathbf{y}_j|\mathbf{y}_i)p(\mathbf{y}_i) = G(\mathbf{y}_i|\mathbf{y}_j)p(\mathbf{y}_j) \tag{9.11}$$

for arbitrary label assignments $\mathbf{y}_i$ and $\mathbf{y}_j$.

To avoid subscript clutter, let us write $\mathbf{y}$ and $\mathbf{y}'$ instead of $\mathbf{y}_i$ and $\mathbf{y}_j$. The label that $\mathbf{y}$ assigns to the k-th node is y_k. Let us write $\bar{\mathbf{y}}_k$ for the rest of $\mathbf{y}$. That is, $\bar{\mathbf{y}}_k$ is just like $\mathbf{y}$, except that it has a "hole" in place of the k-th label. With this notation, the probability $G(\mathbf{y}'|\mathbf{y})$ can be written

$$G(\mathbf{y}'|\mathbf{y}) = \sum_k [\![\bar{\mathbf{y}}'_k = \bar{\mathbf{y}}_k]\!] p(k) p(y'_k|\bar{\mathbf{y}}'_k).$$

In words, the probability of going from $\mathbf{y}$ to $\mathbf{y}'$ is obtained by summing over possible choices k of nodes to relabel. The probability of going from $\mathbf{y}$ to $\mathbf{y}'$ by changing the label of node k is zero unless $\mathbf{y}$ and $\mathbf{y}'$ are identical everywhere except possibly at node k. If that condition is met, then the probability of going from $\mathbf{y}$ to $\mathbf{y}'$ by relabeling node k is equal to the probability of choosing node k for relabeling, times the probability that y'_k is chosen as the new value, given the context $\bar{\mathbf{y}}'_k$.

This allows us to expand the left side of (9.11) as follows:

$$\begin{aligned} G(\mathbf{y}'|\mathbf{y})p(\mathbf{y}) &= \sum_k [\![\bar{\mathbf{y}}'_k = \bar{\mathbf{y}}_k]\!] p(k) p(y'_k|\bar{\mathbf{y}}'_k) p(\mathbf{y}) \\ &= \sum_k [\![\bar{\mathbf{y}}'_k = \bar{\mathbf{y}}_k]\!] p(k) \frac{p(\mathbf{y}')p(\mathbf{y})}{p(\bar{\mathbf{y}}'_k)}. \end{aligned}$$

Similarly, the right side of (9.11) expands as

$$\begin{aligned} G(\mathbf{y}|\mathbf{y}')p(\mathbf{y}') &= \sum_k [\![\bar{\mathbf{y}}_k = \bar{\mathbf{y}}'_k]\!] p(k) p(y_k|\bar{\mathbf{y}}_k) p(\mathbf{y}') \\ &= \sum_k [\![\bar{\mathbf{y}}_k = \bar{\mathbf{y}}'_k]\!] p(k) \frac{p(\mathbf{y})p(\mathbf{y}')}{p(\bar{\mathbf{y}}_k)}. \end{aligned}$$

Noting that $\bar{\mathbf{y}}_k = \bar{\mathbf{y}}'_k$ within the summation, we conclude that the left and right sides of (9.11) are equal, as desired. This shows that the Gibbs sampler is in detailed balance with the true distribution over labelings, and hence that the Gibbs sampler converges to the true distribution.

9.4 Bibliographic notes

For random fields, see Winkler [233]. In particular, he discusses the Ising model (p. 52). The discussion of Gibbs sampling follows the presentation in Russell & Norvig [199], chapter 14.

10

Propagation Methods

We have seen several iterative methods for semisupervised learning: self-training and co-training, McLachlan's algorithm, the EM algorithm, boosting. In section 7.6, we showed that the iterations in self-training and co-training can be viewed as propagating label information through a graph. EM provides a soft version, and the Gibbs sampler can be seen in similar light: each update probabilistically adjusts the label of a node to make it like the labels of its neighbors.

Viewed abstractly enough, just about any classification algorithm can be seen as propagating labels based on similarity. The nearest-neighbor classifier asks directly which training instance a new instance is most similar to, and predicts the label of that most-similar training instance. Other classifiers create more abstract representations of classes and compare a given test instance to the abstract representations, predicting the class whose representation the new instance is most similar to. An abstract class representation, in turn, is based on the features that training instances belonging to the class have in common.

For example, as we have seen, many classifiers construct a weight vector $\mathbf{w}$ and, given a new instance $\mathbf{x}$, predict positive if $\mathbf{x} \cdot \mathbf{w}$ exceeds a threshold and negative otherwise. One can see $\mathbf{w}$ as an abstract representation of the positive class, and the dot product $\mathbf{x} \cdot \mathbf{w}$ as the measure of similarity between $\mathbf{x}$ and $\mathbf{w}$.

A similarity relation defines a weighted graph. The objects that participate in the relation represent nodes in the graph, and there is an edge between two objects just in case they have non-zero similarity. The weight on an edge represents the similarity between the two objects connected by the edge.

We have mentioned two different sorts of graph. In one, nodes represent instances, and features play a role in determining the similarities that are represented as edge weights. In the other sort of graph, both instances and features are represented by nodes. An instance is "similar" to a feature if it possesses the feature.

We are particularly interested in the semisupervised case in which some, but not all, of the nodes are labeled. If the seed consists of labeled data, then the labeled nodes are instances. If the seed consists of a seed classifier, the labeled nodes are features. The labels of nodes corresponding to the seed are fixed, and other labels are allowed to vary. The basic mechanism for assigning labels to unlabeled nodes is to propagate labels from neighboring nodes, in accordance with the weight of the edge connecting the nodes.

In this chapter, we examine the idea of label propagation through a graph in more detail. The picture that emerges is of a graph as a fabric or net, in which instances and features are knots connected by cords to similiar instances and features. The height of the fabric corresponds to the probability of having the positive label: an altitude of 1 indicates that the node is certainly labeled positive, an altitude of 0 indicates that the node is certainly labeled negative, and an altitude of 1/2 indicates that the two labels are equally likely. The labeled instances form the perimeter, at fixed heights of 1 or 0, and the unlabeled instances form the interior. The overall effect is a complex interpolation, with each unlabeled node being pulled to the average height of its neighbors. This yields a labeling of the entire graph, with nodes above altitude 1/2 labeled positive in the end and nodes below altitude 1/2 labeled negative.

10.1 Label propagation

We begin by considering a simple label propagation process. It assumes a weighted undirected graph in which weights are non-negative real numbers. The first ℓ nodes have fixed labels $y_1, \ldots, y_\ell$, with $y_i \in \{-1, +1\}$, and the remaining $n - \ell$ nodes are unlabeled; let us define L to be the set of labeled nodes, and U to be the set of unlabeled nodes. There are n nodes all together.

The label propagation process repeatedly updates node labels by propagating labels from neighbors. To choose a new label for the i-th node, one probabilistically chooses a node j among the neighbors of i, and one propagates j's label to i.

Only one neighbor j is allowed to propagate its label to i. The probability that j is the propagator is defined to be proportional to the weight of the edge connecting j to i. That is, the probability of j propagating its label to i is

$$p_{ij} = \frac{w_{ij}}{\sum_k w_{ik}} \tag{10.1}$$

where w_{ij} is the weight of the edge between i and j, and the summation is over edges incident to i. Note that weights are symmetric ($w_{ij} = w_{ji}$) because the graph is undirected, but it does not follow that $p_{ij} = p_{ji}$, because the denominator in (10.1) may differ for nodes i and j.

Label propagation is a stochastic process. We begin by labeling the unlabeled nodes at random, with probability 1/2 for positive and 1/2 for negative. Then, at each time step, the labels of the originally unlabeled nodes are updated in parallel. The update for the i-th node is

$$y_i^{(t+1)} = y_j^{(t)}$$

where j is a neighbor of i chosen at random according to p_{ij}. A new choice of neighbor is made at each update. Updates are applied only to originally unlabeled nodes; the labels of the originally labeled nodes are indelible.

We cannot define what the new label $y_i^{(t+1)}$ for the i-th node will be; it is not deterministic. But if we know the *probabilities* that the neighbors of the i-th node are labeled positive, then we can determine the probability that i is labeled positive after one label propagation step, and likewise for the probability of being labeled negative. We will give a definition for $g_i^{(t)}$, representing the probability that the i-th node is labeled positive at time t. The probability that it is labeled negative is simply $1 - g_i^{(t)}$.

The probability $g_i^{(t)}$ is trivial for nodes in L, which is to say, for the originally labeled nodes. If node i belongs to L, then for all times t, we have $g_i^{(t)} = 1$ if node i is originally labeled positive, and $g_i^{(t)} = 0$ otherwise.

The interesting case is where node i belongs to U (i is originally unlabeled). The probability $g_i^{(t)}$ for times $t > 0$ is defined recursively. The base case is $t = 0$. Since labels for nodes in U are initially assigned uniformly at random, it follows that $g_i^{(0)} = 1/2$. Now let us assume that $g_k^{(t)}$ is defined for all nodes k at time t. Consider an arbitrary node i, and let j range over the (indices of the) neighbors of node i. Note that the neighbors may be either in L or U. The probability that node i receives the positive label from a particular neighbor j is the probability p_{ij} that j is the propagator, times the probability $g_j^{(t)}$ that j is labeled positive at the beginning of the propagation step. Summing over neighbors yields the probability that node i receives the positive label:

$$g_i^{(t+1)} = \sum_j p_{ij} g_j^{(t)}. \tag{10.2}$$

This leads us to an algorithm. We collect the positive-label probabilities g_i into a vector $\mathbf{g}$. The first ℓ components are fixed, but the remaining $n - \ell$ components are variable, and are updated using (10.2). We write $\mathbf{g}_L$ for the half-vector $[g_1 \dots g_\ell]$, and $\mathbf{g}_U$ for the half-vector $[g_{\ell+1} \dots g_n]$. The probabilities p_{ij} define an $n \times n$ matrix $\mathbf{P}$. Note that the rows of $\mathbf{P}$ sum to unity, though its columns generally do not. If $\mathbf{P}_i$ is the i-th row of $\mathbf{P}$, then the new value g_i for node i defined in (10.2) can be written as $\mathbf{P}_i \cdot \mathbf{g}$. Updating all nodes in parallel, we obtain

$$\mathbf{g} \leftarrow \mathbf{P}\mathbf{g}. \tag{10.3}$$

This is actually not quite correct as written, because it updates the probabilities for the labeled nodes as well as the unlabeled nodes. The algorithm repairs that by restoring the probabilities for $\mathbf{g}_L$ to their original values, after the update.

The complete algorithm is given in figure 10.1. The algorithm runs until convergence, at which point a node in U is deemed to be positive if its probability of being positive exceeds $1/2$, and negative otherwise. Of course, saying that the algorithm runs to convergence presumes that it converges.

procedure `propagateLabels` $(\mathbf{W}, \mathbf{y})$
1 Input $\mathbf{W} = \{w_{ij}\}$ are edge weights for the graph
2 Input $\mathbf{y}$ contains labels for nodes $1, \ldots, \ell$
3 Compute $\mathbf{P}$ from $\mathbf{W}$ using (10.1)
4 Define $\mathbf{g}_L^{(0)}$ with elements $g_i^{(0)} = 1$ if $y_i = +$ and $g_i^{(0)} = 0$ otherwise
5 **set** $\mathbf{g}_L \leftarrow \mathbf{g}_L^{(0)}$, $\mathbf{g}_U \leftarrow (0.5)\mathbf{1}$
6 **until** convergence, **do**
7 **set** $\mathbf{g} \leftarrow \mathbf{Pf}$
8 **set** $\mathbf{g}_L \leftarrow \mathbf{g}_L^{(0)}$
9 **return** a labeling $\hat{\mathbf{y}}$ for originally unlabeled nodes,
10 where $\hat{y}_i = +1$ if $g_i > 1/2$ and $\hat{y}_i = -1$ otherwise.

FIGURE 10.1
A simple label propagation algorithm.

We should rather ask *whether* it converges. If it converges, it will converge to a fixed point of the update (10.3). That is, it will converge to a vector $\mathbf{g}$ satisfying

$$\mathbf{g} = \mathbf{Pg}. \tag{10.4}$$

If such a vector exists, its i-th component g_i provides a natural definition of the intrinsic probability that the i-th node is labeled positive. Note that what equation (10.2) provides is the probability that node i is labeled positive after the t-th step of label propagation. That probability varies from time step to time step. Equation (10.4) is what we expect of a label probability assignment that depends only on the matrix $\mathbf{P}$, and not on time.

It turns out that there is a natural relationship between the intrinsic probability of a node being labeled positive, and a random walk. We consider that connection first, before addressing the existence of $\mathbf{g}$.

10.2 Random walks

The probabilities p_{ij} defined in (10.1) can be viewed as transition probabilities for a Markov chain. We think of the graph as the state graph of the automaton; a computation of the automaton corresponds to a particle pursuing a random walk on the graph. At each moment of time t, the automaton is in some particular state, which is to say, the particle is located at some particular node x_i. The particle chooses an edge out of x_i stochastically according to the distribution p_{ij}, and moves along the chosen edge, arriving at node x_j at discrete time $t+1$.

It is important to notice that the particle moves in the direction *opposite* to the direction of label propagation. The probabilities p_{ij} represent a probability distribution over neighbors j. In terms of label propagation, the distribution determines which neighbor gets to propagate its label to i. In terms of a random walk, the distribution determines which neighbor the particle goes to next. The two processes are identical mathematically, but move in opposite directions. The particle can be thought of as moving backward in time along the same path that the label traversed forward in time.

A particle that begins its random walk in node x will eventually arrive at a node in L. Since the particle's path traces the path of label propagation in reverse, the label of x is determined by the label of the node where the particle's path terminates. We think of the particle as being absorbed at that point. If it is absorbed at a positively labeled node, the label of x is positive, and if it is absorbed at a negatively labeled node, the label of x is negative.

Following a single particle back from the start node x to a node in L tells us only one way that the label propagation might have gone. But since the motion of the particle is governed by the same probabilities that govern label propagation, the probability of the particle arriving at a positively labeled node in L is equal to the probability of x being labeled positive by the propagation process.

Let us define $f(x_i)$ to be the probability that a particle starting in x_i is absorbed at a positively labeled node. If $x_i \in L$, then the particle is immediately absorbed, and $f(x_i) = 1$ if x_i is positively labeled and $f(x_i) = 0$ otherwise. If $x_i \in U$, then the particle will go to some neighboring node x_j with the next step of its random walk. The probability of going to a particular neighbor x_j and subsequently (eventually) being absorbed at a positive node is the product $p_{ij} f(x_j)$. Summing over all neighbors gives the total probability of eventually being absorbed at a positive node:

$$f(x_i) = \sum_j p_{ij} f(x_j). \tag{10.5}$$

Here j ranges over (the indices of) the neighbors of node i, though we can equally well take j to range over all node indices, inasmuch as p_{ij} is zero for any j that is not a neighbor of i.

Equation (10.5) looks very much like equation (10.2), so it is worthwhile to point out the differences explicitly. How could the quantity $g_i^{(t)}$ defined by (10.2) differ from $f(x_i)$? The value $f(x_i)$ is the probability that x_i is labeled positive *on the assumption that* its label was propagated (ultimately) from a node in L. Let us call such labels "genuine" labels, as distinct from "pseudo" labels that came from the random assignment of labels at time $t = 0$. At time $t = 0$, any node in U has a pseudo label, not a genuine label propagated from a node in L. Hence $g_i^{(0)}$ is simply $1/2$, and it is unrelated to $f(x_i)$. Even at times

$t > 0$, there will be a discrepancy between f and g. The probability $f(x_i)$ averages over all paths along which the label of x_i might have propagated, which is to say, all paths along which a particle might travel before reaching a node in L. Some portion of those paths will have a length greater than t, and they are omitted from the calculation of $g_i^{(t)}$. Conversely, even if node i has received a genuine label by time t, the recursive computation of $g_i^{(t)}$ may still reflect indirect influence of pseudo labels, propagated from earlier times. But with time, the paths with length greater than t will represent an ever smaller portion of $f(x_i)$, and the influence of pseudo labels on $g_i^{(t)}$ will decrease as genuine labels are continually propagated from nodes in L. In short, we expect g_i to converge to $f(x_i)$. In the limit, when $g_i = f(x_i)$ for all i, the equation (10.5) is equivalent to the fixed-point equation (10.4).

The fixed-point equation (10.4) naturally brings to mind the stationary distribution of a Markov chain, as in equation (9.8). But in fact the two are not the same. The Markov transition matrix in (9.8) is column-stochastic; that is, the columns sum to unity. The matrix in (10.4) is row-stochastic. Moreover, the vector $\mathbf{x}$ in (9.8) was a probability distribution, but there is no guarantee that the elements of $\mathbf{g}$ in (10.4) sum to unity; in fact, they almost certainly do not. We *can* safely say that $\mathbf{g}$ in (10.4) is an eigenvector of $\mathbf{P}$ with eigenvalue 1, but we do not yet have grounds to assert that an eigenvector with eigenvalue 1 exists. (We also have no grounds to expect that such an eigenvector, if it exists, should agree with $\mathbf{g}_L$.)

10.3 Harmonic functions

The equation (10.5) can be viewed as an "averaging property" of the function f: for any node x_i, the value of f at node x_i is the weighted average of its values at neighboring nodes. A function that possesses this property is known as a **harmonic function**. Harmonic functions arise in a number of contexts. For example, a harmonic function is a natural generalization of linear interpolation. Consider a simple linear graph, such as that in figure 10.2. Assume that the value at the left end of the line is fixed at 0, and the value at the right end is fixed at 1, and assume that the edge weights are all unity. Linear interpolation of values at the interior nodes gives the results shown. Linear interpolation is equivalent to the averaging property (10.5) in this case; and it is natural to take the averaging property as the basis for generalizing linear interpolation to more complicated graphs.

The averaging property also arises if we think of the nodes of the graph as points in an elastic mesh, and values as heights above the ground. Edge weights correspond to the strength of the elastic forces along the edge. At equilibrium, an unclamped node will have an altitude at which the forces

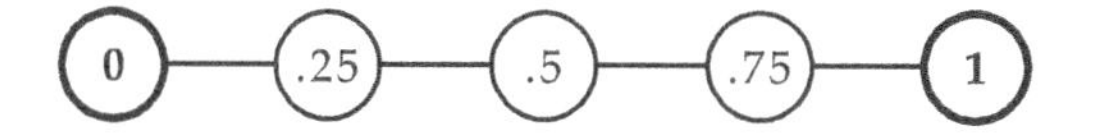

FIGURE 10.2
A simple linear graph. The heavy circles represent nodes whose values are clamped.

pulling it upward are balanced by the forces pulling it downward, which is to say, when the net force on it is zero. The force exerted on node i by a neighboring node j is the strength of the connection between them, w_{ij}, times the difference in their altitudes, $f(x_j) - f(x_i)$. Hence, at equilibrium, we have

$$\sum_j [f(x_j) - f(x_i)]\, w_{ij} = 0.$$

That is

$$\sum_j f(x_j) w_{ij} = f(x_i) \sum_j w_{ij}.$$

Dividing through by $\sum_j w_{ij}$ yields the averaging property (10.5).

Interpreting values as the height of an elastic mesh accords with physical intuition in the example of the linear graph (figure 10.2). It corresponds to a cord stretched between the clamped nodes at either end of the graph. We assume there is no gravity, so that the cord does not sag, but rises linearly from one end to the other.

Figure 10.3a provides a somewhat more complicated example. The corners are the only clamped values. Two corners, diagonally opposite, are clamped at 0 and the other two are clamped at 1. Given the symmetry of the graph, the center node, along with the two axes that intersect there, should have a value halfway between the extremes. The surface as a whole should obviously be saddle-shaped, but intuition does not give clear guidance regarding the shapes of the corners.

The averaging property does give clear guidance, however. It is easy to confirm that the value for each unclamped node in figure 10.3a is the arithmetic mean of its neighbors' values. (We note that the clamped nodes do not

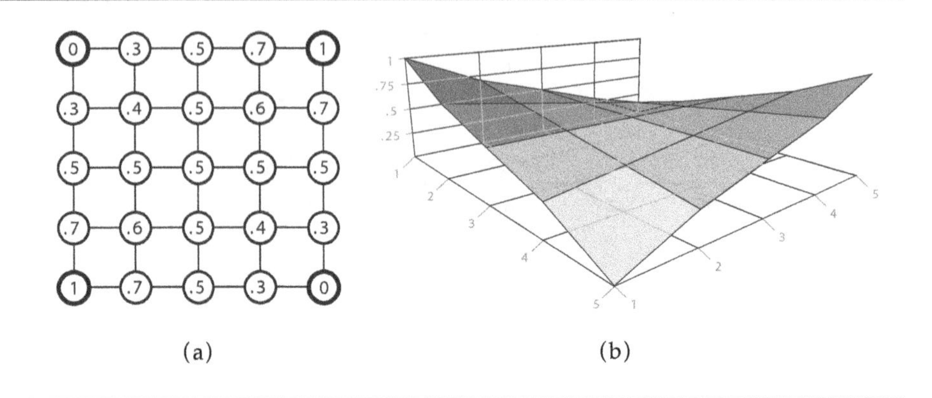

FIGURE 10.3
(a) A square graph with clamped corners. (b) The resulting saddle-shaped surface.

satisfy the averaging property, but then we should not expect them to – they are not free to "seek their level.") The resulting shape, when plotted (figure 10.3b), does accord qualitatively with physical intuition.

Note that the clamped values need not be at the edge of the graph. In figure 10.4, we have a graph with the same shape as the previous one, but this time the corners are clamped at 0 and the centermost node is clamped at 1. The values shown are determined by the averaging property: the result is a "circus tent" shape. In the absence of gravity, there is no sag, but the edges are pulled up by the tension of the "center pole."

The clamped nodes are conventionally called the **boundary**, even though they need not actually lie on the outer edge of the graph, as we have just seen. The unclamped nodes constitute the **interior** of the graph.

There are other physical systems that behave like an elastic mesh, in the sense of satisfying the averaging property. An example that we will return to is pressure of a fluid flowing through a network of pipes. Electricity is another example – indeed, one can treat electricity as a fluid in which voltage represents pressure.

A version of the label-propagation problem arises with harmonic functions. We have exhibited several graphs with value assignments satisfying the averaging property, but how can such a value assignment be computed in general? The problem can be stated as finding a harmonic function that agrees with given values on the boundary nodes.

The solution will be the topic of subsequent sections. In the remainder of this section, we address the preliminary question of whether a unique solution exists. We expect a unique solution to exist, given the physical intuition that a stretched fabric settles quickly into an equilibrium. We will see that this expectation is correct.

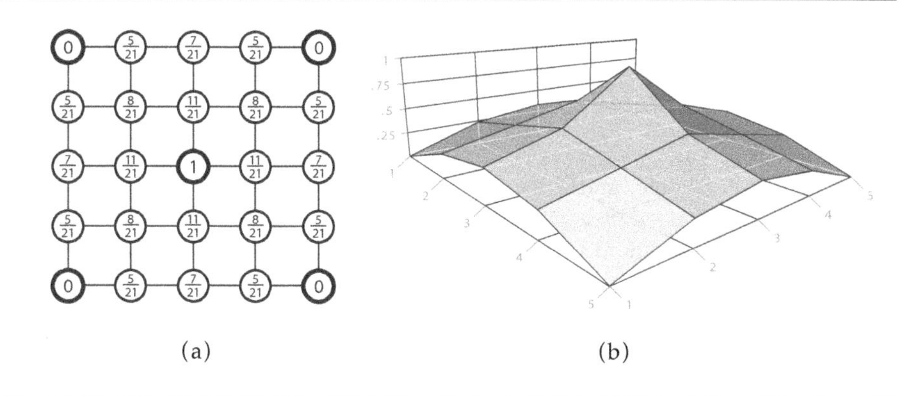

FIGURE 10.4
A clamped value in the middle of the graph.

We start with three simple properties of harmonic functions. The first is not actually needed for our immediate purposes, but this is an appropriate place to mention it.

- Every constant function is harmonic.
- The **superposition principle**: a linear combination of harmonic functions is itself a harmonic function.
- The **extreme value property**: the extreme values of a harmonic function occur in the boundary.

The first property (the harmonicity of any constant function) is easily seen. If $f(x_i) = k$ for all i, then clearly

$$f(x_i) = k = \sum_j p_{ij} k = \sum_j p_{ij} f(x_j).$$

If f is constant, then it is harmonic.

The second property is the superposition principle. To see that it holds, consider any harmonic functions $f_1, f_2, \ldots, f_n$ and any weights $\alpha_1, \alpha_2, \ldots, \alpha_n$. These define the linear combination $g(x_i) = \sum_t \alpha_t f_t(x_i)$. To show that the function g is harmonic, it suffices to show that an arbitrary interior node i satisfies the averaging property:

$$\begin{aligned} g(x_i) &= \sum_t \alpha_t f_t(x_i) \\ &= \sum_t \alpha_t \sum_j p_{ij} f_t(x_j) \end{aligned}$$

$$= \sum_j p_{ij} \sum_t \alpha_t f_t(x_j)$$

$$= \sum_j p_{ij} g(x_j).$$

Hence g is harmonic, and since we assumed nothing about g except that it is a linear combination of harmonic functions, we have shown that every linear combination of harmonic functions is harmonic.

The third property is the extreme value property. It states that, for any harmonic function f, the extreme values of f occur on the boundary. More precisely: a maximum of f occurs at a boundary node, and a minimum of f occurs at a boundary node. To prove it, we consider maxima; the case for minima is symmetric. Suppose $f(x_i)$ is a maximum of f. If x_i is a boundary node, we are done. Suppose instead that x_i is an interior node. If the neighbors of x_i have varying values for f, then $f(x_i)$, being an average of the values of x_i's neighbors, must be less than the maximum (and greater than the minimum). But we supposed $f(x_i)$ to be a maximum of f; hence we must conclude that all neighbors of x_i have the same value for f as x_i does. By the same argument, all the neighbors of neighbors of x_i have the same value for f, and so on across the entire graph. In short, if the maximum is an interior node, then f is a constant function, and all boundary nodes are also maxima of f. In every case, then, a maximum of f occurs on the boundary.

Now we return to the assertion of central interest, the uniqueness of the harmonic solution for interior node values. We will show that, if f is a harmonic function, then its values at interior nodes are uniquely determined by its values at boundary nodes. More precisely, assume that f and g are harmonic functions that agree on the values for boundary nodes. We will show that f and g are identical.

By the superposition principle, the difference $h = f - g$ is also a harmonic function. Since f and g agree on boundary nodes, we know that $h(x) = 0$ for all boundary nodes. By the extreme value property, the minimum and maximum of h occur on boundary nodes. But since $h(x) = 0$ for all boundary nodes, it follows that the minimum and maximum of h are both 0, hence $h(x) = 0$ everywhere, that is, $f(x) = g(x)$ for all nodes.

To recap, we have seen that the averaging property (10.5) accords with physical intuition about an elastic mesh clamped at boundary nodes – at least in cases where the physical intuition is clear. We have also seen that the averaging property, which is to say, harmonicity, determines a unique solution to the problem of assigning values to interior nodes. Since the fixed point equation (10.4) is simply a restatement in linear algebraic terms of the averaging property (10.5), we have just shown that a unique solution $\mathbf{g}$ does in fact exist. The originally labeled nodes L and the graph weights $\mathbf{P}$ determine a unique assignment of labels to the unlabeled nodes U.

10.4 Fluids

Before addressing the question of how to compute the solution **g**, let us consider one more physical analogue, namely, that of a fluid moving through a network. The discussion will lead to a proof that harmonic functions minimize energy, which will in turn be useful in understanding algorithms for computing solutions.

Specifically, let us think of edges in the graphs as pipes, and nodes as junctures. Then the positive-label probabilities $f(x)$ turn out to correspond naturally to pressure. In the interior, the network is closed; the only flow into or out of an interior node is through the pipes of the network. The weight of an edge will turn out to correspond naturally to the capacity of the pipe represented by the edge.

At boundary nodes, pressure is maintained either at 1 or 0 by pumping fluid in or out, as needed. We expect fluid to flow from nodes of higher pressure toward nodes of lower pressure. An interior node that is allowed to seek its own level should intuitively have a pressure somewhere between that of its high-pressure neighbors, from which it receives fluid, and its low-pressure neighbors, to which it sends fluid. Thus it is natural to expect the pressure at an interior node to be an average of the pressures at its neighbors. It is plausible that pressure satisfies the averaging property (10.5). However, rather than assuming that pressure satisfies the averaging property, we will define a concept of *flow* related to pressure, and show that the averaging property arises by minimizing the energy dissipation of the system.

Incidentally, if we interpret the fluid to be electricity, then pressure f corresponds to voltage, the pipe capacity corresponds to conductance (the reciprocal of resistance), and flow corresponds to electrical current. For this reason, we will use the symbol "i" for flow: "i" is the standard symbol for electrical current. That choice means, however, that we cannot continue to use i as a node index. Instead we will use x and y to represent nodes, and write in particular w_{xy} for the weight of the edge connecting x and y. The averaging property (10.5) becomes

$$f(x) = \sum_y p(y|x) f(y)$$

where

$$p(y|x) = \frac{w_{xy}}{\sum_z w_{xz}}.$$

10.4.1 Flow

A flow is a function $i(x, y)$ associated with edges (x, y). If there is no edge connecting x and y, we define $i(x, y)$ to be 0. Flow is directional. Positive

$i(x, y)$ means that the flow is from x to y. Negative $i(x, y)$ means that the flow is from y to x. An immediate consequence is that $i(y, x) = -i(x, y)$.

When the graph is partitioned into a boundary and interior, as we have been assuming, we require that the only flow into or out of the system be at the boundary. Let us define the **(net) external inflow** at a node x to be the total inflows to x from outside the system, less the total outflows from x. If there is net inflow at x from the outside, then the external inflow at x is positive, and if there is net leakage from x to the outside, then the external inflow at x is negative.

The external inflow is whatever is not accounted for by flow from or to neighboring nodes within the system. That is, we define external inflow $i(x)$ as

$$i(x) \equiv \sum_y i(x, y).$$

If the neighbor y is downstream of x, then $i(x, y)$ is positive, and if the neighbor y is upstream of x, then $i(x, y)$ is negative. If the positives are greater than the negatives, then there is more outflow from x than can be accounted for by inflows within the system, so there must be additional inflow from outside the system. Hence positive $i(x)$ represents inflow to the system from the outside. Conversely, negative $i(x)$ represents outflow from the system to the outside. If x has no connections to the outside, then the inflows and outflows within the system must exactly balance, and the sum is 0.

In short, we define a **(proper) flow** to be a function $i(x, y)$ that satisfies three conditions:

$$\begin{array}{lll} \text{(a)} & i(x,y) = 0 & \text{if there is no edge } (x,y) \\ \text{(b)} & i(x,y) = -i(y,x) & \\ \text{(c)} & i(x) = \sum_y i(x,y) = 0 & \text{for any } x \text{ in the interior.} \end{array} \tag{10.6}$$

In the context of electricity, property (10.6c) is known as **Kirchhoff's current law**. It states that there is no "leakage" between the interior of the system and the outside world.

As with harmonic functions, the linear combination of flows is a flow. Define:

$$i(x, y) = \sum_t \alpha_t i_t(x, y).$$

The resulting quantity i satisfies condition (10.6a): if x and y are not neighbors, then $i_t(x, y) = 0$ for all t, so $i(x, y) = 0$. The quantity i also satisfies condition (10.6b):

$$\begin{aligned} i(x, y) &= \sum_t \alpha_t i_t(x, y) \\ &= -\sum_t \alpha_t i_t(y, x) \end{aligned}$$

$$= -i(y, x).$$

Finally, to see that i satisfies condition (10.6c), let x be an arbitrary interior node.

$$\begin{aligned}\sum_y i(x, y) &= \sum_y \sum_t \alpha_t i_t(x, y) \\ &= \sum_t \alpha_t \sum_y i_t(x, y) \\ &= 0.\end{aligned}$$

A notable property of flows is that the total net external inflow $\sum_x i(x)$ is always 0. By definition,

$$\begin{aligned}\sum_x i(x) &= \sum_{x,y} i(x, y) \\ &= \tfrac{1}{2} \sum_{x,y} i(x, y) + \tfrac{1}{2} \sum_{x,y} i(x, y) \\ &= \tfrac{1}{2} \sum_{x,y} i(x, y) - \tfrac{1}{2} \sum_{x,y} i(y, x) \\ &= \tfrac{1}{2} \sum_{x,y} i(x, y) - \tfrac{1}{2} \sum_{y,x} i(x, y) \\ &= 0.\end{aligned}$$

Intuitively, the total inflow to the system equals the total outflow.

10.4.2 Pressure

Physical intuition suggests that the flow from x to y should be proportional to the pressure difference between x and y. If there is no pressure difference, there is no flow, and the flow increases monotonically with increase in the pressure difference. Intuition is not sufficiently crisp to dictate that the monotone increase is exactly proportional, but proportionality is the most natural assumption.

Physical intuition also suggests that flow is proportional to the capacity of the pipe connecting x and y. It is natural to take the edge weight w_{xy} to represent the capacity (cross-sectional area) of the pipe between x and y. Recall that $w_{xy} = w_{yx}$, meaning that the capacity is the same in both directions, in accordance with intuition. If there is no edge between x and y, then $w_{xy} = 0$, which we can paraphrase as "no pipe, no capacity," again in accordance with intuition.

In short, the relation between flow and pressure suggested by physical intuition is

$$i(x, y) = [f(x) - f(y)]w_{xy}. \tag{10.7}$$

In the context of electricity, the relation (10.7) is known as **Ohm's law**. The flow is from high pressure to low. It is positive if $f(x) \geq f(y)$ and negative otherwise.

Equation (10.7) establishes a relationship between properties of i and properties of f. Namely, if functions i and f satisfy Ohm's law (10.7), then we can show that i is a flow if and only if f is a harmonic function. In an Ohmic system, properness of flow is equivalent to harmonicity of pressure.

First, Ohm's law (10.7) implies that i satisfies (10.6a) and (10.6b), whether f is harmonic or not. Specifically, the satisfaction of (10.6a) follows immediately from the fact that $w_{xy} = 0$ if x and y are not connected by an edge. And the satisfaction of (10.6b) follows from the symmetry of w_{xy}:

$$\begin{aligned} i(x,y) &= [f(x) - f(y)]w_{xy} \\ &= -[f(y) - f(x)]w_{yx} \\ &= -i(y,x). \end{aligned}$$

So it remains only to show that Kirchhoff's law (10.6c) is equivalent to harmonicity of f.

Let x be an arbitrary interior node. Then the following equations are all equivalent:

$$\sum_y i(x,y) = 0 \tag{10.8}$$

$$\sum_y [f(x) - f(y)]w_{xy} = 0$$

$$\sum_y [f(x) - f(y)]p(y|x) = 0$$

$$\sum_y p(y|x)f(x) - \sum_y p(y|x)f(y) = 0$$

$$f(x) = \sum_y p(y|x)f(y). \tag{10.9}$$

That is, Kirchhoff's law (10.8) is indeed equivalent to the harmonicity of f (10.9), and we have completed the proof that properness of flow is equivalent to harmonicity of pressure, if Ohm's law holds.

We can view Ohm's law as defining $i(x, y)$ as a function of f and the weights w_{xy}. Take w_{xy} as fixed. If f is a harmonic function, let us call the flow

$$i(x,y) = [f(x) - f(y)]w_{xy},$$

the **Ohmic flow generated by** f. We know that $i(x, y)$ is a flow, because f is harmonic and harmonic pressure implies a proper flow, whenever Ohm's law holds.

Not all flows are Ohmic. Figure 10.5 shows a simple example of a non-Ohmic flow. All edges have weight one, so the flow $i(x, y)$ is simply the

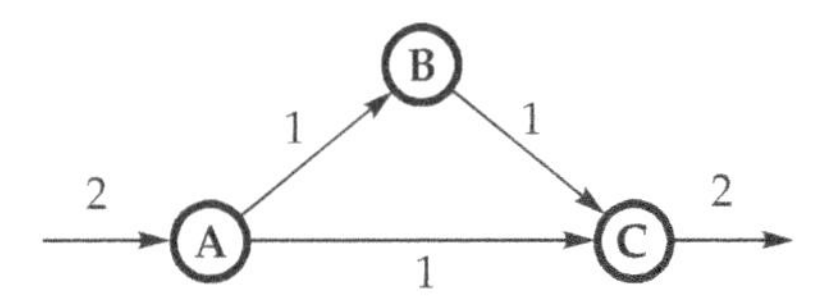

FIGURE 10.5
A non-Ohmic flow. Nodes A and C are boundary nodes, with net inflow of 2 at A and net outflow of 2 at C.

pressure difference $f(x) - f(y)$. If there were a harmonic pressure function f, the edge AC requires that $f(C) = f(A) - 1$, but the edges AB and BC require that $f(C) = f(B) - 1 = f(A) - 2$, a contradiction.

We can also reverse Ohm's law to define f in terms of i:

$$f(x) - f(y) = i(x, y)/w_{xy}. \tag{10.10}$$

Unfortunately, (10.10) does not determine a unique function f. If $f(x)$ satisfies (10.10) for a given i, then so does every function $g(x) = f(x) + k$ for any constant k.

However, a given flow i does determine family of harmonic functions $\{f(x) = f_0(x) + k | k \in \mathbf{R}\}$ where f_0 is any arbitrary member of the family. Namely, if two functions f and g both stand in relation (10.10) to a fixed i, then

$$f(x) - f(y) = g(x) - g(y)$$

for all edges (x, y). (For non-edges, w_{xy} is 0.) Rearranging terms yields

$$f(x) - g(x) = f(y) - g(y).$$

That is, if x and y are connected by an edge, there is a constant k such that $f(x) - g(x) = k$ and $f(y) - g(y) = k$. Assuming the graph is connected, we can follow edges to show that f and g differ by k everywhere. A given Ohmic flow i determines a family of harmonic functions that have the same "shape," and differ only by an additive constant. We can choose one of them to serve as f_0.

Let us say that a harmonic function f is in **standard form** if

$$\min_x f(x) = 0.$$

A standard-form harmonic function is non-negative everywhere, and has a zero value for at least one point. Any Ohmic flow i determines a unique standard-form harmonic function: if i is related to any function f by (10.10), simply translate f to set its minimum to 0. We know that the standard-form function is unique, because if f and g are two standard-form harmonic functions in the same family, then they have the same minimum point x_0 and $f(x_0) = 0 = g(x_0)$, hence $k = 0$ and f and g are the same everywhere.

We note in passing that, by the extreme value property, we can take the minimum point x_0 to be a boundary point.

There are a couple of useful facts to note about Ohmic flows. First, the linear combination of Ohmic flows is Ohmic. We already know that the linear combination of flows is a flow; now we note that it is generated by a harmonic function:

$$\begin{aligned} i(x,y) &= \sum_t \alpha_t i_t(x,y) \\ &= \sum_t \alpha_t [f_t(x) + k_t - f_t(y) - k_t] w_{xy} \\ &= \left[\sum_t \alpha_t f_t(x) - \sum_t \alpha_t f_t(y) \right] w_{xy}. \end{aligned}$$

We know that $\sum_t \alpha_t f_t$ is harmonic because linear combinations of harmonic functions are harmonic.

Another useful property is that a constant pressure generates a null flow – as intuition would lead us to expect. Assume that $f(x) = f(y)$ for all x and y. Then $i(x,y) = [f(x) - f(y)]w_{xy}$ is 0 for all edges, hence $i(x) = 0$ everywhere. Conversely, if i is the null flow, then f is a constant pressure function. Namely, we know that a constant pressure function f_0 exists that generates i. Hence any pressure function f that generates i differs from f_0 by a constant; and since f_0 is constant, it follows that $f = f_0 + k$ is constant.

We established earlier that fixing the weights w_{xy} and values $f(x)$ for boundary nodes x determines a unique harmonic function f. In terms of fluids, fixing capacities and boundary pressures determines a unique harmonic pressure function f.

That is unfortunately not true of flows, in general. Fixing the values $i(x)$ for boundary nodes x, does *not* determine a unique flow. Fixing the boundary flows does trivially determine a unique *net external* inflow, inasmuch as we already know that $i(x) = 0$ for all interior nodes. But many different flows $i(x,y)$ are consistent with a given boundary flow $i(x)$ for $x \in B$.

However, it *is* true of Ohmic flows. For fixed capacities w_{xy} and fixed boundary flows $i(x)$, there is a unique Ohmic flow $i(x,y)$. To see this, suppose that i and j are both Ohmic flows, and that $i(x) = j(x)$ for all boundary nodes. Define $d(x,y) = j(x,y) - i(x,y)$. Since d is a linear combination of Ohmic flows, it is Ohmic. But $d(x) = 0$ for all boundary nodes, since $i(x) = j(x)$ on the boundary, and $d(x) = 0$ for all interior nodes by Kirchhoff's law. That is, d is the null flow, hence $d(x, y) = 0$ for all edges, meaning that i and j are identical.

10.4.3 Conservation of energy

Energy is the rate at which work is done, and the unit of work in this context is a single unit of fluid m moving through a single unit of pressure difference d. That is, work $W = md$. Energy is the rate of work: $H = W/t = md/t$. Intuitively, flow is the rate at which fluid is moving, m/t, so it follows that energy is flow times pressure difference.

That is, the reasoning of the previous paragraph suggests that we define the energy associated with edge (x, y) as follows:

$$H_{xy}(f, i) \equiv [f(x) - f(y)]i(x, y). \tag{10.11}$$

If the flow is "uphill," from low pressure to high, then the energy is being applied to move the fluid against the stream, and if the flow is downhill, the energy is being released by the fluid. In downhill flow, either $f(x) > f(y)$ and $i(x, y)$ is positive, or $f(x) < f(y)$ and $i(x, y)$ is negative. In either case, the value $H_{xy}(f, i)$ is positive. So we should think of $H_{xy}(f, i)$ as energy release: the conversion of potential energy to kinetic energy. If the energy dissipation is negative, then the system consumes energy.

We have motivated the definition for H_{xy} by physical intuition, but for the moment, let us not rely on those physical intuitions, but simply consider H_{xy} as an abstract quantity defined by equation (10.11). It is defined for any functions f and i.

If i is a flow, then H_{xy} is symmetric in x and y:

$$\begin{aligned} H_{yx}(f, i) &= [f(y) - f(x)]i(y, x) \\ &= [f(x) - f(y)]i(x, y) \\ &= H_{xy}(f, i). \end{aligned}$$

The only "directionality" of the energy is the intrinsic directionality of the pressure difference. When we compute the energy dissipation of the whole system, it is conventional to multiply by 1/2 in order to correct for "double-counting" of edges:

$$H(f, i) \equiv \tfrac{1}{2} \sum_{x,y} [f(x) - f(y)]i(x, y).$$

If i is a flow, then we can apportion the energy dissipation among the nodes of the graph:

$$H_x(f, i) \equiv f(x)i(x).$$

We confirm that summing H_x over all nodes yields H:

$$\begin{aligned}
\sum_x H_x(f,i) &= \sum_x f(x)i(x) \\
&= \sum_{x,y} f(x)i(x,y) \\
&= \tfrac{1}{2}\sum_{x,y} f(x)i(x,y) + \tfrac{1}{2}\sum_{x,y} f(x)i(x,y) \\
&= \tfrac{1}{2}\sum_{x,y} f(x)i(x,y) - \tfrac{1}{2}\sum_{x,y} f(x)i(y,x) \\
&= \tfrac{1}{2}\sum_{x,y} f(x)i(x,y) - \tfrac{1}{2}\sum_{y,x} f(y)i(x,y) \\
&= \tfrac{1}{2}\sum_{x,y} [f(x) - f(y)]i(x,y) \\
&= H(f,i).
\end{aligned}$$

We note that what we have said so far about H has relied only on properties (10.6a) and (10.6b) of i. Nothing yet has depended on any properties of f except that it is a function defined on the nodes of the graph. The Kirchhoffian property of i (10.6c) allows us to show that energy is conserved in the system, in the sense that the only energy dissipation is at the boundary.

THEOREM 10.1 (Conservation of Energy)
For any function f on the graph, and any flow i, the net energy dissipation in the interior of the graph is zero. That is, where B is the set of boundary nodes:

$$H(f,i) = \sum_{x \in B} f(x)i(x).$$

PROOF. We have just seen that $H(f,i) = \sum_x f(x)i(x)$. By Kirchhoff's law (10.6c), $i(x) = 0$ for all interior nodes x, and the theorem follows immediately.

10.4.4 Thomson's principle

In the previous section, we made no use of Ohm's law (10.7) to relate flow and pressure. Everything we said about H is true for any flow i and any function f on the nodes of the graph.

Suppose i is an Ohmic flow generated by a harmonic function f. Then we can express $H(f,i)$ as a function of i only, or as a function of f only:

$$I(i) = \tfrac{1}{2}\sum_{x,y} i(x,y)^2/w_{xy}$$

$$J(f) = \frac{1}{2} \sum_{x,y} [f(x) - f(y)]^2 w_{xy}. \tag{10.12}$$

If i is generated by f, then $H(f, i) = I(i) = J(f)$.

But even if i is not Ohmic, $I(i)$ is well-defined, and even if f is not harmonic, $J(f)$ is well-defined. That is, I and J provide natural generalizations of H to non-Ohmic flows and non-harmonic pressure functions.

Interestingly, Ohmic flows and harmonic functions still have a special status with respect to I and J.

THEOREM 10.2 (Thomson's principle, primal form)
For fixed boundary flows, and fixed capacities w_{xy}, the flow that minimizes "I" energy is the Ohmic flow determined by the given boundary values.

PROOF. Let $k(x)$ for $x \in B$ represent fixed net boundary flow values, and let j be an arbitrary flow that satisfies $j(x) = k(x)$ for all boundary nodes. Let i be the Ohmic flow that satisfies $i(x) = k(x)$ on the boundary. (We saw at the end of section 10.4.2 that it is unique.)

Define $d(x, y) = j(x, y) - i(x, y)$. Since a linear combination of flows is a flow, we know that d is a flow:

$$\begin{aligned} 2I(j) &= \sum_{x,y} j(x,y)^2/w_{xy} \\ &= \sum_{x,y} i(x,y)^2/w_{xy} + 2\sum_{x,y} i(x,y)d(x,y)/w_{xy} + \sum_{x,y} d(x,y)^2/w_{xy}. \end{aligned}$$

Let us consider the middle term. Since i is Ohmic, $i(x, y)$ can be expressed as $[f(x) - f(y)]w_{xy}$ for some function f. That is, the middle term can be expressed as

$$2\sum_{x,y} [f(x) - f(y)]d(x,y) = 4H(f, d).$$

By conservation of energy, this is equal to

$$4\sum_{x \in B} f(x)d(x).$$

This equals zero, since $d(x) = j(x) - i(x)$ and $j(x) = i(x)$ on the boundary. It follows that

$$\begin{aligned} 2I(j) &= \sum_{x,y} i(x,y)^2/w_{xy} + \sum_{x,y} d(x,y)^2/w_{xy} \\ &\geq 2I(i). \end{aligned}$$

We have equality if and only if $d(x, y) = 0$ everywhere, which is to say, if and only if $j = i$. This proves that i is unique.

THEOREM 10.3 (Thomson's principle, dual form)
For fixed capacities and fixed boundary pressures, there is a unique pressure function that minimizes "J" energy, and it is harmonic.

PROOF. The proof is similar to that for the primal form. Let $k(x)$ represent fixed boundary pressures, and let g be an arbitrary function on the nodes of the graph with $g(x) = k(x)$ for all $x \in B$. Let f be any harmonic function with $f(x) = k(x)$ for all $x \in B$, and define $d(x) = g(x) - f(x)$. Then

$$\begin{aligned} 2J(g) &= \sum_{x,y} [g(x) - g(y)]^2 w_{xy} \\ &= \sum_{x,y} [f(x) + d(x) - f(y) - d(y)]^2 w_{xy} \\ &= \sum_{x,y} [f(x) - f(y)]^2 w_{xy} + 2 \sum_{x,y} [f(x) - f(y)][d(x) - d(y)] w_{xy} \\ &\qquad + \sum_{x,y} [d(x) - d(y)]^2 w_{xy}. \end{aligned}$$

Again we consider the middle term. Define $i(x, y) = [f(x) - f(y)] w_{xy}$. The middle term is equal to

$$2 \sum_{x,y} [d(x) - d(y)] i(x, y) = 4H(d, i).$$

Since f is a harmonic function, we know that i is a flow, hence conservation of energy applies to H. (Conservation of energy makes no requirement of d.) Hence the middle term equals

$$4 \sum_{x \in B} d(x) i(x) = 4 \sum_{x \in B} [g(x) - f(x)] i(x).$$

Since $g(x)$ and $f(x)$ agree at boundary nodes, the middle term reduces to zero. It follows that

$$\begin{aligned} 2J(g) &= \sum_{x,y} [f(x) - f(y)]^2 w_{xy} + \sum_{x,y} [d(x) - d(y)]^2 w_{xy} \\ &\geq 2J(f). \end{aligned}$$

Equality holds just in case $d(x) - d(y) = g(x) - f(x) - g(y) + f(y) = 0$ for all edges (x, y), which is to say, just in case f and g differ only by an additive constant. Since f and g agree on the boundary, there is at least one point with $f(x) = g(x)$, so the constant is 0. In short, equality holds just in case $f = g$, so we know that f is unique.

10.5 Computing the solution

We turn now to the question of how to compute a harmonic function f, given values for boundary nodes.

The connection to random walks provides one conceptually simple way to compute f, namely, by doing a Monte Carlo simulation. Recall that $f(x)$ is interpretable as the probability that a particle starting in node x is ultimately absorbed at a positively labeled boundary node. To determine that probability, we simulate the motion of a particle. We start it at node x, and at each step of the simulation, we choose a neighbor at random in accordance with $p(y|x)$. Eventually the particle reaches a boundary node, and we record the label of that boundary node, positive or negative. If we run many simulated particles and count the number of times that the particle is absorbed in positive versus negative boundary nodes, we can estimate $f(x)$ as the proportion of positive absorptions out of total runs. The reasoning is valid, and the computer code is easy to write, but it takes a very large number of simulated trials to get much accuracy in the estimates.

A second method is called the **method of relaxations**. It is in fact equivalent to the label propagation algorithm of figure 10.1. We start off with the boundary nodes clamped at their known values, but randomly assigned labels in the interior. Then we cycle through the interior nodes. At each interior node, we assume for the moment that the neighbors' values are correct, and we replace the current node's value with the value required by the averaging property:

$$f(x) = \sum_y p(y|x)f(y).$$

If x already satisfies the averaging property, we skip it. We keep iterating until all nodes satisfy the averaging property, or at least until the changes become smaller than some threshold.

Notice that the update just described, namely, setting a node's value to make it satisfy the averaging property, is identical to the update (10.2) used by the label propagation algorithm. In fact, the only difference between the algorithms is that the method of relaxations updates nodes one at a time, whereas the label propagation algorithm updates nodes in parallel. That difference is not significant. Either algorithm is exact in the limit, though they must be considered approximate in practice, as one must stop after some finite number of iterations.

The dual form of Thomson's principle gives us a proof that the method of relaxations finds a harmonic function with the given boundary values. Suppose f is not harmonic, and let

$$J = \frac{1}{2}\sum_{x,y}[f(x) - f(y)]^2 w_{xy}$$

be its energy value. In a single "relaxation," we pick a single point x and change the value $f(x)$. Consider the subgraph consisting of just x and its neighbors. The overall energy J is a summation over "edge energies" $[f(x) - f(y)]w_{xy}$. Thus, the only change in J consists in changes to "edge energies" within the subgraph. We can view the subgraph as a graph in its own right in which all nodes except x are boundary nodes with fixed values. We change $f(x)$ so as to make f harmonic on the subgraph. By Thomson's principle, that change minimizes J on the subgraph, and since J is held constant outside of the subgraph, the change cannot increase J overall. Since we skip nodes whose subgraph is already harmonic, the change in fact strictly decreases J. Hence the method of relaxations produces a strictly decreasing sequence of J values, terminating only when J reaches its minimum, at which point f is harmonic.

A third method is to use matrix operations. This method is exact and non-iterative. Consider the matrix $\mathbf{P}$ and vector $\mathbf{g}$ used in the label propagation algorithm. Recall that we partitioned $\mathbf{g}$ into half-vectors $\mathbf{g}_L$ and $\mathbf{g}_U$, and we can partition $\mathbf{P}$ similarly. The fixed-point equation (10.4) becomes:

$$\begin{bmatrix} \mathbf{g}_L \\ \mathbf{g}_U \end{bmatrix} = \begin{bmatrix} \mathbf{P}_{LL} & \mathbf{P}_{LU} \\ \mathbf{P}_{UL} & \mathbf{P}_{UU} \end{bmatrix} \begin{bmatrix} \mathbf{g}_L \\ \mathbf{g}_U \end{bmatrix}. \tag{10.13}$$

The half-vector $\mathbf{g}_L$ is constrained to match $\mathbf{g}_L^{(0)}$, so our interest is to determine the value of $\mathbf{g}_U$:

$$\mathbf{g}_U = \mathbf{P}_{UL}\mathbf{g}_L + \mathbf{P}_{UU}\mathbf{g}_U$$

which yields

$$\mathbf{g}_U = (\mathbf{I} - \mathbf{P}_{UU})^{-1}\mathbf{P}_{UL}\mathbf{g}_L.$$

Substituting the known value for $\mathbf{g}_L$, we conclude that

$$\mathbf{g}_U = (\mathbf{I} - \mathbf{P}_{UU})^{-1}\mathbf{P}_{UL}\mathbf{g}_L^{(0)}. \tag{10.14}$$

The solution exists only if $\mathbf{I} - \mathbf{P}_{UU}$ is invertible, but since we know from our previous discussion that the solution does exist, we can conclude that $\mathbf{I} - \mathbf{P}_{UU}$ is invertible.

Actually, we do have to make a minimal assumption about the graph, namely, that each connected component contains at least one labeled node. We are quite happy, in fact, to make stronger assumptions. If the graph consists of multiple connected components, we might as well treat each as a separate learning problem, so we can without loss of generality assume that the graph consists of a single connected component. Moreover, unless there is at least one positive node and one negative node, the learning problem is trivial, so we assume that the set of labeled nodes is not only nonempty, but contains at least one positive and one negative node.

10.6 Graph mincuts revisited

In section 7.5 we introduced the idea of clustering by cutting a graph into two connected components, choosing the cut so as to minimize the total weight of the edges that are severed. Though not directly related to harmonic functions, minimum cuts do turn out to have a strong connection to flows. In particular, there is a well-known algorithm to find a mincut given a weighted graph, called the Ford-Fulkerson algorithm, which is predicated on an equivalence between minimum cuts and maximum flows through a capacitated graph. We describe the algorithm in this section.

In previous sections, we thought of edge weights as capacities, in the sense of cross-sectional areas of pipes. Those capacities imposed no maximum on flow, however – multiplying any flow by a constant greater than one yields a larger flow. It is easy to confirm that $ki(x, y)$ satisfies conditions (10.6) if $i(x, y)$ does.

In a physical system, though, there are limits on how much flow can be achieved through a particular pipe. If nothing else, the flow is limited by the power of the available pumps and the strength of the walls of the pipe. Define a **capacitated graph** to be a weighted graph in which the weights w_{xy} are interpreted as maximum permissible flow. Again, this is a different notion of "capacity" than before – not cross-sectional area, but maximum flow. The **maximum flow problem** is to find the largest overall flow that does not exceed any edge capacity.

This raises the question of how we measure the total flow. After all, summing $i(x)$ for all nodes always yields 0. The maximum flow problem as standardly formulated assumes a graph with only two boundary nodes, a **source** s and a **sink** t. Let us call such a graph an **s-t graph**. If s and t are the only boundary nodes, then $i(s) = -i(t)$; we define the source to be the one with positive flow. The measure of total flow is simply $i(s)$.

Finding the maximum flow might seem trivial: simply fill all the pipes to capacity. But that is not usually possible. Figure 10.6a gives a trivial example. We cannot fill pipe AB to capacity, because pipe BC has a smaller capacity. Figure 10.6b gives a more interesting example. We cannot send any more flow down the pipe AB, because it is at capacity. We cannot send more flow down the path ACD, because CD is at capacity, and we cannot send any more flow down ACEF, because EF is at capacity. It would seem that the flow is the maximum possible. But it turns out there is a way to increase it, namely, by reversing the flow along the edge DE and rerouting it along DF; additional flow can then be sent along the path ACEDF. We discuss this in more detail below.

Granted that finding the maximum flow is not trivial, how does it help us find a mincut? What is the relationship between flow and cut? Let us consider a couple of cuts through the graph of figure 10.6b, as shown in figure

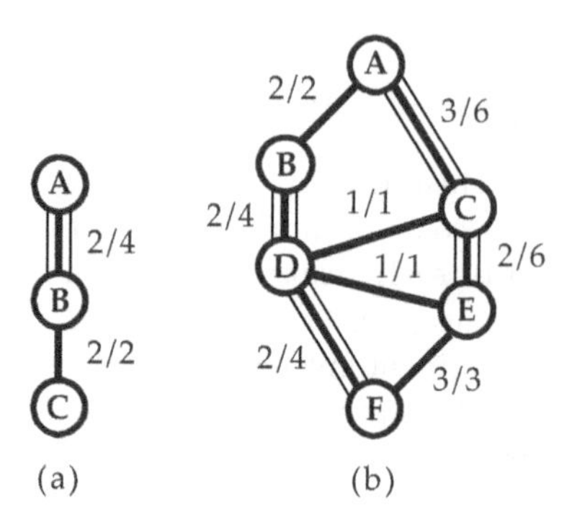

FIGURE 10.6
One cannot always fill all pipes.

10.7. If we add up the flow going through the edges that are cut, we find that it adds up to 5 in all three cases. It should be clear that this will always be the case. There is a fixed amount of flow going in, and by Kirchhoff's law, the total flow going into a node is the same as the total flow going out. If we cut the graph a little lower or higher, we are still cutting the same total flow, though it may be apportioned a little differently among the edges.

Note that this is true even if we consider wiggly cuts as in figure 10.7b. At least, it is true if we consider the cut to have an "uphill" side and a "downhill" side, and flows that cross the cut in the wrong direction count as negative flows.

So we see that the total flow across a cut is the same for all cuts. But the value of a cut is not the total flow that it cuts, but rather, the total *capacity* that it cuts. Different cuts necessarily transect the same amount of flow, but they need not transect the same amount of capacity. Consider any two cuts, like the two cuts in figure 10.7a. Since both cuts transect the same amount of flow, we can subtract the amount of transected flow from the value of each without changing their relative values. What we are comparing then is the *excess capacity* of the transected edges. To find the cut with minimum value, we should look for the cut that transects the minimum amount of excess capacity. The upper cut has $(4-2)+(6-3)=5$ units of excess capacity, whereas the lower cut has $(4-2)+(1-1)+(6-2)=6$ units of excess capacity; the upper cut is better.

Flow plus excess capacity equals capacity. Unlike flow, capacity is not dependent on the direction in which one follows an edge. When one follows an edge "upstream," flow is negative, and as a result excess capacity actually exceeds capacity.

One can increase the flow through a graph just in case one can find a path from source to sink in which every edge has excess capacity. For example,

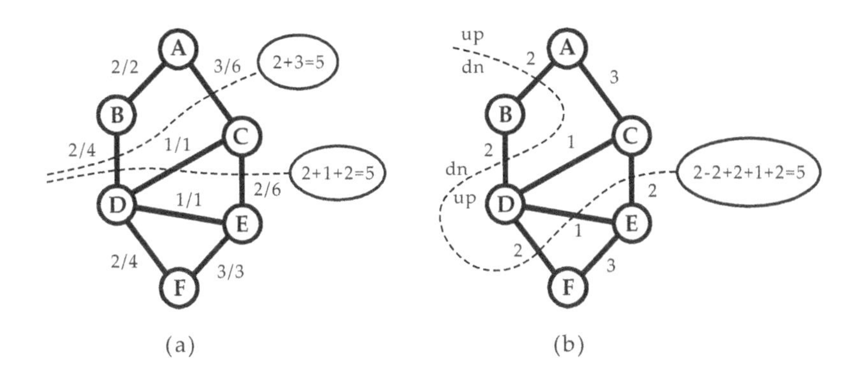

FIGURE 10.7
Some cuts.

consider figure 10.8. In the marked path, every edge has excess capacity. We can increase flow down the marked path by one unit – that is the minimum of the excess capacities on edges in the path. Thereby, the total flow is increased by one unit.

One can think of edges that have no excess capacity as "blocked." We are seeking a path from source to sink that is not blocked. Since excess capacity depends on the direction in which one follows the edge, a blocked edge actually represents a one-way barrier. For example, in figure 10.6b, all the strictly downhill paths are blocked, but the path ACEDF is open – by reversing the flow along DE, we can increase flow along ACEDF while maintaining the rest of the graph unchanged. In the downhill direction, that is, the direction in which the flow is currently traversing the edge, DE has one unit of excess capacity, but in the uphill direction (ED), it has three units of excess capacity. In the path ACEDF, the edge with the least excess capacity is DF; it has two units of excess capacity. Hence we can increase flow along the marked path by two units. Note that "increasing flow" along the uphill edge ED means changing the flow from -1 to $+1$. What we are actually doing is shunting the flow that goes along CDE to CDF, and making up for it with increased flow from E to D. One can confirm that the overall result is an increase of two units in total flow, as shown in figure 10.9.

We are placing nodes in the figures so that flow is oriented from top to bottom. Edges with 0 flow are horizontal. Let us mark blocked edges as in figure 10.10. In seeking a path, one is not allowed to cross barriers in a downhill direction, but one *is* allowed to cross them in an uphill direction.

We can seek a path by systematically marking nodes that we can reach from the source. If we can reach a given node, and it has an unblocked edge to another node, that other node is reachable as well. We mark the reachable nodes by circling them. When there is a path, the sink is reachable. When

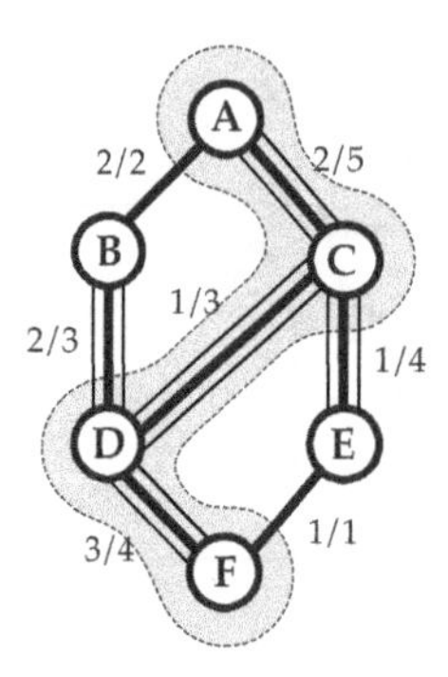

FIGURE 10.8
Flow can be increased along the marked path.

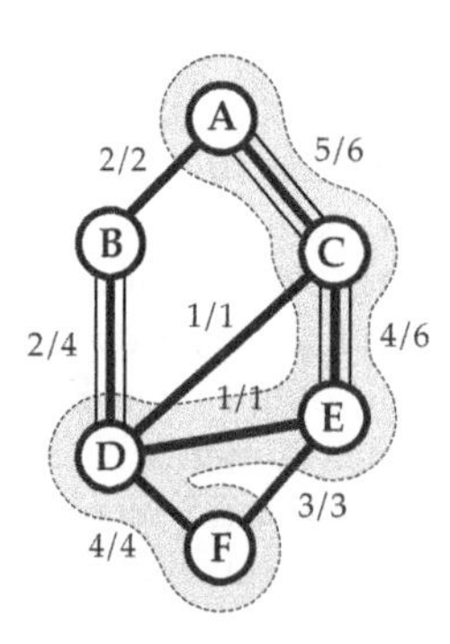

FIGURE 10.9
The outcome of increasing flow along the path ACEDF by reversing the flow along the edge ED.

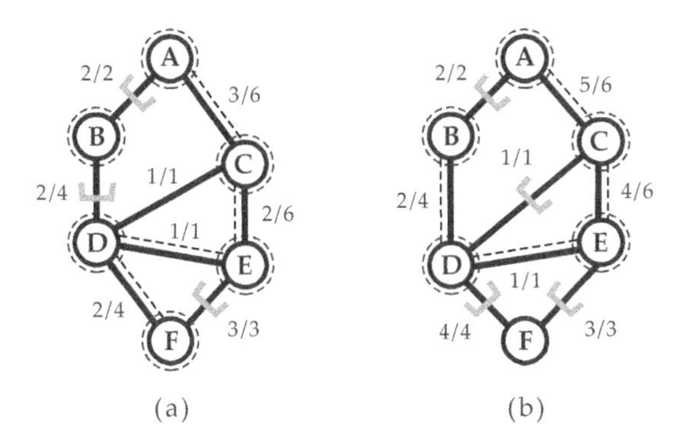

FIGURE 10.10
Blocked edges are marked, reachable nodes are circled. (a) Flow can be increased along dotted path; sink is reachable. (b) Sink is not reachable; flow cannot be increased.

the sink is not reachable, we have found the maximum flow. Figure 10.10b illustrates.

When we have found a maximum flow, every edge that connects a reachable node to an unreachable node is blocked. That set of edges constitutes a cut, separating the graph into two sets – the reachable nodes and the unreachable nodes. Moreover, the value of that cut is the capacity of the transected edges, which is equal to the total flow, since none of the edges has excess capacity.

This, then, is a sketch of the algorithm for finding a mincut. Propagate reachability through the graph. If node x is reachable with i units of additional flow, and there is an edge with excess capacity j connecting x to another node y, then y is reachable with $\min(i, j)$ units of additional flow. There may be more than one edge by which y is reachable. Keep track of which edge maximizes the additional flow that reaches y – call that the "best" edge leading into y.

If the sink is reachable with i units of additional flow, then trace backward along the path of "best" edges and increase flow along that path by i units. If the sink is not reachable, then find all the edges that connect reachable nodes to unreachable nodes. The resulting set of edges is the mincut, and the sum of their capacities is its value.

One might be concerned that requiring a single source and a single sink limits us to having two labeled instances. But Blum & Chawla [20] propose a straightforward way of representing a data set with multiple labeled instances (as well as unlabeled instances) as an s-t graph. Each instance is represented by a vertex, as we have assumed all along. But the source and sink vertices do not correspond to instances; rather, they represent the two class labels. The source represents the positive class and the sink represents the negative class. The source and sink are connected to the labeled instances. The source is connected to each positively labeled instance, and the sink is connected to each negatively labeled instance. Those edges represent the labeling, and they all have infinite weight. Edges connecting one instance to another have finite weight; this guarantees that the mincut includes only edges connecting instances, not edges representing labeling.

10.7 Bibliographic notes

This chapter, particularly the sections on harmonic functions, random walks, and fluids, relies heavily on the excellent book *Random Walks and Electric Networks* by Doyle & Snell [79]. The presentation of the label propagation algorithm (section 10.1) follows Zhu [244]. The Ford-Fulkerson algorithm is described in standard introductions to algorithms such as Cormen [51] or Sedgewick [205].

11

Mathematics for Spectral Methods

11.1 Some basic concepts

11.1.1 The norm of a vector

A vector is defined by a **direction** and a **length**. Fixing the length r and allowing the direction to vary generates a family of vectors that describe a sphere of radius r, centered at the origin. Of more interest for linear algebra is the family of vectors that have the same direction, but vary in length. The unit vector $\mathbf{x}$ in the given direction can be taken as the canonical representative of the family. The other members of the family have the form $r\mathbf{x}$ for some scalar r, which is incidentally the length of the vector in question.

The **norm** of a vector is simply its length. Usually, we define the length of $\mathbf{x}$ to be

$$\sqrt{\mathbf{x} \cdot \mathbf{x}} = \left(\sum_i x_i^2 \right)^{\frac{1}{2}}.$$

This is the **Euclidean norm**. Other definitions of norm are possible. For example, the Euclidean norm is a member of the family of p-norms $(\sum_i x_i^p)^{1/p}$, each value of p defining a different norm. Technically, a norm is any function that satisfies the following three conditions:

$$\begin{aligned} & f(\mathbf{x}) \geq 0 && \text{with equality iff } \mathbf{x} = \mathbf{0} \\ & f(\mathbf{x}+\mathbf{y}) \leq f(\mathbf{x}) + f(\mathbf{y}) && \text{(the triangle inequality)} \\ & f(\alpha\mathbf{x}) = |\alpha| f(\mathbf{x}) && \text{for all scalars } \alpha. \end{aligned} \tag{11.1}$$

The notation $\|\mathbf{x}\|$ represents the norm of $\mathbf{x}$. We will limit our attention to the Euclidean norm.

Normalizing a vector $\mathbf{x}$ means finding the unit vector that has the same direction as $\mathbf{x}$, which is

$$\frac{\mathbf{x}}{\|\mathbf{x}\|}.$$

This vector represents the direction of $\mathbf{x}$.

11.1.2 Matrices as linear operators

Usually one is trained to think about matrix multiplication in terms of the individual entries of the product matrix: the (i, j)-th entry of AB is the dot product of the i-th row of A with the j-th column of B. But it is often better to consider the matrix-vector product as the more basic operation:

$$\mathbf{y} = A\mathbf{x} = \begin{bmatrix} \mathbf{a}_1 \dots \mathbf{a}_n \end{bmatrix} \mathbf{x} = \sum_i x_i \mathbf{a}_i.$$

Here $\mathbf{y}$, $\mathbf{x}$, and $\mathbf{a}_1, \dots, \mathbf{a}_n$ are all column vectors. The output vector $\mathbf{y}$ is a linear combination of the columns $\mathbf{a}_i$, the coefficients being the components x_i of $\mathbf{x}$.

One can think of a matrix as a **linear operator**. The product $A\mathbf{x}$ represents the application of the operator A to the input vector $\mathbf{x}$, and the output is $\mathbf{y} = A\mathbf{x}$. The matrix product AB arises as the composition of the operators A and B. The i-th column of AB is $A\mathbf{b}_i$, where $\mathbf{b}_i$ is the i-th column of B.

11.1.3 The column space

The set of all possible linear combinations $\sum_i x_i \mathbf{a}_i$ of the columns of a matrix A is called the **column space** of A. It is the range of outputs of A, treating A as an operator. We also say that the column space is the space **spanned by** the columns of A. The dimensionality of the column space is the **rank** of A. A rank of one means that all the outputs lie in a line (they are all multiples of one vector). A rank of two means that all the outputs lie in a plane, and so on.

Suppose that a column of A, say $\mathbf{a}_i$, is a linear combination of the other columns $\mathbf{a}_j$ for $j \neq i$. That column is said to be **linearly dependent** on the other columns. If $\mathbf{a}_i$ is linearly dependent on the other columns, then any vector in the column space that we form using $\mathbf{a}_i$ could just as easily be formed by using the other columns directly, without using $\mathbf{a}_i$. In this sense, $\mathbf{a}_i$ is "dispensable" – it can be omitted from the matrix without changing the column space. (Actually, it is never possible to identify a unique dispensable column – if any column is dispensable, there is always more than one choice of which one to eliminate. Said another way, linear dependence is a property of a set of vectors, not an individual vector.) A matrix with no dispensable columns is said to be of **full column rank**. A matrix has full column rank just in case the number of columns is equal to the rank (which is equal to the dimensionality of the column space).

The rank cannot exceed the number of rows or the number of columns, whichever is less. A matrix can have full column rank only if it is either square or "tall and skinny" (more rows than columns). If it is "short and fat" (more columns than rows) it cannot have full column rank.

The columns of the matrix form a **basis**, which is to say, a set of coordinate axes, for the column space. This is true by definition: every vector in the column space is a linear combination of the columns of the matrix, and a set of vectors forms a basis just in case every vector in the space is a linear combination of the basis vectors.

A basis is not generally unique. For example, a one-dimensional space is a line and a two-dimensional space is a plane. Any nonzero vector that lies in a given line provides a basis for the line. Any pair of independent (non-parallel) nonzero vectors lying in a plane provides a basis for the plane. In general, any set of n independent vectors provides a basis for an n-dimensional space.

Two vectors can be independent without being perpendicular. The set of vectors forming a basis must be independent, but they need not be mutually perpendicular. If they *are* mutually perpendicular, they form an **orthogonal basis**. An orthogonal basis is what we usually have in mind when we think of a set of coordinate axes. A matrix whose columns are mutually perpendicular and of unit length is an **orthonormal matrix**. The columns of an orthonormal matrix provide an orthogonal basis for its column space.

As an operator, an orthonormal matrix represents a rigid rotation. In particular, the operation of an orthonormal matrix preserves vector lengths, and applying an orthonormal matrix to perpendicular vectors yields perpendicular vectors. Both of those properties are consequences of a more general fact: the operation of an orthonormal matrix preserves dot products. Writing $\mathbf{m}_i$ for the i-th column of M, and for any vectors $\mathbf{x}$ and $\mathbf{y}$, we have

$$\begin{aligned} M\mathbf{x} \cdot M\mathbf{y} &= \left(\sum_i x_i \mathbf{m}_i\right) \cdot \left(\sum_j y_j \mathbf{m}_j\right) \\ &= \sum_{ij} x_i y_j (\mathbf{m}_i \cdot \mathbf{m}_j) \\ &= \sum_i x_i y_i (\mathbf{m}_i \cdot \mathbf{m}_i) + \sum_{i \neq j} x_i y_j (\mathbf{m}_i \cdot \mathbf{m}_j) \\ &= \mathbf{x} \cdot \mathbf{y}. \end{aligned}$$

The last line follows because $\mathbf{m}_i \cdot \mathbf{m}_i = 1$ for all i, and $\mathbf{m}_i \cdot \mathbf{m}_j = 0$ for $j \neq i$. If $\mathbf{x}$ and $\mathbf{y}$ are perpendicular (their dot product is zero), it follows immediately that $M\mathbf{x}$ and $M\mathbf{y}$ are perpendicular, and vice versa. Also, taking $\mathbf{y} = \mathbf{x}$, we deduce that the operation of M preserves vector length:

$$\|M\mathbf{x}\|^2 = M\mathbf{x} \cdot M\mathbf{x} = \mathbf{x} \cdot \mathbf{x} = \|\mathbf{x}\|^2.$$

In the same way that two vectors can be perpendicular, two spaces can also be perpendicular. Consider an orthonormal matrix M whose columns provide an orthogonal basis for an n-dimensional space. Let us choose k of the n columns. The chosen columns define a k-dimensional space $\mathcal{A}$, and the remaining columns define an $(n-k)$-dimensional space $\mathcal{B}$. Consider an arbitrary vector $\mathbf{a} \in \mathcal{A}$ and an arbitrary vector $\mathbf{b} \in \mathcal{B}$. The vector $\mathbf{a}$ represents a linear combination of the first k columns of M, meaning that $\mathbf{a}$ can be written as $M\mathbf{x}$ where the last $n - k$ components of $\mathbf{x}$ are zero. Similarly, $\mathbf{b}$ can be written as $M\mathbf{y}$ where the first k components of $\mathbf{y}$ are zero. Since the dot product of $\mathbf{x}$ and $\mathbf{y}$ is zero, they are perpendicular; and since M is orthonormal, it follows that $\mathbf{a} = M\mathbf{x}$ and $\mathbf{b} = M\mathbf{y}$ are perpendicular. But $\mathbf{a}$ and $\mathbf{b}$ were chosen arbitrarily, so it follows that every vector in the space $\mathcal{A}$ is perpendicular to every vector in the space $\mathcal{B}$. $\mathcal{A}$ and $\mathcal{B}$ are **orthogonal spaces**. Taking the n-dimensional column space of M as the universe, $\mathcal{B}$ is the **orthogonal complement** of $\mathcal{A}$:

$$\mathcal{B} = \mathcal{A}^{\perp}. \tag{11.2}$$

For a concrete example, consider a 3×3 matrix whose columns are independent but not necessarily perpendicular. Its first two columns define a plane (that is subspace $\mathcal{A}$) and its third column defines a line intersecting the plane (that is subspace $\mathcal{B}$). In general, the line forms an oblique angle to the plane. In that case, a vector $\mathbf{b}$ in the line has a nonzero dot product with vectors in the plane: there is a nonzero component of $\mathbf{b}$ that is oriented in the direction of the plane. The special case in which the line is perpendicular to the plane is the case in which $\mathcal{A}$ and $\mathcal{B}$ are orthogonal subspaces.

11.2 Eigenvalues and eigenvectors

We have already encountered eigenvalues and eigenvectors, along with the concept of diagonalization, but we give a more systematic and detailed development here.

11.2.1 Definition of eigenvalues and eigenvectors

A natural question to ask about an operator is whether it has **fixed points**, that is, inputs that are left unchanged by the operation. Now since a vector can be thought of as the combination of a direction and a magnitude, a matrix actually has two effects: it scales the vector (changes its length) and rotates it (changes its direction). For matrices, the question of a fixed point turns out to be less interesting than a somewhat weaker question: are there vector *directions* that are left unchanged by the operation of the matrix?

Two vectors share the same direction if one is a scalar multiple of the other, that is, if $\mathbf{v} = \lambda\mathbf{u}$ for some scalar λ. An **eigenvector** of a matrix A is a vector $\mathbf{u}$ whose direction is not changed by the operation of the matrix:

$$A\mathbf{u} = \lambda\mathbf{u}. \tag{11.3}$$

The number λ is the **eigenvalue** corresponding to the eigenvector $\mathbf{u}$.

To emphasize, only the direction of an eigenvector matters, not its length. If $\mathbf{u}$ is an eigenvector with eigenvalue λ, then so is $c\mathbf{u}$ for any nonzero constant c:

$$A(c\mathbf{u}) = cA\mathbf{u} = c\lambda\mathbf{u} = \lambda(c\mathbf{u}). \tag{11.4}$$

An eigenvector is actually a family of vectors sharing the same direction.

However, the eigenvalue λ is fixed. Notice that, when we change the length of the eigenvector by choosing a different value for c in (11.4), the eigenvalue λ does *not* change. To say it another way, we cannot eliminate λ by adjusting the length of $\mathbf{u}$.

There is one property of eigenvalues that one needs to be aware of: they may be imaginary or complex, even if the original matrix contains only real numbers.

11.2.2 Diagonalization

Consider a matrix A with eigenvalues $\lambda_1, \ldots, \lambda_n$ and corresponding eigenvectors $\mathbf{u}_1, \ldots, \mathbf{u}_n$. We assume A to have full column rank. Define S to be a matrix whose columns are the eigenvectors:

$$S = \begin{bmatrix} \mathbf{u}_1 \ldots \mathbf{u}_n \end{bmatrix}.$$

Since the columns of S are the eigenvectors of A, premultiplying any column of S by A is the same as multiplying by the corresponding eigenvalue. That is,

$$AS = \begin{bmatrix} A\mathbf{u}_1 \ldots A\mathbf{u}_n \end{bmatrix} = \begin{bmatrix} \lambda_1\mathbf{u}_1 \ldots \lambda_n\mathbf{u}_n \end{bmatrix} = \begin{bmatrix} \mathbf{u}_1 \ldots \mathbf{u}_n \end{bmatrix} \begin{bmatrix} \lambda_1 & & \\ & \ddots & \\ & & \lambda_n \end{bmatrix}.$$

Let us write Λ for the latter diagonal matrix containing eigenvalues. We have just derived

$$AS = S\Lambda.$$

Postmultiplying by S^{-1} yields

$$A = S\Lambda S^{-1}. \tag{11.5}$$

Factoring A into this form – the matrix of eigenvectors, a diagonal matrix of eigenvalues, and the inverse of the eigenvector matrix – is called **diagonalization**. It shows how any matrix can be factored into its eigenvectors and eigenvalues.

The factorization $A = S\Lambda S^{-1}$ is actually not unique. Remember that any multiple of an eigenvector is also an eigenvector, as discussed in connection with equation (11.4). We can scale the lengths of the eigenvectors any way we like (as long as we do not set them to zero), and we still have $A = S\Lambda S^{-1}$.

Diagonalization shows that S and A have the same column space. To see this, let $\mathbf{a}_i$ be the i-th column of A and let $\mathbf{x}_i$ be the i-th column of the matrix ΛS^{-1}. We can write (11.5) as

$$\begin{bmatrix}\mathbf{a}_1 \dots \mathbf{a}_n\end{bmatrix} = S \begin{bmatrix}\mathbf{x}_1 \dots \mathbf{x}_n\end{bmatrix}.$$

This says that each vector $\mathbf{a}_i$ is a linear combination of the columns of S. (Namely, $\mathbf{a}_i = S\mathbf{x}_i$.) Since every column of A is a linear combination of the columns of S, every vector in the column space of A (every linear combination of the columns of A) is also in the column space of S. The reverse implication follows from $AS\Lambda^{-1} = S$, which is

$$\begin{bmatrix}\mathbf{u}_1 \dots \mathbf{u}_n\end{bmatrix} = A \begin{bmatrix}\mathbf{y}_1 \dots \mathbf{y}_n\end{bmatrix}$$

where $\mathbf{u}_i$ is the i-th column of S and $\mathbf{y}_i$ is the i-th column of $S\Lambda^{-1}$.

11.2.3 Orthogonal diagonalization

When A is a symmetric matrix, the matrix S in the diagonalization (11.5) can be chosen to be orthonormal. We have noted the fact previously; we will show now why it is true, at least in the case when all eigenvalues of A are distinct. For the more general case the reader is referred to a text on linear algebra, for example Strang [213], pp. 318ff.

First, when A is symmetric, we can show that the columns of S (the eigenvectors of A) are perpendicular. Let $\mathbf{u}_i$ and $\mathbf{u}_j$ be any two eigenvectors with eigenvalues $\lambda_i \neq \lambda_j$. Then:

$$\begin{aligned}
\lambda_i \mathbf{u}_i^{\mathrm{T}} \mathbf{u}_j &= (A\mathbf{u}_i)^{\mathrm{T}} \mathbf{u}_j \\
&= \mathbf{u}_i^{\mathrm{T}} A^{\mathrm{T}} \mathbf{u}_j \\
&= \mathbf{u}_i^{\mathrm{T}} A \mathbf{u}_j \qquad \text{by symmetry of } A \\
&= \lambda_j \mathbf{u}_i^{\mathrm{T}} \mathbf{u}_j.
\end{aligned}$$

Since $\lambda_i \neq \lambda_j$, the only way to have $\lambda_i \mathbf{u}_i^{\mathrm{T}} \mathbf{u}_j = \lambda_j \mathbf{u}_i^{\mathrm{T}} \mathbf{u}_j$ is if $\mathbf{u}_i^{\mathrm{T}} \mathbf{u}_j = 0$. That is, the eigenvectors are orthogonal.

Second, we recall that any multiple of an eigenvector is also an eigenvector, as discussed in connection with equation (11.4). In particular, we can normalize the columns of S without affecting diagonalization. That is, define a matrix Q whose columns are proportional to the columns of S, but their length (as vectors) is one. Then AQ is

$$A \begin{bmatrix}\frac{\mathbf{u}_1}{\|\mathbf{u}_1\|} \cdots \frac{\mathbf{u}_n}{\|\mathbf{u}_n\|}\end{bmatrix} = \begin{bmatrix}\lambda_1 \frac{\mathbf{u}_1}{\|\mathbf{u}_1\|} \cdots \lambda_n \frac{\mathbf{u}_n}{\|\mathbf{u}_n\|}\end{bmatrix} = Q\Lambda.$$

Hence

$$A = Q\Lambda Q^{-1}.$$

Because the columns of Q have length one, taking the dot product of a column with itself yields value one. Because any two distinct columns of Q are perpendicular, the dot product of two distinct columns has value zero. Succinctly, $Q^{\mathrm{T}}Q = I$, the identity matrix. Postmultiplying by Q^{-1} yields $Q^{\mathrm{T}} = Q^{-1}$. In short, if A is symmetric, its diagonalization can be expressed as

$$A = Q\Lambda Q^{\mathrm{T}} \tag{11.6}$$

with Q orthonormal.

Orthogonal diagonalization is significant because, as we noted previously, A and Q have the same column space. Hence the columns of Q provide a basis for the column space of A that has the properties we are accustomed to in coordinate axes: the basis vectors are all unit length and mutually perpendicular.

In addition, we have seen that the operation of an orthonormal matrix like Q is a rigid rotation. The operation of a diagonal matrix like Λ is a scaling without rotation: if $\mathbf{x}_i$ is a coordinate axis of the input space, having 1 in the i-th position and 0 everywhere else, then $\Lambda\mathbf{x}_i = \lambda_i\mathbf{x}_i$. That is, Λ stretches the axes of the input space, but does not change their directions. Hence $A = Q\Lambda Q^{\mathrm{T}}$ decomposes the action of A into a scaling Λ and a rigid rotation Q. This is only the beginning of the story, though, as we will see shortly.

11.3 Eigenvalues and the scaling effects of a matrix

11.3.1 Matrix norms

We discussed vector norms earlier. Norms are also defined for matrices, and they must satisfy the same three properties (11.1), replacing vectors $\mathbf{x}$ and $\mathbf{y}$ with matrices X and Y. The norm of particular interest is again the Euclidean norm. As already mentioned, the operation of a matrix can be viewed as a combination of a scaling effect and a rotation effect. The Euclidean norm is the maximum scaling effect. A formal definition requires a little explanation.

A matrix's scaling effect is measured by the ratio of the output and input vector lengths. We define the **scaling function** of matrix A to be

$$\rho(\mathbf{x}) = \frac{\|A\mathbf{x}\|}{\|\mathbf{x}\|}. \tag{11.7}$$

The scaling function $\rho(\mathbf{x})$ is sensitive to the orientation of the input vector $\mathbf{x}$, but not its length. The sensitivity of ρ to orientation is easily seen by considering the example

$$A = \begin{bmatrix} 1\,0 \\ 0\,0 \end{bmatrix} \quad \mathbf{x}_1 = \begin{bmatrix} 1 \\ 0 \end{bmatrix} \quad \mathbf{x}_2 = \begin{bmatrix} 0 \\ 1 \end{bmatrix}.$$

The vectors $\mathbf{x}_1$ and $\mathbf{x}_2$ differ in direction but not length. (They both have length one.) The scaling effect of A is sensitive to that difference: A leaves $\mathbf{x}_1$'s length unchanged ($\|A\mathbf{x}_1\| = 1$) but reduces $\mathbf{x}_2$'s length to zero ($\|A\mathbf{x}_2\| = 0$).

But the scaling function is not sensitive to the original length of the input vector. Consider any two vectors $\mathbf{x}_1$ and $\mathbf{x}_2$ with the same direction but different lengths. Let c be the ratio of their lengths, that is, let $\mathbf{x}_2 = c\mathbf{x}_1$. The effect of A on $\mathbf{x}_2$ is

$$\rho(\mathbf{x}_2) = \frac{\|A(c\mathbf{x}_1)\|}{\|c\mathbf{x}_1\|} = \frac{c\|A\mathbf{x}_1\|}{c\|\mathbf{x}_1\|} = \rho(\mathbf{x}_1).$$

All vectors in a family of vectors sharing the same direction are scaled in the same way by A.

When we say that the Euclidean norm $\|A\|$ of a matrix A is its maximum scaling effect, we mean that

$$\|A\| = \max_{\mathbf{x} \neq \mathbf{0}} \frac{\|A\mathbf{x}\|}{\|\mathbf{x}\|}. \tag{11.8}$$

Since all vectors that share orientation are scaled the same way, we can take the unit vector with a given orientation as the canonical representative of its family. That is, (11.8) is equivalent to

$$\|A\| = \max_{\|\mathbf{x}\|=1} \|A\mathbf{x}\|. \tag{11.9}$$

Note that the definandum, $\|A\|$, is an array norm, and it is being defined in terms of $\|A\mathbf{x}\|$, a vector norm.

The next question is what we can say about the orientation that maximizes the scaling effect of A.

11.3.2 The Rayleigh quotient

Since the scale factor ρ is a ratio of nonnegative numbers (lengths are never negative!), it is monotonically related to its square:

$$R(\mathbf{x}) = \rho^2(\mathbf{x}) = \frac{\|A\mathbf{x}\|^2}{\|\mathbf{x}\|^2} = \frac{(A\mathbf{x})^{\mathrm{T}}(A\mathbf{x})}{\mathbf{x}^{\mathrm{T}}\mathbf{x}} = \frac{\mathbf{x}^{\mathrm{T}}A^{\mathrm{T}}A\mathbf{x}}{\mathbf{x}^{\mathrm{T}}\mathbf{x}}. \tag{11.10}$$

The ratio $\mathbf{x}^{\mathrm{T}}A^{\mathrm{T}}A\mathbf{x}/\mathbf{x}^{\mathrm{T}}\mathbf{x}$ is known as the **Rayleigh quotient**.

The vector direction that maximizes the Rayleigh quotient of $A^{\mathrm{T}}A$ is the same one that maximizes the scaling function of A. The value of the scaling function in that direction is the norm of A. (It should be clear that the Rayleigh quotient, like the matrix norm, is insensitive to the length of $\mathbf{x}$.) To learn something about the vector that maximizes R, let us take its derivative and set it to zero. Using the rules developed in section 5.2.3, the derivative is

$$\mathbf{D}_{\mathbf{x}}R = \mathbf{x}^{\mathrm{T}}A^{\mathrm{T}}A\mathbf{x}\mathbf{D}_{\mathbf{x}}(\mathbf{x}^{\mathrm{T}}\mathbf{x})^{-1} + (\mathbf{x}^{\mathrm{T}}\mathbf{x})^{-1}\mathbf{D}_{\mathbf{x}}\mathbf{x}^{\mathrm{T}}A^{\mathrm{T}}A\mathbf{x}$$

$$= -\mathbf{x}^{\mathrm{T}} A^{\mathrm{T}} A\mathbf{x}(\mathbf{x}^{\mathrm{T}}\mathbf{x})^{-2}\mathbf{D}_{\mathbf{x}}\mathbf{x}^{\mathrm{T}}\mathbf{x} + (\mathbf{x}^{\mathrm{T}}\mathbf{x})^{-1}\mathbf{x}^{\mathrm{T}}(A^{\mathrm{T}}A + (A^{\mathrm{T}}A)^{\mathrm{T}})\mathbf{D}_{\mathbf{x}}\mathbf{x}$$
$$= -\mathbf{x}^{\mathrm{T}} A^{\mathrm{T}} A\mathbf{x}(\mathbf{x}^{\mathrm{T}}\mathbf{x})^{-2}(2\mathbf{x}^{\mathrm{T}}) + (\mathbf{x}^{\mathrm{T}}\mathbf{x})^{-1}\mathbf{x}^{\mathrm{T}}(2A^{\mathrm{T}}A)$$
$$= \frac{2}{(\mathbf{x}^{\mathrm{T}}\mathbf{x})^2}\left(\|\mathbf{x}\|^2\mathbf{x}^{\mathrm{T}}A^{\mathrm{T}}A - \|A\mathbf{x}\|^2\mathbf{x}^{\mathrm{T}}\right).$$

Setting this expression to zero, and transposing both sides in order to convert the row vectors into column vectors, we obtain:

$$A^{\mathrm{T}}A\mathbf{x} = \frac{\|A\mathbf{x}\|^2}{\|\mathbf{x}\|^2}\mathbf{x}. \tag{11.11}$$

What we learn is that the critical points of the Rayleigh quotient, and hence the critical points of the scaling function, are eigenvectors of $A^{\mathrm{T}}A$, whose corresponding eigenvalues have the form of the Rayleigh quotient itself. This can be strengthened to a biconditional. Consider an arbitrary eigenvalue λ of $A^{\mathrm{T}}A$:

$$A^{\mathrm{T}}A\mathbf{x} = \lambda\mathbf{x}.$$

Premultiplying both sides by $\mathbf{x}^{\mathrm{T}}$ yields

$$\mathbf{x}^{\mathrm{T}}A^{\mathrm{T}}A\mathbf{x} = \lambda\mathbf{x}^{\mathrm{T}}\mathbf{x}$$
$$\lambda = \frac{\|A\mathbf{x}\|^2}{\|\mathbf{x}\|^2}.$$

The eigenvalues of $A^{\mathrm{T}}A$ all satisfy the Rayleigh quotient of $A^{\mathrm{T}}A$, and they are the only vectors that do so.

We have already observed that the value of the Rayleigh quotient $R(\mathbf{x})$ is insensitive to the length of $\mathbf{x}$, and we recall from the discussion surrounding equation (11.4) that the property of being an eigenvector with a given eigenvalue λ is insensitive to length. Hence, without loss of generality, we can take the vector $\mathbf{x}$ in (11.11) to be a unit vector, and we can write:

$$A^{\mathrm{T}}A\mathbf{x} = \|A\mathbf{x}\|^2\mathbf{x}$$

if and only if $\mathbf{x}$ is a critical point of the Rayleigh quotient.

One of the (unit-length) vectors that satisfy (11.11) maximizes the Rayleigh quotient, and hence also maximizes the scaling function ρ; call that vector $\mathbf{z}$. Because $\mathbf{z}$ maximizes ρ, it follows by definition of the matrix norm (11.8) that $\|A\mathbf{z}\| = \|A\|$, so we can write:

$$A^{\mathrm{T}}A\mathbf{z} = \|A\|^2\mathbf{z}. \tag{11.12}$$

This tells us that the squared norm of matrix A is the eigenvalue of $A^{\mathrm{T}}A$ with the largest absolute value.

Since the vectors satisfying (11.11) are the critical points of the Rayleigh quotient, one of them also *minimizes* the Rayleigh quotient. It will turn out to be $\mathbf{u}_2$, the eigenvector corresponding to the second smallest eigenvalue of

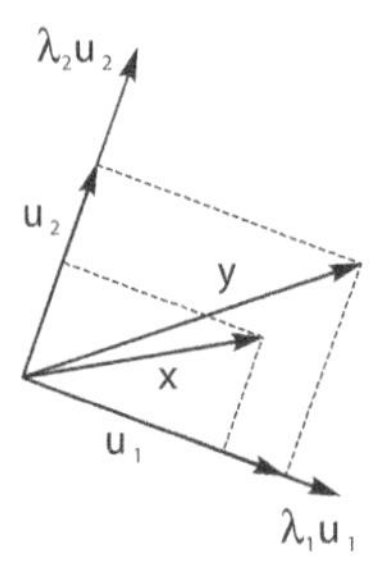

FIGURE 11.1
Eigenvectors of a 2×2 matrix.

$A^{\mathrm{T}}A$. That will bring us back to the mincut problem. Before we go there, however, let us look further at the geometry of eigenvectors and the scaling function.

We began the discussion of matrix norms by defining the norm of a matrix as its maximum scaling effect. We also pointed out that the scaling function is sensitive to the direction but not to the original magnitude of the scaled vector. Now (11.12) shows that the maximum squared scaling effect $\|A\|^2$ is an eigenvalue of $A^{\mathrm{T}}A$, and the direction of maximum scaling is the corresponding eigenvector.

11.3.3 The 2×2 case

Some geometric intuition can be obtained as follows. Let us consider a symmetric 2×2 matrix A with two eigenvectors $\mathbf{u}_1$ and $\mathbf{u}_2$. We can assume, without loss of generality, that the eigenvectors $\mathbf{u}_1$ and $\mathbf{u}_2$ are unit vectors, and the symmetry of A implies that $\mathbf{u}_1$ and $\mathbf{u}_2$ are perpendicular. Let the vector $\mathbf{x}$ represent an arbitrary input direction, and let the vector $\mathbf{y} = A\mathbf{x}$ be the result of applying the matrix A to $\mathbf{x}$. These four vectors are shown in figure 11.1. Since we are interested only in the scaling effect of A, we take $\mathbf{x}$ to be a unit vector. The output vector $\mathbf{y}$ is not generally a unit vector; indeed, the scaling effect of A on $\mathbf{x}$ is

$$\rho(\mathbf{x}) = \frac{\|A\mathbf{x}\|}{\|\mathbf{x}\|} = \|\mathbf{y}\|. \tag{11.13}$$

As orthogonal unit vectors, $\mathbf{u}_1$ and $\mathbf{u}_2$ can be viewed as axes of a coordinate space. The input vector $\mathbf{x}$ can be represented by its coordinates in "eigenspace":

$$\mathbf{x} = (\cos\theta)\mathbf{u}_1 + (\sin\theta)\mathbf{u}_2$$

where θ is the angle between $\mathbf{x}$ and $\mathbf{u}_1$. The value $\cos\theta$ is the length of the projection of $\mathbf{x}$ onto the $\mathbf{u}_1$ axis, and $\sin\theta$ is the length of its projection onto the $\mathbf{u}_2$ axis. The dotted lines from the end of $\mathbf{x}$ in figure 11.1 represent those projections. The coordinates of $\mathbf{x}$ in the eigenspace constitute a new vector $\mathbf{x}'$:

$$\mathbf{x}' = \begin{bmatrix} \cos\theta \\ \sin\theta \end{bmatrix}.$$

Defining Q to be a matrix whose columns are $\mathbf{u}_1$ and $\mathbf{u}_2$, notice that

$$Q\mathbf{x}' = \begin{bmatrix} \mathbf{u}_1 \ \mathbf{u}_2 \end{bmatrix} \begin{bmatrix} \cos\theta \\ \sin\theta \end{bmatrix} = (\cos\theta)\mathbf{u}_1 + (\sin\theta)\mathbf{u}_2 = \mathbf{x}.$$

That is, the conversion from $\mathbf{x}$ to $\mathbf{x}'$ can be represented as $\mathbf{x}' = Q^{-1}\mathbf{x}$.

We can compute $\mathbf{y}$ in terms of this decomposition:

$$\mathbf{y} = A\mathbf{x} = A[(\cos\theta)\mathbf{u}_1 + (\sin\theta)\mathbf{u}_2] = (\cos\theta)\lambda_1\mathbf{u}_1 + (\sin\theta)\lambda_2\mathbf{u}_2.$$

In eigenspace coordinates, $\mathbf{y}$ corresponds to the vector $\mathbf{y}'$:

$$\mathbf{y}' = \begin{bmatrix} \lambda_1\cos\theta \\ \lambda_2\sin\theta \end{bmatrix} = \begin{bmatrix} \lambda_1 & 0 \\ 0 & \lambda_2 \end{bmatrix} \mathbf{x}'.$$

When we apply A to vector $\mathbf{u}_1$, the result is $\lambda_1\mathbf{u}_1$, and $A\mathbf{u}_2$ is $\lambda_2\mathbf{u}_2$. The coordinates of $\mathbf{x}$ on the eigenspace axes, namely $\cos\theta$ on the $\mathbf{u}_1$ axis and $\sin\theta$ on the $\mathbf{u}_2$ axis, are scaled similarly by A, yielding $\mathbf{y}' = (\lambda_1 x'_1, \lambda_2 x'_2)$. Projecting back from the eigenspace coordinates to the original coordinates yields $\mathbf{y}$. Those reverse projections are the dotted lines converging on the end of $\mathbf{y}$ in the figure. They can also be represented by a matrix multiplication:

$$\mathbf{y} = (\lambda_1\cos\theta)\mathbf{u}_1 + (\lambda_2\sin\theta)\mathbf{u}_2 = Q\mathbf{y}'.$$

The scaling effect of A on $\mathbf{u}_1$ and $\mathbf{u}_2$ is obviously λ_1 and λ_2, respectively. Assume that they are not identical; suppose $\lambda_1 < \lambda_2$, as in the figure. Let us compute the scaling effect of A on $\mathbf{x}$. We saw in equation (11.13) that $\rho(\mathbf{x})$ is simply the length of $\mathbf{y}$:

$$\rho^2(\mathbf{x}) = \|\mathbf{y}\|^2 = \|\mathbf{y}'\|^2 = (\cos^2\theta)\lambda_1^2 + (\sin^2\theta)\lambda_2^2.$$

Since $\cos^2\theta$ and $\sin^2\theta$ are nonnegative and $\cos^2\theta + \sin^2\theta = 1$, we observe that $\rho^2(\mathbf{x})$ is a weighted average, which means that $\rho^2(\mathbf{x})$ lies between λ_1^2 and λ_2^2. That is, recalling the expression for ρ^2 given in (11.10):

$$\lambda_1^2 \le \frac{\mathbf{x}^{\mathrm{T}}A^{\mathrm{T}}A\mathbf{x}}{\mathbf{x}^{\mathrm{T}}\mathbf{x}} \le \lambda_2^2. \tag{11.14}$$

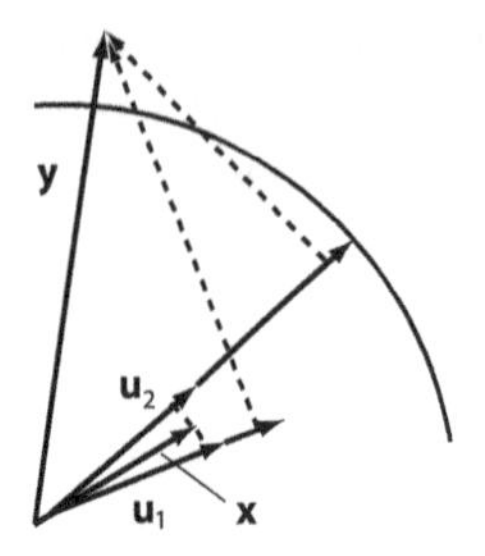

FIGURE 11.2
Non-orthogonal eigenvectors. Input vector $\mathbf{x}$ is projected onto $\mathbf{u}_1$ and $\mathbf{u}_2$, yielding coordinates $x'_1 \approx 1$ and $x'_2 \approx 1$ on the $\mathbf{u}_1$ and $\mathbf{u}_2$ "axes." The eigenvectors are stretched by the operation of matrix A, and the coordinates in the stretched system, $\lambda_1 x'_1 \approx \lambda_1$ and $\lambda_2 x'_2 \approx \lambda_2$, are projected back (the dotted lines) to determine the location of $\mathbf{y}$. In this case, $\mathbf{y}$ is longer than either of the stretched eigenvectors.

Hence $|\lambda_1| \leq |\rho(\mathbf{x})| \leq |\lambda_2|$.

In short, for a symmetric 2×2 matrix, the largest scaling effect is in the direction of the eigenvector with the largest eigenvalue, and the smallest scaling effect is in the direction of the other eigenvector. All other scaling effects are mixtures of those two.

We mention in passing that the assumption of symmetry is important. Without symmetry, the eigenvectors may fail to be orthogonal, and if the eigenvectors are not orthogonal, the scaling factor for $\mathbf{x}$ may actually exceed $|\lambda_2|$. Figure 11.2 illustrates.

11.3.4 The general case

The discussion generalizes readily to multiple dimensions. Assuming that A is symmetric, it can be expressed as

$$A = Q \Lambda Q^{\mathrm{T}} \tag{11.15}$$

with Q orthonormal, as in (11.6). The application of A to an arbitrary unit vector $\mathbf{x}$ takes the form:

$$\mathbf{y} = Q \Lambda Q^{\mathrm{T}} \mathbf{x}.$$

Though it may not be immediately obvious, this decomposition reflects exactly the steps we just went through in the two-dimensional case. Working from right to left, we first express $\mathbf{x}$ as a linear combination of the eigenvectors; this yields its representation in the coordinate space whose axes are the eigenvectors:

$$\mathbf{x} = Q\mathbf{x}'$$

which is to say

$$\mathbf{x}' = Q^{\mathrm{T}}\mathbf{x}$$

since $Q^{-1} = Q^{\mathrm{T}}$ for orthonormal matrices. Next we scale each axis by the corresponding eigenvalue:

$$\mathbf{y}' = \Lambda\mathbf{x}' = \Lambda Q^{\mathrm{T}}\mathbf{x}.$$

Finally, we also represent $\mathbf{y}$ as a linear combination of eigenvectors:

$$\mathbf{y} = Q\mathbf{y}'.$$

Applying Q to $\mathbf{y}'$ converts it back to the original coordinate space, yielding $\mathbf{y}$:

$$\mathbf{y} = Q\Lambda Q^{\mathrm{T}}\mathbf{x} = A\mathbf{x}.$$

Now let us order the eigenvalues by increasing squared value:

$$\lambda_1^2 \leq \lambda_2^2 \leq \ldots \leq \lambda_n^2 \tag{11.16}$$

and let us order the eigenvectors $\mathbf{u}_1, \ldots, \mathbf{u}_n$ accordingly. Let the first two eigenvectors $\mathbf{u}_1$ and $\mathbf{u}_2$ be the eigenvectors in the two-dimensional space of figure 11.1. Let us call this space $\mathcal{Q}_2$; it is the plane spanned by the first two eigenvectors. The discussion of the previous section (section 11.3.3) leads to the following conclusions. Any unit vector $\mathbf{x}$ in the plane $\mathcal{Q}_2$ can be expressed as a linear combination of $\mathbf{u}_1$ and $\mathbf{u}_2$, and the squared scaling effect of A on $\mathbf{x}$ is a weighted average of its squared scaling effects on $\mathbf{u}_1$ and $\mathbf{u}_2$. Hence

$$\lambda_1^2 \leq \frac{\mathbf{x}^{\mathrm{T}}A^{\mathrm{T}}A\mathbf{x}}{\mathbf{x}^{\mathrm{T}}\mathbf{x}} \leq \lambda_2^2.$$

Now suppose we add the third eigenvector $\mathbf{u}_3$ to the space. The resulting eigenspace $\mathcal{Q}_3$ has three dimensions instead of two, but otherwise nothing changes. Let $\mathbf{x}$ be an arbitrary vector in $\mathcal{Q}_3$, and let $\mathbf{x}'$ represent its coordinates in $\mathcal{Q}_3$:

$$\mathbf{x}' = \begin{bmatrix} x_1' \\ x_2' \\ x_3' \end{bmatrix}$$

where x_i' is the length of the projection of $\mathbf{x}$ onto $\mathbf{u}_i$. The squared scaling effect is the length of $A\mathbf{x}$, namely:

$$\rho^2(\mathbf{x}) = \|\mathbf{y}\|^2 = (x_1')^2\lambda_1^2 + (x_2')^2\lambda_2^2 + (x_3')^2\lambda_3^2. \tag{11.17}$$

The coefficients $(x'_i)^2$ in (11.17) are obviously nonnegative, and their sum is the squared length of $\mathbf{x}'$, that is, they sum to one. Hence $\rho^2(\mathbf{x})$ is a weighted average of the squared eigenvalues. Since λ_3^2 is the largest of the values being averaged, we know that

$$\rho^2(\mathbf{x}) = \frac{\mathbf{x}^{\mathrm{T}} A^{\mathrm{T}} A \mathbf{x}}{\mathbf{x}^{\mathrm{T}} \mathbf{x}} \leq \lambda_3^2$$

for any vector $\mathbf{x}$ in the space $\mathcal{Q}_3$ spanned by the first 3 eigenvectors. We have equality if we take $\mathbf{x}$ equal to $\mathbf{u}_3$, so

$$\max_{\mathbf{x} \in \mathcal{Q}_3} \frac{\mathbf{x}^{\mathrm{T}} A^{\mathrm{T}} A \mathbf{x}}{\mathbf{x}^{\mathrm{T}} \mathbf{x}} = \lambda_3^2.$$

Applying this reasoning recursively, we conclude that

$$\max_{\mathbf{x} \in \mathcal{Q}_k} \frac{\mathbf{x}^{\mathrm{T}} A^{\mathrm{T}} A \mathbf{x}}{\mathbf{x}^{\mathrm{T}} \mathbf{x}} = \lambda_k^2 \qquad \arg\max_{\mathbf{x} \in \mathcal{Q}_k} \frac{\mathbf{x}^{\mathrm{T}} A^{\mathrm{T}} A \mathbf{x}}{\mathbf{x}^{\mathrm{T}} \mathbf{x}} = \mathbf{u}_k. \tag{11.18}$$

We can also make a statement about minima. Let us consider the subspace spanned by the eigenvectors *excluding* the first k, which is to say, $\mathcal{Q}_k^{\perp}$ (11.2). That is, $\mathcal{Q}_{n-1}^{\perp}$ is the line spanned by $\mathbf{u}_n$, $\mathcal{Q}_{n-2}^{\perp}$ is the plane spanned by $\mathbf{u}_{n-1}$ and $\mathbf{u}_n$, and so on. In general, the dimensionality of the space $\mathcal{Q}_k^{\perp}$ is $n-k$. The eigenvectors spanning $\mathcal{Q}_k^{\perp}$ correspond to the $n-k$ eigenvalues with the *largest* squared values: $\lambda_{k+1}, \ldots, \lambda_n$. Clearly, if we restrict attention to $\mathbf{x}$ in the space $\mathcal{Q}_k^{\perp}$, the squared scaling effect $\rho^2(\mathbf{x})$ is minimized by choosing the smallest available eigenvector direction. That is

$$\min_{\mathbf{x} \in \mathcal{Q}_{k-1}^{\perp}} \frac{\mathbf{x}^{\mathrm{T}} A^{\mathrm{T}} A \mathbf{x}}{\mathbf{x}^{\mathrm{T}} \mathbf{x}} = \lambda_k^2 \qquad \arg\min_{\mathbf{x} \in \mathcal{Q}_{k-1}^{\perp}} \frac{\mathbf{x}^{\mathrm{T}} A^{\mathrm{T}} A \mathbf{x}}{\mathbf{x}^{\mathrm{T}} \mathbf{x}} = \mathbf{u}_k. \tag{11.19}$$

The equations use $\mathcal{Q}_{k-1}^{\perp}$ instead of $\mathcal{Q}_k^{\perp}$ in order to have k range from 1 to n, for consistency with equations (11.18).

11.3.5 The Courant-Fischer minimax theorem

One further simplification of (11.18) and (11.19) is possible. Continuing to assume that A is symmetric and hence has the decomposition $Q\Lambda Q^{\mathrm{T}}$, we observe that

$$A^2 = (Q\Lambda Q^{\mathrm{T}})(Q\Lambda Q^{\mathrm{T}}) = Q\Lambda^2 Q^{\mathrm{T}}$$

since $Q^{\mathrm{T}}Q = I$. This is useful because Λ^2 has a particularly simple form. Being a diagonal matrix, it is squared elementwise:

$$\Lambda^2 = \begin{bmatrix} \lambda_1^2 & & \\ & \ddots & \\ & & \lambda_n^2 \end{bmatrix}.$$

Since the columns of Q are eigenvectors and the entries of Λ are eigenvalues, we see that A^2 has the same eigenvectors as A, and that the eigenvalues of A^2 are the squares of the eigenvalues of A.

We can also take the square root of A in the same fashion:

$$A^{1/2} = Q\Lambda^{1/2}Q^{\mathrm{T}} \tag{11.20}$$

where $\Lambda^{1/2}$ is a diagonal matrix whose i-th entry is $\sqrt{\lambda_i}$. We confirm

$$A^{1/2}A^{1/2} = Q\Lambda^{1/2}Q^{\mathrm{T}}Q\Lambda^{1/2}Q^{\mathrm{T}} = Q\Lambda Q^{\mathrm{T}} = A.$$

The equation (11.18) is true of any matrix that can be decomposed in the form (11.15), with Q orthonormal and Λ diagonal. $A^{1/2}$ does have such a decomposition, though a subtlety must be pointed out. The factorization (11.15) is assured for symmetric *real* matrices, but if any of the eigenvalues in Λ are negative, then $\Lambda^{1/2}$ contains imaginary values. However, we do not need to show that $A^{1/2}$ is symmetric and real in order to know that it has the desired factorization: we already know that it satisfies (11.20), and we know independently that Q is orthonormal. Hence $A^{1/2}$ does have a factorization of the required form, and equation (11.18) applies to $A^{1/2}$.

Since A and $A^{1/2}$ have the same eigenvectors, the eigenspaces $\mathcal{Q}_k$ are the same for both, but we replace the eigenvalues λ_k with the eigenvalues of $A^{1/2}$, namely, $\sqrt{\lambda_k}$:

$$\max_{\mathbf{x}\in\mathcal{Q}_k} \frac{\mathbf{x}^{\mathrm{T}}(A^{1/2})^{\mathrm{T}}(A^{1/2})\mathbf{x}}{\mathbf{x}^{\mathrm{T}}\mathbf{x}} = (\sqrt{\lambda_k})^2.$$

That is, for $1 \leq k \leq n$,

$$\max_{\mathbf{x}\in\mathcal{Q}_k} \frac{\mathbf{x}^{\mathrm{T}}A\mathbf{x}}{\mathbf{x}^{\mathrm{T}}\mathbf{x}} = \lambda_k \qquad \arg\max_{\mathbf{x}\in\mathcal{Q}_k} \frac{\mathbf{x}^{\mathrm{T}}A\mathbf{x}}{\mathbf{x}^{\mathrm{T}}\mathbf{x}} = \mathbf{u}_k. \tag{11.21}$$

Applying the same reasoning to (11.19) yields, for $0 \leq k \leq n-1$,

$$\min_{\mathbf{x}\in\mathcal{Q}_k^{\perp}} \frac{\mathbf{x}^{\mathrm{T}}A\mathbf{x}}{\mathbf{x}^{\mathrm{T}}\mathbf{x}} = \lambda_{k+1} \qquad \arg\min_{\mathbf{x}\in\mathcal{Q}_k^{\perp}} \frac{\mathbf{x}^{\mathrm{T}}A\mathbf{x}}{\mathbf{x}^{\mathrm{T}}\mathbf{x}} = \mathbf{u}_{k+1}. \tag{11.22}$$

Finally, recall that we ordered the eigenvalues by increasing squared value (11.16). With the replacement of $\sqrt{\lambda_k}$ for λ_k, the definition of the ordering becomes

$$(\sqrt{\lambda_1})^2 \leq (\sqrt{\lambda_2})^2 \leq \ldots \leq (\sqrt{\lambda_n})^2.$$

That is, we now order eigenvalues simply by increasing size:

$$\lambda_1 \leq \lambda_2 \leq \ldots \leq \lambda_n.$$

This ordering is valid even if some of the eigenvalues are negative. If so, some of the values $\sqrt{\lambda_k}$ are imaginary, but we still have $(\sqrt{\lambda_k})^2 = \lambda_k$.

The equations (11.21) and (11.22) represent a special case of the **Courant-Fischer minimax theorem**. The general theorem states that (11.22) is true not just for the space $\mathcal{Q}_k^\perp$, but for *any* space with the same dimensionality as $\mathcal{Q}_k^\perp$, hence in particular for the space that maximizes the value on the left-hand side of (11.22):

$$\max_{S|\dim(S)=n-k} \min_{\mathbf{x}\in S} \frac{\mathbf{x}^{\mathrm{T}} A\mathbf{x}}{\mathbf{x}^{\mathrm{T}}\mathbf{x}} = \lambda_{k+1}.$$

The equation (11.21) can be similarly generalized. A proof for the general case can be found in Golub & Loan [97], p. 394.

11.4 Bibliographic notes

For the reader who really has no more background than I have assumed, I cannot recommend Strang [213] highly enough. For discussion of more advanced topics, including matrix norms and the Courant-Fischer theorem, see Horn & Johnson [113] and Golub & Loan [97].

12

Spectral Methods

At first glance, it is not at all obvious what form spectral methods for learning might take. The connection between spectra and semisupervised learning can perhaps most readily be seen by thinking back to label propagation, and specifically our picture of label propagation as doing an interpolation of label information from boundaries across the interior of the graph. With spectral methods, the idea is to construct an interpolation in the form of a "standing wave." Accordingly, we begin with a discussion of wave-shaped functions, specifically *harmonics* (not to be confused with the harmonic functions of chapter 10) that arise from simple harmonic motion.

12.1 Simple harmonic motion

12.1.1 Harmonics

Most readers will be familiar with the idea that a musical tone is a mixture of harmonics. The perceived pitch is the **fundamental frequency**, and the timbre – what makes a violin playing a particular note sound different from a flute playing the same note – is determined by the relative strengths of **overtones**. The perception of pitch and overtones is possible because the ear breaks the sound into its component frequencies. Similarly, the idea that white light is a mixture of frequencies, and is broken into its spectrum by a prism, is surely also familiar. A spectrum is the same as timbre – it is an association of coefficients with frequencies, specifying the relative weight of different frequencies in the mix.

A "pure tone" consisting of a single frequency is represented by a harmonic. Technically, a **harmonic** is any function

$$y = A\sin(\omega x + \psi). \tag{12.1}$$

The parameters of a harmonic are its **amplitude** A, its **phase** ψ, and the **angular velocity** ω. Angular velocity is measured in radians per unit time, and is related to **frequency** (cycles per unit time) by

$$F = \frac{\omega}{2\pi}.$$

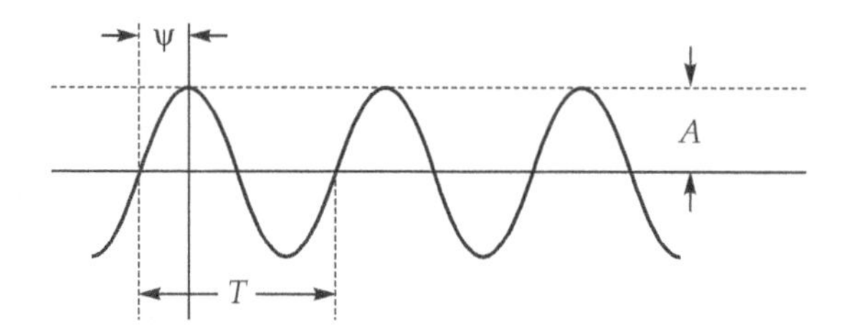

FIGURE 12.1
A harmonic. A is amplitude, ψ is (sine) phase, and T is period.

The inverse of frequency is the **period**:

$$T = \frac{2\pi}{\omega}.$$

See figure 12.1 for the graphical correlates of amplitude, phase, and period.

Circular motion and oscillating linear motion are two simple physical systems that are described by a harmonic. Suppose a particle is moving counterclockwise around a circle of radius A. The circular motion can be (physically) translated to oscillating linear motion by replacing the particle with a peg on an arm, inserted into a slot in a horizontal bar that is constrained to move only up and down (figure 12.2). Let θ represent the angular location of the particle, that is, the angle it forms with the horizontal axis. If the particle moves around the unit circle, the angle θ in radians is the distance it has traveled; for a circle of radius A, the distance traveled is $A\theta$. The vertical position of the particle is $y = A \sin \theta$. Let x represent time. If the particle starts at angle θ_0 at time zero, and travels at angular velocity ω radians per second, then its angular position at time x is $\theta_0 + \omega x$, and its vertical position is

$$y = A \sin(\omega x + \theta_0).$$

This is a harmonic with amplitude A, angular velocity ω, and phase θ_0. The particle completes one revolution at the time T such that $\omega T = 2\pi$. That is, the time required for one revolution of the physical system $(2\pi/\omega)$ is equal to the period of the harmonic. If x is time in seconds, the frequency of the harmonic is the number of revolutions per second.

Since $\cos(\theta) = \sin(\theta + \pi/2)$, the harmonic (12.1) can be expressed equivalently as (12.2):

$$y = A \cos(\omega x + \phi) \tag{12.2}$$

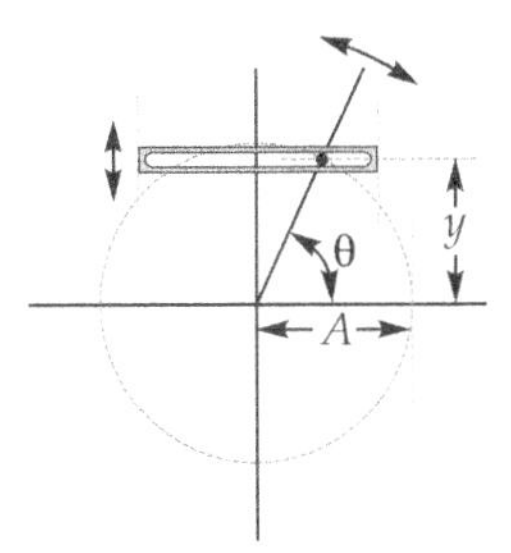

FIGURE 12.2
Circular motion and oscillating linear motion. The horizontal bar is constrained to move only up and down.

where

$$\phi = \psi - \pi/2.$$

Observe that ϕ plays the same role in (12.2) as ψ does in (12.1). For that reason, ϕ is called the **cosine phase** of the harmonic, and ψ is more specifically called the **sine phase** to distinguish it from cosine phase. Since (12.1) and (12.2) are equivalent, the choice between them is a matter of convenience. Note that $\sin\psi = \cos\phi$.

12.1.2 Mixtures of harmonics

Applying the addition formula for sine to (12.1), we obtain:

$$y = A\cos\omega x\sin\psi + A\sin\omega x\cos\psi.$$

Writing

$$a = A\sin\psi \qquad b = A\cos\psi$$

the harmonic (12.1) can be expressed as

$$y = a\cos\omega x + b\sin\omega x. \tag{12.3}$$

In this form, we see that the harmonic is the dot product of two vectors:

$$y = \mathbf{u}\cdot\mathbf{v} = \begin{bmatrix} A\sin\psi \\ A\cos\psi \end{bmatrix}^{\mathrm{T}} \begin{bmatrix} \cos\omega x \\ \sin\omega x \end{bmatrix}.$$

This representation is convenient because it separates frequency from phase and amplitude. The vector $\mathbf{u}$ depends only on phase ψ and amplitude A, and $\mathbf{v}$ depends only on frequency $F = \omega/(2\pi)$, not on phase or amplitude. The vector $\mathbf{v}$ can be thought of as a harmonic in standard form (amplitude one and phase zero), and $\mathbf{u}$ is a coefficient vector providing the phase and amplitude. For lack of a better term, I will call $\mathbf{u}$ the "phase-amplitude" of the harmonic.

We already noted that sound waves and light waves consist of mixtures of harmonics. Remarkably, that turns out to be true not only for sound and light waves, but for *any* piecewise continuous function that is either periodic with period T, or else defined only on a bounded interval $[0, T]$. Unfortunately, we will not be able to prove the theorem here – the reader is referred to more specialized works, such as that of Tolstov [225], for a proof. Omitting some details, the theorem states that any piecewise continuous function $f(x)$ defined on a bounded interval $[0, T]$ can be expressed as an infinite sum of harmonics:

$$f(x) = \frac{a_0}{2} + \sum_{k=1}^{\infty} \left(a_k \cos k\frac{2\pi}{T}x + b_k \sin k\frac{2\pi}{T}x \right). \tag{12.4}$$

That is, for any given function f, there exists a choice of coefficients (a_0, a_1, b_1, a_2, b_2, ...) that makes (12.4) true. The coefficient a_0 sets the baseline. The remaining coefficients come in pairs (a_k, b_k) that represent the phase-amplitudes of the component harmonics. Observe that the expression inside the summation in (12.4) is a harmonic in the form (12.3), where $\omega_k = 2\pi k/T$. The frequency of the k-th harmonic is

$$F_k = \frac{\omega_k}{2\pi} = \frac{k}{T}.$$

The value $F_1 = 1/T$ represents the **fundamental frequency**, and the frequencies of the harmonics are integer multiples of the fundamental. (The more common notation for fundamental frequency is "F_0," but we will prefer consistency over convention and write "F_1.") The **spectrum** of the function f is the list of coefficients:

$$(a_0, a_1, b_1, a_2, b_2, \ldots).$$

The k-th pair (a_k, b_k) represents the weight, in phase-amplitude form, of the k-th harmonic in the weighted sum (12.4).

A more convenient form is obtained by defining

$$t = \frac{2\pi}{T}x.$$

Then (12.4) can be written as

$$f(t) = \frac{a_0}{2} + \sum_{k=1}^{\infty} (a_k \cos kt + b_k \sin kt). \tag{12.5}$$

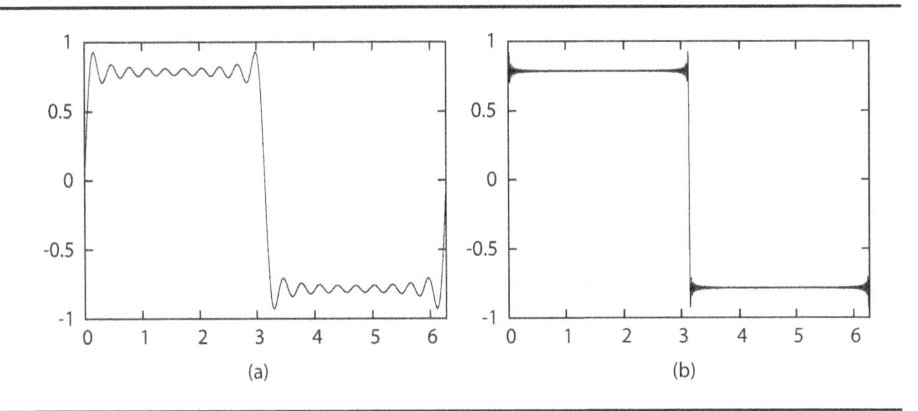

FIGURE 12.3
Approximation of a piecewise linear function by a sum of sines: (a) first ten terms of the Fourier series, (b) first 100 terms.

This is known as the **Fourier expansion** of the function.

An impression of how an arbitrary function can be represented as a sum of harmonics can be obtained by considering the examples in figure 12.3. The plots show the series

$$f_n(x) = \sum_{k=1}^{n} \frac{1}{k} \sin kx$$

for $n = 10$ and $n = 100$. These are the first n terms of a Fourier series with coefficients $a_0 = a_k = 0$, $b_k = 1/k$. The limit is a step function.

12.1.3 An oscillating particle

Let us return to the physical example of oscillating linear motion. Imagine a particle moving in one dimension, under the influence of an attractive force at the origin. Let us assume that the force is an elastic force, behaving like an ideal spring. Its magnitude is proportional to the amount of stretching, the constant of proportionality being the elastic constant K. If the position of the particle is y, the force exerted on it is

$$F = -Ky.$$

When the particle is above the origin ($y > 0$), the force is directed downward, and when the particle is below the origin, the force is directed upward. The magnitude of the force increases linearly as the particle moves away from the origin.

By Newton's second law of motion, force is equal to mass times acceleration, and, by definition, acceleration is the second derivative of position with respect to time:

$$m\frac{d^2y}{dt^2} = -Ky.$$

That is

$$\frac{d^2}{dt^2}y = -\frac{K}{m}y. \tag{12.6}$$

This is a differential equation. To solve it, we require a function that is proportional to its own second derivative. That property should bring the exponential function immediately to mind:

$$\frac{d}{dt}Ce^{\lambda t} = \lambda Ce^{\lambda t}$$

$$\frac{d^2}{dt^2}Ce^{\lambda t} = \lambda^2 Ce^{\lambda t}$$

for any constants C and λ. This is exactly the form we require for (12.6), with

$$y = Ce^{\lambda t} \qquad \lambda^2 = -K/m.$$

Since $y = C$ when $t = 0$, let us rename C as y_0. Since K and m are both positive, $\lambda = \sqrt{-K/m}$ is an imaginary number; let us replace λ with $i\omega$. Here ω is an arbitrary value, but the choice of variable is not accidental; the connection to the ω of (12.1) will emerge shortly. In sum, the solution to (12.6) is

$$y = y_0 e^{i\omega t} \qquad \text{where } \omega = \sqrt{K/m}. \tag{12.7}$$

The reader should confirm that (12.7) implies (12.6).

The solution (12.7) is not fully general. The negative square root gives a second solution: $y = y_0 e^{-i\omega t}$. In fact, any linear combination of these two solutions is also a solution:

$$y = Ce^{i\omega t} + De^{-i\omega t} \qquad \text{where } \omega = \sqrt{K/m}. \tag{12.8}$$

The function (12.8) satisfies (12.6) for any choice of C and D. When $t = 0$, we have $y_0 = C + D$.

Now other functions that are proportional to their own second derivative lie near to hand: recall that the derivative of sine is cosine, and the derivative of cosine is (the negative of) sine again. In fact, we can use harmonics more generally. The form (12.3) will prove most convenient:

$$\begin{aligned}
\frac{d}{dt}(a\cos\omega t + b\sin\omega t) &= -\omega(a\sin\omega t - b\cos\omega t) \\
\frac{d^2}{dt^2}(a\cos\omega t + b\sin\omega t) &= -\omega^2(a\cos\omega t + b\sin\omega t).
\end{aligned}$$

This has the form (12.6) with

$$y = a\cos\omega t + b\sin\omega t \qquad \omega = \sqrt{K/m}. \tag{12.9}$$

The function (12.9) is a solution for (12.6) for any choice of a and b.

It turns out to be no accident that sine, cosine, and exponential all have the property of being proportional to their own second derivative. In fact, the functions (12.8) and (12.9) are one and the same. This follows from Euler's formula:

$$e^{i\theta} = \cos\theta + i\sin\theta.$$

Here is the derivation. Immediate consequences of Euler's formula are:

$$\begin{aligned} e^{i\theta} + e^{-i\theta} &= 2\cos\theta \\ e^{i\theta} - e^{-i\theta} &= 2i\sin\theta. \end{aligned}$$

Making the substitution $\theta = \omega t$ and multiplying through by $a/2$ and $b/2i$, respectively, yields

$$\begin{aligned} \frac{a}{2}e^{i\omega t} + \frac{a}{2}e^{-i\omega t} &= a\cos\omega t \\ \frac{b}{2i}e^{i\omega t} - \frac{b}{2i}e^{-i\omega t} &= b\sin\omega t. \end{aligned}$$

Adding these two equations, we obtain

$$a\cos\omega t + b\sin\omega t = \left(\frac{a - ib}{2}\right)e^{i\omega t} + \left(\frac{a + ib}{2}\right)e^{-i\omega t} \tag{12.10}$$

which is to say, (12.9) is equal to (12.8) with

$$C = \frac{a - ib}{2} \qquad D = \frac{a + ib}{2}. \tag{12.11}$$

Conversely, given a solution in the form (12.8) with arbitrary complex constants C and D, Euler's formula gives us

$$\begin{aligned} Ce^{i\omega t} + De^{-i\omega t} &= C\cos\omega t + iC\sin\omega t + D\cos\omega t - iD\sin\omega t \\ &= (C + D)\cos\omega t + (iC - iD)\sin\omega t \end{aligned}$$

which gives us a solution in the form (12.9) with

$$a = C + D \qquad b = i(C - D).$$

12.1.4 A vibrating string

In the previous section, we considered a single particle moving in one dimension under the influence of a restoring force, which is to say, a force that pulls the mass back toward the origin with magnitude proportional to distance from the origin. We saw that such a physical system is described by a single harmonic.

Now let us consider a physical system that gives rise to a mixture of harmonics, namely, a vibrating string. Let us assume that the string has length

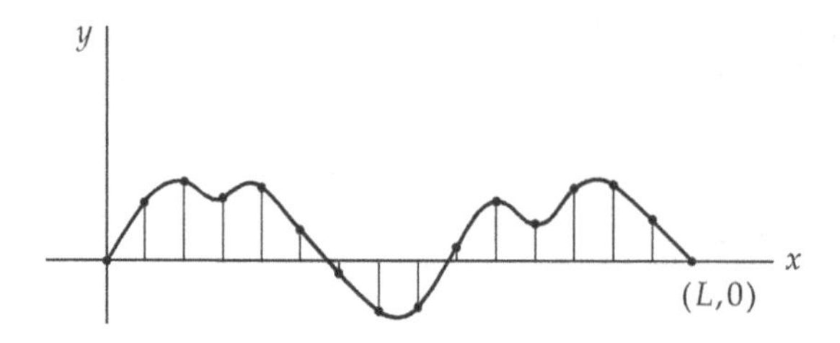

FIGURE 12.4
The shape of a vibrating string is given by $y = f(x)$. It is approximated by a sequence of particles, connected by elastic forces, each particle constrained to move only vertically.

L, and that both ends are fixed. The value y represents displacement. For convenience, let us take the string to be oriented in the direction of the x-axis, with endpoints fixed at the origin and $(L, 0)$. Then the shape of the string at a given point in time is simply the graph of a function $f(x)$, with the constraints that $f(0) = f(L) = 0$. (See figure 12.4.)

We can approximate the string by a sequence of particles connected by elastic forces. We assume that the particles move only vertically, not horizontally. Two forces are applied to any given particle, one from each neighbor. Since the forces are elastic, the force imposed on a particle by its neighbor is

$$F = -K\ell$$

where ℓ is the distance between the particles (figure 12.5). The force is the sum of a horizontal component $-K\Delta x$ and a vertical component $-K\Delta y$. We assume that the string is homogeneous, so that K is the same everywhere. If we also assume that the particles are equally spaced horizontally, then Δx and hence $K\Delta x$ is the same for both neighbors of a given particle, with the consequence that the horizontal forces exerted by the two neighbors of a given particle cancel out – if the initial velocities for all particles are purely vertical, then they will remain purely vertical. At any rate, we consider only the vertical force components.

The net vertical force F_j applied to the j-th particle is

$$\begin{aligned} F_j &= -K(y_j - y_{j-1}) - K(y_j - y_{j+1}) \\ &= -K(-y_{j-1} + 2y_j - y_{j+1}) \end{aligned} \tag{12.12}$$

where y_j is the vertical position of the j-th particle. If the neighbors are lower than the particle, the force acting on the particle is directed downward

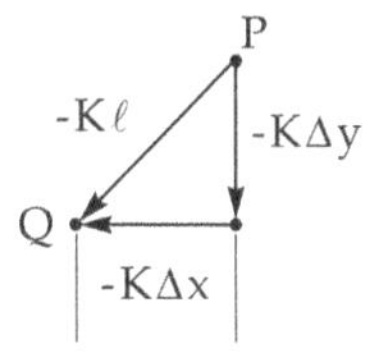

FIGURE 12.5
The force imposed on particle P by its neighbor Q.

(negative), and if the neighbors are higher, the force is positive. The first and last masses are exceptional. The two forces on the first particle are applied by the endpoint, fixed at height 0, and by the second particle:

$$\begin{aligned} F_1 &= -Ky_1 - K(y_1 - y_2) \\ &= -K(2y_1 - y_2). \end{aligned} \tag{12.13}$$

Similarly, for the last particle:

$$\begin{aligned} F_n &= -K(y_n - y_{n-1}) - Ky_n \\ &= -K(-y_{n-1} + 2y_n). \end{aligned} \tag{12.14}$$

The entire string of forces can be represented as a vector $\mathbf{F} = (F_1, \dots, F_n)$, and the vertical positions as a vector $\mathbf{y} = (y_1, \dots, y_n)$. The relationship between $\mathbf{F}$ and $\mathbf{y}$ is given by the system of equations (12.12), (12.13), and (12.14). It can be stated concisely as

$$\mathbf{F} = -KA\mathbf{y}$$

where A is a matrix that has twos on the diagonal and negative ones in the bands just above and below the diagonal. For example, for the case of five particles, the equation is

$$\begin{bmatrix} F_1 \\ F_2 \\ F_3 \\ F_4 \\ F_5 \end{bmatrix} = -K \begin{bmatrix} 2 & -1 & 0 & 0 & 0 \\ -1 & 2 & -1 & 0 & 0 \\ 0 & -1 & 2 & -1 & 0 \\ 0 & 0 & -1 & 2 & -1 \\ 0 & 0 & 0 & -1 & 2 \end{bmatrix} \begin{bmatrix} y_1 \\ y_2 \\ y_3 \\ y_4 \\ y_5 \end{bmatrix}. \tag{12.15}$$

The value for F_j is $-K$ times the dot product of the j-th row of A with the vector of y_j's. For example,

$$F_2 = -K(-1, 2, -1, 0, 0) \cdot (y_1, y_2, y_3, y_4, y_5)$$

in accordance with (12.12).

To determine how $f(x)$ changes with time, we again apply Newton's second law of motion:

$$m\frac{d^2}{dt^2}\mathbf{y} = -KA\mathbf{y}. \tag{12.16}$$

What is meant by $d^2\mathbf{y}/dt^2$ is simply the vector of second derivatives

$$\left(\frac{d^2}{dt^2}y_1, \ldots, \frac{d^2}{dt^2}y_n\right).$$

Dividing (12.16) through by m gives us a system of differential equations of the form (12.6). Unfortunately, we cannot simply apply the solution that we derived in the previous section for the one-particle case to each particle separately: the particles interact.

The differential equation we wish to solve is

$$\frac{d^2}{dt^2}\mathbf{y} = -\frac{K}{m}A\mathbf{y}. \tag{12.17}$$

Define $B = -(K/m)A$, and suppose that we can identify an eigenvector $\mathbf{u}$ of B with eigenvalue λ. Then

$$\mathbf{y} = \mathbf{u}e^{i\omega t} \qquad \text{where } \omega = \sqrt{-\lambda} \tag{12.18}$$

is a solution for (12.17). (Matrices of the form we will be concerned with often have negative eigenvalues, so, despite appearances, ω is not necessarily imaginary.) Let us confirm that (12.18) is a solution to (12.17):

$$\begin{aligned}
\frac{d}{dt}\mathbf{y} &= i\omega\mathbf{u}e^{i\omega t} \\
\frac{d^2}{dt^2}\mathbf{y} &= -\omega^2\mathbf{u}e^{i\omega t} \\
&= \lambda\mathbf{u}e^{i\omega t} \\
&= B\mathbf{u}e^{i\omega t}.
\end{aligned}$$

The last step follows because λ is the eigenvalue associated with eigenvector $\mathbf{u}$ of B, hence $\lambda\mathbf{u} = B\mathbf{u}$. Substituting $\mathbf{y}$ for $\mathbf{u}e^{i\omega t}$ yields (12.17). QED.

We will make no use of the elastic constant K or the particle mass m, so we will henceforth assume for convenience that

$$-\frac{K}{m} = 1.$$

The differential equation (12.17) becomes

$$\frac{d^2}{dt^2}\mathbf{y} = A\mathbf{y} \tag{12.19}$$

and in the solution (12.18), $\mathbf{u}$ is an eigenvector of A itself, with eigenvalue λ.

It is interesting to note that the solution (12.18) to the multi-particle case is just like the solution (12.7) to the single-particle case, except that the eigenvector $\mathbf{u}$ takes the place of the initial value y_0.

As in the single-particle case, (12.18) is only one solution. First, as in the single-particle case, the negative square root provides a second solution:

$$\mathbf{y} = \mathbf{u}e^{-i\omega t}.$$

Second, there is in general more than one eigenvalue-eigenvector pair for a matrix; in fact, there may be as many eigenvalues $\lambda_1, \ldots, \lambda_n$ and corresponding eigenvectors $\mathbf{u}_1, \ldots, \mathbf{u}_n$ as there are columns in A. Each pair provides a different special solution – or rather, each pair provides a pair of solutions, including both positive and negative square roots. The general solution is the linear combination of the special solutions:

$$\mathbf{y} = \sum_{k=1}^{n} \left(C_k \mathbf{u}_k e^{i\omega_k t} + D_k \mathbf{u}_k e^{-i\omega_k t}\right) \qquad \text{where } \omega_k = \sqrt{-\lambda_k}. \tag{12.20}$$

We can confirm that (12.20) is a solution in much the same way as we confirmed that (12.18) is a solution; it is left as an exercise for the reader.

The expression inside the summation in (12.20) has the same form as (12.8). So, recalling the discussion at the end of section 12.1.3, the family of solutions (12.20) with arbitrary coefficients C_k and D_k is equivalent to the following family of solutions with arbitrary coefficients a_k and b_k:

$$\mathbf{y} = \sum_{k=1}^{n} \mathbf{u}_k (a_k \cos\omega_k t + b_k \sin\omega_k t) \qquad \text{where } \omega_k = \sqrt{-\lambda_k}. \tag{12.21}$$

The expression inside the parentheses is a harmonic; let us call it c_k:

$$c_k(t) = a_k \cos(\omega_k t) + b_k \sin(\omega_k t) \qquad \text{for } \omega_k = \sqrt{-\lambda_k}.$$

Its angular velocity ω_k is determined by the matrix A, but its amplitude and phase are the length and angle, respectively, of the vector (a_k, b_k). Equation (12.21) becomes

$$\mathbf{y} = \sum_{k=1}^{n} c_k(t)\mathbf{u}_k. \tag{12.22}$$

The coefficients a_k and b_k of the harmonic $c_k(t)$ are determined by the initial positions and velocities of the particles. At time $t = 0$, we have

$$\mathbf{y} = \sum_{k=1}^{n} a_k \mathbf{u}_k.$$

That is, the coefficients a_k are chosen so as to match the shape of the string at time zero. The particle velocities are

$$\frac{d}{dt}\mathbf{y} = \sum_{k=1}^{n} \mathbf{u}_k(b_k \cos \omega_k t - a_k \sin \omega_k t)$$

which at time $t = 0$ becomes

$$\frac{d}{dt}\mathbf{y} = \sum_{k=1}^{n} b_k \mathbf{u}_k.$$

The coefficients b_k are chosen so as to match the initial velocities of the particles.

The vector equation (12.22) represents a system of equations:

$$y_j = \sum_{k=1}^{n} u_{kj} c_k(t). \tag{12.23}$$

It states that the motion of each particle individually is described by a mixture of harmonics. It does *not* say that the shape of the string at any point in time is described by a mixture of harmonics.

Nonetheless, it is true that the shape of the string is also a mixture of harmonics. Equation (12.22) says that, at any given time t, the shape of the string $\mathbf{y}$ is a weighted sum of the eigenvectors $\mathbf{u}_k$. As we will show now, the shape of the k-th eigenvector $\mathbf{u}_k$ is a harmonic. Specifically, $\mathbf{u}_k$ contains the values of a sine harmonic with period $2L/k$ and phase zero, sampled at points equally spaced in the interval $(0, L)$ (figure 12.6). Note that a period of $2L/k$ implies an angular velocity of $k\pi/L$. That is, we claim that the k-th eigenvector is defined as

$$\mathbf{u}_k = \begin{bmatrix} \sin((k\pi/L)x_1) \\ \vdots \\ \sin((k\pi/L)x_n) \end{bmatrix} \tag{12.24}$$

where

$$x_j = j\Delta x \qquad \Delta x = \frac{L}{n+1}$$

and the k-th eigenvalue is

$$\lambda_k = 2 - 2\cos((k\pi/L)\Delta x). \tag{12.25}$$

To verify our claim that the $\mathbf{u}_k$ and λ_k thus defined are eigenvectors and eigenvalues of A, we must show that

$$A\mathbf{u}_k = \lambda_k \mathbf{u}_k$$

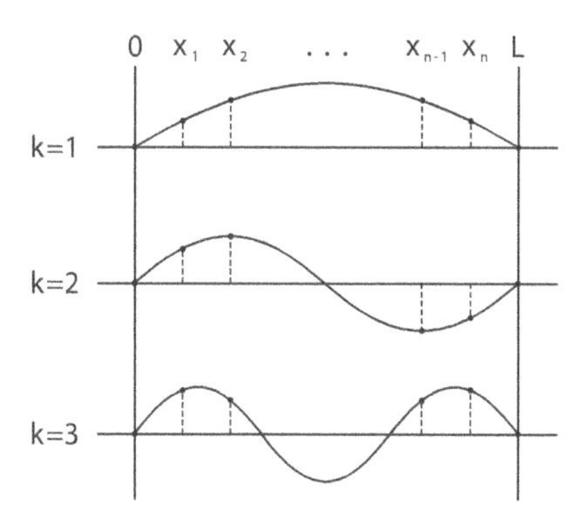

FIGURE 12.6
Eigenvector $\mathbf{u}_k$ contains the values $u_j = \sin((k\pi/L)x_j)$ with the x_j distributed uniformly in the interval $(0, L)$. The period corresponding to angular velocity $k\pi/L$ is $2L/k$.

for all k, which we do by showing that it is true for a fixed but arbitrary k. To reduce clutter, we drop the "k" subscripts and write simply $\mathbf{u}$ and λ for $\mathbf{u}_k$ and λ_k. It will also be convenient to define $\Delta\theta = k\pi/(n+1)$, so that

$$\mathbf{u} = \begin{bmatrix} \sin(1\Delta\theta) \\ \vdots \\ \sin(n\Delta\theta) \end{bmatrix} \qquad \lambda = 2 - 2\cos(\Delta\theta).$$

We must show that the vectors $A\mathbf{u}$ and $\lambda\mathbf{u}$ are elementwise equal, that is, that

$$(A\mathbf{u})_j = (\lambda\mathbf{u})_j$$

for all j.

Recalling the shape of the matrix A, given in (12.15), and writing u_j for the j-th element of $\mathbf{u}$, we have

$$\begin{aligned}(A\mathbf{u})_j &= -u_{j-1} + 2u_j - u_{j+1} \qquad \text{for } j \neq 1, n \\ (A\mathbf{u})_1 &= 2u_1 - u_2 \\ (A\mathbf{u})_n &= 2u_n - u_{n-1}.\end{aligned}$$

We consider the cases in the order given, beginning with the middle elements u_j where j is neither 1 nor n.

We can expand u_{j-1} as

$$u_{j-1} = \sin((j-1)\Delta\theta) = \sin(j\Delta\theta)\cos(\Delta\theta) - \cos(j\Delta\theta)\sin(\Delta\theta).$$

Similarly, u_{j+1} expands as

$$u_{j+1} = \sin(j\Delta\theta)\cos(\Delta\theta) + \cos(j\Delta\theta)\sin(\Delta\theta).$$

Their sum is

$$u_{j-1} + u_{j+1} = 2\sin(j\Delta\theta)\cos(\Delta\theta) = 2u_j\cos(\Delta\theta).$$

Hence

$$(A\mathbf{u})_j = 2u_j - 2u_j\cos(\Delta\theta) = (\lambda\mathbf{u})_j$$

as desired.

This leaves the first and last elements. A preliminary fact:

$$u_2 = \sin(2\Delta\theta) = 2\sin(\Delta\theta)\cos(\Delta\theta) = 2u_1\cos(\Delta\theta). \tag{12.26}$$

Then the derivation for the first element is straightforward:

$$\begin{aligned}(A\mathbf{u})_1 &= 2u_1 - u_2 \\ &= 2u_1 - 2u_1\cos(\Delta\theta) \\ &= (\lambda\mathbf{u})_1.\end{aligned}$$

As for the last element, observe that, for any k, the k-th harmonic crosses the x-axis at x_{n+1}, which equals L (figure 12.6). Hence, stepping backward from x_{n+1} is equivalent to stepping forward from x_0, except that there is a change of sign when k is even. That is, defining $\sigma = +1$ if k is odd and $\sigma = -1$ if k is even,

$$u_n = \sigma u_1$$

$$u_{n-1} = \sigma u_2.$$

Hence

$$\begin{aligned}(A\mathbf{u})_n &= 2u_n - u_{n-1} \\ &= 2u_n - \sigma u_2 \\ &= 2u_n - 2\sigma u_1\cos(\Delta\theta) && \text{by (12.26)} \\ &= 2u_n - 2u_n\cos(\Delta\theta) \\ &= (\lambda\mathbf{u})_n.\end{aligned}$$

In short, the shape of the string at a fixed time t is given by (12.22). It is a mixture of eigenvectors, where each eigenvector is a sampled harmonic (12.24). If we take the limit as the number of sample points n goes to infinity, the eigenvector $\mathbf{u}_k$ becomes an **eigenfunction** f_k. Instead of $u_{kj} = \sin((k\pi/L)x_j)$ at the discrete sample points x_j, we have

$$f_k(x) = \sin((k\pi/L)x)$$

continuously. The vector $\mathbf{y}$ of (12.22) also becomes a function, namely

$$f(x) = \sum_{k=1}^{\infty} c_k(t) f_k(x).$$

This is the Fourier expansion of the function f – it has the form (12.4). For this reason, our solution (12.22) is known as a **discrete Fourier expansion**.

Finally, the solution (12.22) can be rewritten in the form of a matrix times a vector:

$$\begin{aligned} \mathbf{y} &= \sum_k c_k \mathbf{u}_k \\ &= \begin{bmatrix} \mathbf{u}_1 \dots \mathbf{u}_n \end{bmatrix} \begin{bmatrix} c_1(t) \\ \vdots \\ c_n(t) \end{bmatrix} \\ &= S\mathbf{c}(t). \end{aligned}$$

The last line defines the matrix S and vector $\mathbf{c}$. S is a matrix whose columns are the eigenvectors of A, and $\mathbf{c}$ is a vector whose elements are the coefficients c_k determined by the initial positions and velocities of the particles.

The fact that the columns of S are the eigenvectors of A should call diagonalization to mind. Recall (section 11.2.2) that A can be expressed as

$$A = S\Lambda S^{-1}. \tag{12.27}$$

Substituting this expression into the original equation (12.17), we have

$$\frac{d^2}{dt^2}\mathbf{y} = A\mathbf{y} = S\Lambda S^{-1} S\mathbf{c}(t) = S\Lambda\mathbf{c}(t).$$

In short, the particle accelerations can be expressed as a product of a set of harmonics S, their weights Λ, and a vector of harmonics $\mathbf{c}(t)$ whose parameters are determined by initial conditions.

12.2 Spectra of matrices and graphs

The diagonalization (12.27) arose rather incidentally in the previous section, but it plays a central role in the definition of the spectra of matrices and graphs.

12.2.1 The spectrum of a matrix

Let S continue to represent the matrix whose columns $\mathbf{u}_k$ are the eigenvectors of A. Define the row vector $\mathbf{v}_k^{\mathrm{T}}$ to be the k-th row of S^{-1}. Then the diagonalization (12.27) can be written as

$$A = \begin{bmatrix} \mathbf{u}_1 \dots \mathbf{u}_n \end{bmatrix} \begin{bmatrix} \lambda_1 & & \\ & \ddots & \\ & & \lambda_n \end{bmatrix} \begin{bmatrix} \mathbf{v}_1^{\mathrm{T}} \\ \vdots \\ \mathbf{v}_n^{\mathrm{T}} \end{bmatrix}.$$

This can be expressed as a sum of outer products:

$$A = \lambda_1 \mathbf{u}_1 \mathbf{v}_1^{\mathrm{T}} + \dots + \lambda_n \mathbf{u}_n \mathbf{v}_n^{\mathrm{T}}.$$

The outer product $\mathbf{u}_k \mathbf{v}_k^{\mathrm{T}}$ is a matrix. Each column is $\mathbf{u}_k$ multiplied by v_k, the k-th element of $\mathbf{v}_k^{\mathrm{T}}$, hence each column of $\mathbf{u}_k \mathbf{v}_k^{\mathrm{T}}$ is a harmonic with the same frequency and phase as $\mathbf{u}_k$, though possibly with a different amplitude. Hence it is natural to think of $\mathbf{u}_k \mathbf{v}_k^{\mathrm{T}}$ as a matrix analogue of a harmonic, and A becomes a mixture of harmonics with weights $\lambda_1, \dots, \lambda_n$. For this reason, the **spectrum of the matrix** is defined to be the list of its eigenvalues, in order of increasing size

$$\lambda_1 \leq \dots \leq \lambda_n.$$

When A is symmetric, as it was in (12.15), its diagonalization can be expressed as

$$A = Q \Lambda Q^{\mathrm{T}}$$

with Q orthonormal, and the spectral decomposition is

$$A = \sum_k \lambda_k \mathbf{u}_k \mathbf{u}_k^{\mathrm{T}}.$$

In this form, the outer product $\mathbf{u}_k \mathbf{u}_k^{\mathrm{T}}$ is symmetric. Its k-th column is identical to its k-th row, and both have the form $u_{kk} \mathbf{u}_k$. It is a discrete sampling of a two-dimensional surface. Each row and column represents an axis-parallel cross section through the surface, and is the harmonic $\mathbf{u}_k$, scaled by a constant u_{kk}. The motivation for considering $\mathbf{u}_k \mathbf{u}_k^{\mathrm{T}}$ to be the matrix analogue of a harmonic is compelling.

In the particular case of the matrix A of the previous section, the k-th eigenvalue λ_k is equal to $2 - 2\cos(k\pi/(n+1))$; see equation (12.25). Since $\cos(k\pi/(n+1))$ decreases monotonically from $+1$ at $k = 0$ to -1 at $k = n+1$, we see that the eigenvalues $\lambda_1, \dots, \lambda_n$ are listed from smallest to largest. The angular velocity of the k-th eigenvector $\mathbf{u}_k$ is $k\pi/L$, so the eigenvectors $\mathbf{u}_1, \dots, \mathbf{u}_n$ are also sorted in order of increasing angular velocity, or decreasing period, as shown in figure 12.6.

12.2.2 Relating matrices and graphs

We have already seen the connection between the matrix A and a graph, though we did not draw attention to it at the time. In our discussion of the vibrating string, we modeled the string as a sequence of particles connected by elastic forces. The particles constitute nodes in a graph, and the elastic forces that connect pairs of adjacent nodes represent edges. The graph is linear, in this case, but it is easy to see how the generalization goes. For example, one might model a vibrating fabric with particles on a grid, each connected to four neighbors. Less regular, and even nonplanar, graphs are certainly possible.

In general, the displacement vector $\mathbf{y}$ is an assignment of values to nodes. The force applied to node i by a neighbor y_j is $F_{ij} = -K(y_i - y_j)$, and the net force applied to node i is the sum $\sum_{j\in\mathcal{N}_i} F_{ij}$, where $\mathcal{N}_i$ is the set of neighbors of node i. It can be written as

$$F_i = -K\left(|\mathcal{N}_i|y_i - \sum_{j\in\mathcal{N}_i} y_j\right). \tag{12.28}$$

We take this equation as defining an entry in the column vector $\mathbf{F}$:

$$\mathbf{F} = -K\begin{bmatrix} |\mathcal{N}_1| & & \\ & \ddots & \\ & & |\mathcal{N}_n| \end{bmatrix}\begin{bmatrix} y_1 \\ \vdots \\ y_n \end{bmatrix} - \begin{bmatrix} \mathbf{e}_1^{\mathrm{T}} \\ \vdots \\ \mathbf{e}_n^{\mathrm{T}} \end{bmatrix}\begin{bmatrix} y_1 \\ \vdots \\ y_n \end{bmatrix} = -K(D-E)\mathbf{y} \tag{12.29}$$

where $\mathbf{e}_i^{\mathrm{T}}$, the i-th row of the matrix E, contains a one in the j-th column just in case $j \in \mathcal{N}_i$. The matrix E is called the **adjacency matrix** of the graph, and the matrix D is called its **degree matrix**, inasmuch as its i-th entry, $|\mathcal{N}_i|$, is the degree of the i-th node.

For example, consider the linear graph in figure 12.7. We have

$$\mathbf{F} = -K\left(\begin{bmatrix} 1\,0\,0\,0 \\ 0\,2\,0\,0 \\ 0\,0\,2\,0 \\ 0\,0\,0\,1 \end{bmatrix} - \begin{bmatrix} 0\,1\,0\,0 \\ 1\,0\,1\,0 \\ 0\,1\,0\,1 \\ 0\,0\,1\,0 \end{bmatrix}\right)\begin{bmatrix} y_1 \\ y_2 \\ y_3 \\ y_4 \end{bmatrix} \tag{12.30}$$

which produces the system of equations

$$\begin{aligned} F_1 &= -K(y_1 - y_2) \\ F_2 &= -K(-y_1 + 2y_2 - y_3) \\ F_3 &= -K(-y_2 + 2y_3 - y_4) \\ F_4 &= -K(-y_3 + y_4) \end{aligned}$$

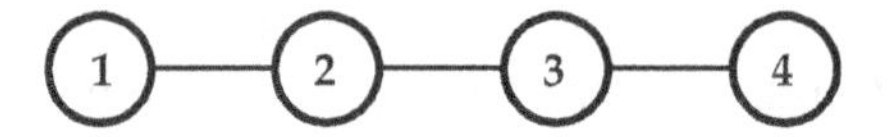

FIGURE 12.7
A small linear graph.

in accordance with (12.28).

Rewriting **F** as mass times acceleration, as before, gives

$$\frac{d^2}{dt^2}\mathbf{y} = -\frac{K}{m}B\mathbf{y} \tag{12.31}$$

where

$$B = D - E = \begin{bmatrix} 1 & -1 & 0 & 0 \\ -1 & 2 & -1 & 0 \\ 0 & -1 & 2 & -1 \\ 0 & 0 & -1 & 1 \end{bmatrix}.$$

Note that B is the parenthesized expression in (12.30). And as before, the solution to the differential equation (12.31) is any linear combination of the special solutions:

$$\mathbf{y} = \mathbf{u}e^{\pm i\omega t} \qquad \text{for } \omega = \sqrt{-\lambda}$$

where $\mathbf{u}$ is an eigenvector of $(-K/m)B$ and λ is the corresponding eigenvalue. As before, nothing rides on the identity of K and m, so we take $-K/m = 1$, in which case $\mathbf{u}$ is an eigenvector of B itself.

The matrix B is very similar to the matrix A of (12.15), differing only in the first and last rows. B arises more naturally than A when one takes a graph as point of departure, as we have just seen, but its physical interpretation is somewhat more exotic than that of A. The physical interpretation of B, as for A, is a string of particles, connected by elastic forces, each constrained to move only vertically, not horizontally. But in the case of B, unlike A, the ends of the string are not fixed. They are free to move vertically just like any other particle. One must imagine a string floating weightlessly and

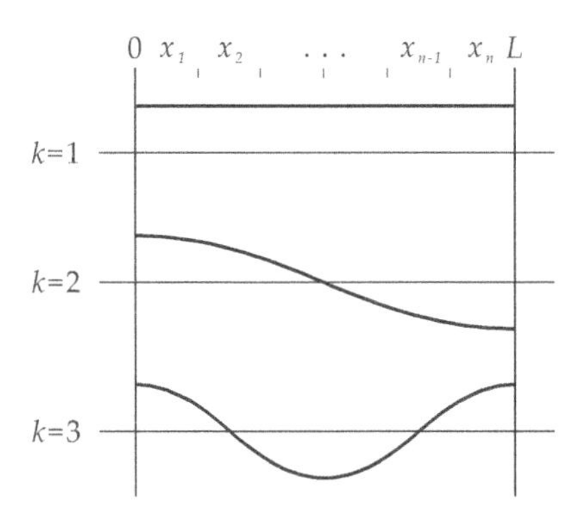

FIGURE 12.8
The first three eigenvectors of the matrix B.

undulating, with its ends moving freely in the vertical direction, but with its ends (and consequently all interior particles) held in a fixed horizontal position by a frictionless constraint.

As in the case of A, the eigenvectors of B are harmonics. But because of the difference in boundary conditions (the first and last rows), the harmonics are different. For B, the k-th eigenvector is

$$\mathbf{u} = \begin{bmatrix} \cos((1-\frac{1}{2})\Delta\theta) \\ \vdots \\ \cos((n-\frac{1}{2})\Delta\theta) \end{bmatrix} \qquad \text{where } \Delta\theta = \tfrac{(k-1)\pi}{n}$$

and the k-th eigenvalue is

$$\lambda = 2 - 2\cos(\Delta\theta).$$

The first three eigenvectors are shown in figure 12.8. In this case, the length of the string L is divided into n equal segments, and the sample positions are at the midpoints of the segments. We have $x = L$ when $j = n + \frac{1}{2}$, which corresponds to an angle of

$$\theta = ((n+\tfrac{1}{2}) - \tfrac{1}{2})\Delta\theta = (k-1)\pi.$$

Since the period of the cosine function is 2π, the k-th eigenvector samples from $k-1$ half-periods of the cosine function. In each half-period, the function either falls from $+1$ to -1, or rises from -1 to $+1$. The case $k = 1$ deserves a comment: in that case, the angle is always zero, and the eigenvector samples from a constant function with value one.

The demonstration that $\mathbf{u}$ and λ thus defined are indeed eigenvectors and eigenvalues of B is similar to the demonstration for A. Briefly, the three cases are:

$$\begin{aligned} Bu_j &= -u_{j-1} + 2u_j - u_{j+1} && \text{for } j \neq 1, n \\ Bu_1 &= u_1 - u_2 \\ Bu_n &= u_n - u_{n-1}. \end{aligned}$$

Showing that $Bu_j = \lambda u_j$ is straightforward for $j \neq 1, n$. First note that

$$\begin{aligned} u_{j-1} &= \cos((j - \tfrac{1}{2})\Delta\theta - \Delta\theta) \\ &= \cos((j - \tfrac{1}{2})\Delta\theta)\cos(\Delta\theta) - \sin((j - \tfrac{1}{2})\Delta\theta)\sin(\Delta\theta) \\ &= u_j \cos(\Delta\theta) - \sin((j - \tfrac{1}{2})\Delta\theta)\sin(\Delta\theta) \end{aligned}$$

and similarly

$$u_{j+1} = u_j \cos(\Delta\theta) + \sin((j - \tfrac{1}{2})\Delta\theta)\sin(\Delta\theta).$$

Hence

$$Bu_j = 2u_j - u_{j-1} - u_{j+1} = 2u_j - 2u_j\cos(\Delta\theta) = (2 - 2\cos(\Delta\theta))u_j$$

which is λu_j, as desired.

For $j = 1$, using the identities $\cos 2x = \cos^2 x - \sin^2 x$ and $\cos^2 x = 1 - \sin^2 x$ yields

$$\lambda = 4\sin^2(\tfrac{1}{2}\Delta\theta)$$

and

$$u_2 = u_1[1 - 4\sin^2(\tfrac{1}{2}\Delta\theta)].$$

The equation $Bu_1 = \lambda u_1$ follows.

As for Bu_n, we observe that

$$\cos((j - \tfrac{1}{2})\Delta\theta) = \cos((j - \tfrac{1}{2})(k - 1)\pi/n)$$

is a multiple of π when $j = \frac{1}{2}$ and again when $j = n + \frac{1}{2}$. It follows that u_n is the same as u_1 except possibly for a change of sign, and similarly u_{n-1} and u_2. Hence, the equality $Bu_n = \lambda u_n$ follows as a consequence of $Bu_1 = \lambda u_1$.

12.2.3 The Laplacian matrix and graph spectrum

The matrix B is known as the **Laplacian matrix** of the graph in figure 12.7. The Laplacian matrix is also sometimes called the *graph Laplacian*. The Laplacian matrix is defined as the difference

$$L = D - E$$

where D is the diagonal degree matrix and E is the adjacency matrix. The Laplacian matrix has positive numbers on its main diagonal, since D is a diagonal matrix containing positive real values and E has zeros on its main diagonal. (We assume that the graph has no edges that start and end at the same node.) It has negative ones off-diagonal, since E has positive ones just in those cells (i, j) where there is an edge from node i to node j, and D has zeros off-diagonal. Notice that the sum of entries in any row or column is zero: the i-th entry in D is the total number of edges that the i-th node participates in, and E contains a one in the i-th row (or column, as it is symmetric) for each of those edges.

The **spectrum of the graph** is defined to be the eigenvalues

$$\lambda_1 \leq \ldots \leq \lambda_n$$

of the Laplacian matrix. As before, the eigenvalues are ordered from smallest to largest, as are the eigenvector frequencies. These definitions hold for any graph, not just linear graphs.

The Laplacian can be generalized to a weighted graph:

$$L = D - W.$$

It is assumed that W is symmetric, that its entries are nonnegative, and that $w_{ii} = 0$ for all i. With a weighted graph, the degree of the i-th node is defined as

$$d_i = \sum_j w_{ij}.$$

We encountered this generalization of degree previously, in (7.9).

12.3 Spectral clustering

12.3.1 The second smallest eigenvector of the Laplacian

The basic idea of spectral clustering is now easily stated. We consider the eigenvectors of the Laplacian. In the case of a linear graph, the eigenvector $\mathbf{u}$ corresponding to the second smallest eigenvalue goes through a single half-period of the cosine function (figure 12.8). It divides the graph into two equal-sized, connected subgraphs. The sign of u_j can be interpreted as a cluster label: it is positive for nodes in the first subgraph, and negative for the remaining nodes.

An example with a non-linear graph is given in figure 12.9. The Laplacian

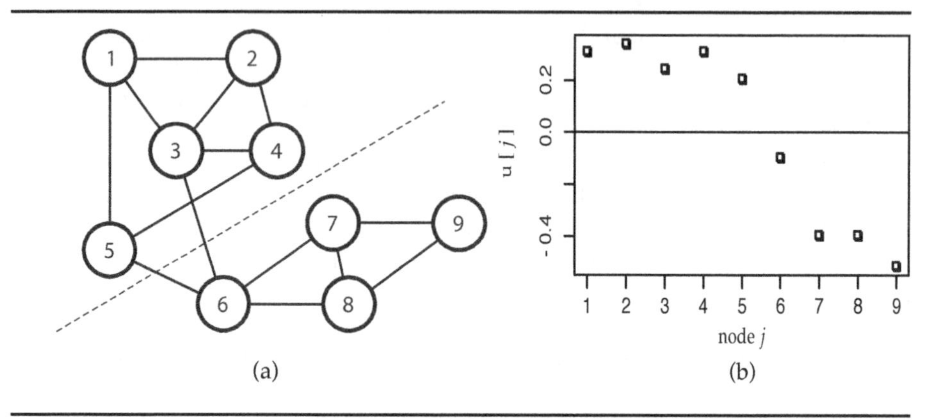

FIGURE 12.9
(a) A graph, and (b) its second smallest eigenvector.

of the graph in figure 12.9a is

	1	*2*	*3*	*4*	*5*	*6*	*7*	*8*	*9*
1	3	−1	−1	0	−1	0	0	0	0
2	−1	3	−1	−1	0	0	0	0	0
3	−1	−1	4	−1	0	−1	0	0	0
4	0	−1	−1	3	−1	0	0	0	0
5	−1	0	0	−1	3	−1	0	0	0
6	0	0	−1	0	−1	4	−1	−1	0
7	0	0	0	0	0	−1	3	−1	−1
8	0	0	0	0	0	−1	−1	3	−1
9	0	0	0	0	0	0	−1	−1	2

.

(The italic numbers are of course not part of the matrix, but are node numbers included for ease of reference.) The eigenvector **u** with the second smallest eigenvalue is shown in figure 12.9b. The function fails to be monotonic because the numbering of the nodes only approximately matches the intrinsic structure of the graph. Sorting the nodes by decreasing value u_j gives an optimal "linearization" of the graph, as shown in figure 12.10. In this case, one extremity is node 2, and the other is node 9. The dotted line in figure 12.10 is the "contour" where **u** equals zero, and represents the cluster boundary; the same boundary is also marked in figure 12.9a.

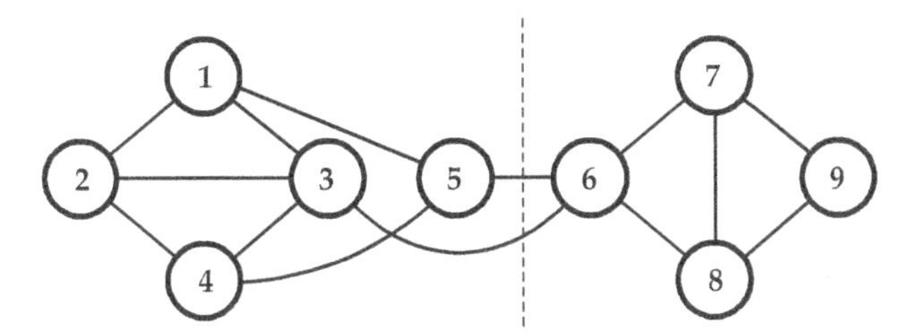

FIGURE 12.10
The nodes of the previous graph, sorted by decreasing value u_j.

12.3.2 The cut size and the Laplacian

It turns out that there is a strong connection between the division of the graph obtained using the second eigenvector, as just discussed, and the graph cuts that we discussed in sections 7.5 and 10.6. In section 7.5, we defined $\partial S = E(S, \bar{S})$ as the set of edges that have one endpoint in S and one endpoint outside of S, and we defined

$$\text{cut}(S, \bar{S}) = \text{size}(\partial S)$$

where the size of a set of edges is the sum of their weights (which is equal to cardinality in the case of an unweighted graph).

A cut dividing the graph into sets S and $\bar{S}$ can be represented by a vector $\mathbf{y}$ in which element y_i is equal to $+1$ if the i-th node belongs to S, and $y_i = -1$ otherwise. Thinking of S and $\bar{S}$ as classes, y_i is the class label of the i-th node. Let us call such a vector $\mathbf{y}$ a *cut vector* for S.

If nodes i and j both belong to the same class (both are in S or both are in $\bar{S}$), then $y_i - y_j = 0$ and hence $(y_i - y_j)^2 = 0$. On the other hand, if nodes i and j belong to different classes, then $y_i - y_j$ is either $+2$ or -2, hence $(y_i - y_j)^2 = 4$. Thus

$$\sum_{ij} \frac{(y_i - y_j)^2}{4} w_{ij} = \sum_{ij} [\![y_i \neq y_j]\!] w_{ij} = 2 \cdot \text{cut}(S, \bar{S}).$$

The factor of two results because each edge gets counted twice, once when we have $i \in S, j \in \bar{S}$, and once when we have $i \in \bar{S}, j \in S$. Define

$$g(\mathbf{y}) = \sum_{ij} w_{ij} \frac{(y_i - y_j)^2}{8} \tag{12.32}$$

$$= \frac{1}{8}\left[\sum_i \left(\sum_j w_{ij}\right) y_i^2 + \sum_j \left(\sum_i w_{ij}\right) y_j^2 - 2\sum_{ij} w_{ij} y_i y_j\right].$$

Assuming symmetric weights, the first two terms are equal, and the sum $\sum_j w_{ij} = \sum_i w_{ij}$ is equal to the degree. Hence

$$g(\mathbf{y}) = \frac{1}{4}\left[\sum_i d_i y_i^2 - \sum_{ij} w_{ij} y_i y_j\right].$$

Writing D for the diagonal degree matrix and W for the weight matrix, this becomes

$$g(\mathbf{y}) = \frac{1}{4}\left[\mathbf{y}^{\mathrm{T}} D\mathbf{y} - \mathbf{y}^{\mathrm{T}} W\mathbf{y}\right] = \frac{1}{4}\mathbf{y}^{\mathrm{T}} L\mathbf{y} \tag{12.33}$$

where L is the Laplacian matrix. When $\mathbf{y}$ is a cut vector for S, then $\mathrm{cut}(S, \bar{S}) = g(\mathbf{y})$. This provides the basic connection between cuts and the Laplacian.

The choice of $+1$ and -1 as class labels in the cut vector is arbitrary. We can actually choose any two distinct real numbers P and N. We can generalize the definition of cut vector in the obvious way: $\mathbf{y}$ is a **cut vector** for S if $y_i = P$ whenever the i-th node belongs to S and $y_i = N$ otherwise. Let α be the halfway point between P and N, and let β be half the difference $P - N$, so that $P = \alpha + \beta$ and $N = \alpha - \beta$. If nodes i and j have the same label, then $y_i - y_j$ is zero, and if i and j have different labels, then $y_i - y_j = \pm[(\alpha + \beta) - (\alpha - \beta)] = \pm 2\beta$, and $(y_i - y_j)^2 = 4\beta^2$. The function $g(\mathbf{x})$ becomes

$$g(\mathbf{y}) = \sum_{ij} w_{ij} \frac{(y_i - y_j)^2}{4\beta^2} = \frac{1}{4\beta^2}\mathbf{y}^{\mathrm{T}} L\mathbf{y}. \tag{12.34}$$

If $\mathbf{y}$ is a cut vector for S then $\mathrm{cut}(S, \bar{S}) = g(\mathbf{y})$ with $\beta = (P - N)/2$. Notice that g is insensitive to α. For the moment, we assume $\alpha = 0$.

12.3.3 Approximating cut size

We arrived at the function g by consideration of vectors $\mathbf{y}$ whose elements are all either $+1$ or -1. Geometrically, such vectors are the corners of a hypercube $\mathbf{C}$ in n-dimensional space, for n the number of nodes in the graph.

It can be shown that minimizing $g(\mathbf{y})$ under the constraint that $\mathbf{y}$ be one of the corners of $\mathbf{C}$ is NP-hard. An exact solution is intractable. But if we relax the constraint that $\mathbf{y}$ lie at one of the corners of the hypercube, and allow $\mathbf{y}$ to vary continuously, an approximate solution can be found.

An obvious idea is to find the vector $\mathbf{x}$ that minimizes $g(\mathbf{x})$, and then find the corner $\mathbf{y}$ that is nearest to $\mathbf{x}$. But there is a problem. The function $g(\mathbf{x})$ varies quadratically with the length of $\mathbf{x}$, and it is trivially minimized by the vector $\mathbf{0}$, which is equidistant from all corners of $\mathbf{C}$.

The fact that all corners are equidistant from the origin suggests a better approach. The distance from the origin to any corner is $\sqrt{n}$. That is, the corners lie on the surface of a sphere, centered at the origin, with radius $\sqrt{n}$. Let us fix the length of $\mathbf{x}$ to be $\sqrt{n}$, so that it ranges over the sphere, and search for the vector *direction* that minimizes $g(\mathbf{x})$ on the sphere. The exact solutions (that is, the corners) lie in diagonal directions. We preserve the constraint that $\|\mathbf{y}\| = \sqrt{n}$, but relax the restriction to diagonal directions, allowing ourselves to consider any direction.

Minimizing g will be easier if we shrink the hypercube a bit. We can do that because the size of the hypercube is determined by the cluster labels, and the cluster labels are arbitrary. When we choose $+1$ and -1, which is to say, $\alpha = 0$ and $\beta = 1$, then the center of $\mathbf{C}$ is at the origin, and its sides are all of length 2. Notice that any side of the hypercube connects two endpoints that differ in only one coordinate; one has $\alpha + \beta$ and the other has $\alpha - \beta$. Hence the sides are all of length 2β. Since α is added to each coordinate, varying α moves the center of the hypercube along the main diagonal $\mathbf{1} = (1, \ldots, 1)$. Each corner of the hypercube can be expressed as $\beta\mathbf{x} + \alpha\mathbf{1}$, where the coordinates of $\mathbf{x}$ are all either $+1$ or -1.

For convenience, then, let us choose β to make the sphere a unit sphere, namely, $\beta = 1/\sqrt{n}$. We continue to assume $\alpha = 0$, so that the center of the hypercube is at the origin. Now the corners lie on the unit sphere. Each corner $\mathbf{y}$ has $\|\mathbf{y}\| = 1$, since

$$\|\mathbf{y}\|^2 = \mathbf{y}^{\mathrm{T}}\mathbf{y} = \sum_{i=1}^{n} (\pm 1/\sqrt{n})^2 = \sum_{i=1}^{n} 1/n = 1.$$

To minimize g under the constraint that $\|\mathbf{y}\| = 1$, we construct a function f whose value agrees with g on the unit sphere, but is insensitive to the length of $\mathbf{y}$. Then we do an unconstrained minimization of f, obtaining a vector $\mathbf{y}_0$, and normalize $\mathbf{y}_0$ to obtain a vector $\mathbf{y}$ on the unit sphere in the same direction as $\mathbf{y}_0$. Since f is insensitive to vector length, we know $f(\mathbf{y}) = f(\mathbf{y}_0)$, hence $\mathbf{y}$ is also a minimum of f. And since $f = g$ on the unit sphere, we know that $\mathbf{y}$ minimizes g on the sphere.

Actually, it is sufficient if $cf(\mathbf{y}) = g(\mathbf{y})$ for some nonzero constant c. We define $f(\mathbf{y})$ to be equal to $\mathbf{y}^{\mathrm{T}} L\mathbf{y}$, which is $g(\mathbf{y})$ omitting the leading constant, but we make it insensitive to the length of $\mathbf{y}$ by normalizing $\mathbf{y}$ beforehand:

$$f(\mathbf{y}) = \left(\frac{\mathbf{y}}{\|\mathbf{y}\|}\right)^{\mathrm{T}} L \left(\frac{\mathbf{y}}{\|\mathbf{y}\|}\right) = \frac{\mathbf{y}^{\mathrm{T}} L\mathbf{y}}{\|\mathbf{y}\|^2} = \frac{\mathbf{y}^{\mathrm{T}} L\mathbf{y}}{\mathbf{y}^{\mathrm{T}}\mathbf{y}}. \tag{12.35}$$

Doing an unconstrained minimization of f and normalizing the resulting vector gives us a vector $\mathbf{x}$ that minimizes g on the unit sphere. At the corners of the hypercube, g is equal to the cut size. The nearest corner to the minimizer is our approximation to the best cut.

12.3.4 Minimizing cut size

We previously introduced the idea of using the second eigenvector to cut the graph – recall figure 12.9. That made intuitive sense for the example given. Now we will show that it has good motivation in general.

The function f in (12.35) is immediately recognizable as a Rayleigh quotient. Recall equation (11.22), restated here for convenience:

$$\min_{\mathbf{x}\in\mathcal{Q}_{k-1}^{\perp}} \frac{\mathbf{x}^{\mathrm{T}}A\mathbf{x}}{\mathbf{x}^{\mathrm{T}}\mathbf{x}} = \lambda_k \qquad \arg\min_{\mathbf{x}\in\mathcal{Q}_{k-1}^{\perp}} \frac{\mathbf{x}^{\mathrm{T}}A\mathbf{x}}{\mathbf{x}^{\mathrm{T}}\mathbf{x}} = \mathbf{u}_k.$$

In words, if we restrict attention to the space $\mathcal{Q}_k^{\perp}$ spanned by the eigenvectors $\mathbf{u}_k, \ldots, \mathbf{u}_n$ corresponding to the $n-k+1$ largest eigenvalues $\lambda_k, \ldots, \lambda_n$ of a symmetric matrix A, then the vector that minimizes the Rayleigh quotient is $\mathbf{u}_k$, and the resulting minimum value is λ_k.

This equation is valid for any symmetric matrix A. Since the Laplacian matrix L is symmetric, the equation is valid in particular if we set $A = L$. Hence

$$\min_{\mathbf{y}\,\in\,\mathcal{Q}_{k-1}^{\perp}} f(\mathbf{y}) = \lambda_k(L)$$

where $\lambda_k(L)$ is the k-th smallest eigenvalue of L. This is true for any value of k from 1 to n. Recall that $\mathcal{Q}_{k-1}$ is the space spanned by the $n-k+1$ largest eigenvalues of L. $\mathcal{Q}_0^{\perp}$ is the entire space spanned by the eigenvalues of L, hence

$$\min_{\mathbf{y}} f(\mathbf{y}) = \lambda_1(L).$$

However, as we have already seen, the eigenvector $\mathbf{u}_1$ corresponding to the smallest eigenvalue is the constant vector. It corresponds to the trivial case in which all nodes are assigned to the same cluster. The resulting cut value is obviously 0, but hardly of interest.

If we exclude the trivial solution from consideration, the next least-restrictive solution is to consider the space spanned by all eigenvectors except $\mathbf{u}_1$:

$$\min_{\mathbf{y}\in\mathcal{Q}_1^{\perp}} f(\mathbf{y}) = \lambda_2(L) \qquad \arg\min_{\mathbf{y}\in\mathcal{Q}_1^{\perp}} f(\mathbf{y}) = \mathbf{u}_2(L). \tag{12.36}$$

The nontrivial cut with the smallest value is the one corresponding to the second smallest eigenvalue of the graph Laplacian.

But we must remember that $\mathbf{u}_2$ is a *relaxed* solution. That is, it minimizes f, but it does not necessarily lie at one of the corners of the hypercube $\mathbf{C}$ representing the discrete clusterings of the nodes of the graph. The nearest exact solution is the corner nearest to $\mathbf{u}_2$, and that is the solution we will adopt. It is reasonable to do so, but there is no guarantee that the nearest corner actually minimizes cut size among the corners.

Which corner is nearest? Recall that cut size (12.35) is sensitive only to vector direction, not vector length. Even though the minimization is unconstrained, we are effectively choosing a best direction on the unit sphere. Hence the natural measure of distance between vectors is cosine, which on the unit sphere is identical to dot product. So we would like to find the nearest corner to $\mathbf{u}_2$, where nearness is understood in the sense of dot product. A corner vector is one whose elements are all $\pm 1/\sqrt{n}$, and since every combination is valid (except the two in which all elements are positive or all elements are negative), we can minimize distance elementwise: for each element u_i of the relaxed solution $\mathbf{u} = \mathbf{u}_2$, we choose positive or negative $1/\sqrt{n}$, whichever is closer. In short, the nearest corner to $\mathbf{u}$ is the vector $\mathbf{x}$ in which the i-th element x_i has the same sign as u_i, but magnitude $1/\sqrt{n}$. The resulting vector $\mathbf{x}$ is one of the corners of the hypercube, hence an exact solution.

We described (12.36) by saying that $\mathbf{u}_2$ represents the nontrivial cut with the smallest value. Actually, that is not quite true. The eigenvector $\mathbf{u}_2$ represents the nontrivial cut *perpendicular to* $\mathbf{u}_1$ that has the smallest value. But linear combinations of $\mathbf{u}_1$ and $\mathbf{u}_2$ also represent nontrivial cuts, and their values are *bounded above* by λ_2 (the value of $\mathbf{u}_2$).

A linear combination of $\mathbf{u}_1$ and $\mathbf{u}_2$ can be thought of as adding $\alpha\mathbf{u}_1$ to $\mathbf{u}_2$ and then scaling the result. The scaling has no effect on either the cut value or the nearest corner, so we can ignore it; what matters is the addition of $\alpha\mathbf{u}_1$. Recall that $\mathbf{u}_1$ is the constant vector $\mathbf{1}$, so adding $\alpha\mathbf{u}_1$ is the same as adding α to each element of $\mathbf{u}_2$. The effect is shown in figure 12.11. Adding $\alpha\mathbf{u}_1$ effectively moves the zero line, and changes the cluster label for any nodes whose sign changes as a result. If α is large enough, all nodes are labeled positive. The value α may also be negative, in which case the zero line is raised. Effectively, $\mathbf{u}_2$ defines an ordering on the nodes, and one may choose any point in that ordering to be the dividing line between positives and negatives. This yields a set of candidate exact solutions $\mathbf{x}$, and a simple approach is to compute $\text{cut}(\mathbf{x})$ for each of those candidates, and keep the one with the smallest value. There are at most $n - 1$ candidate solutions (excluding the trivial ones), so tractability is not an issue.

To summarize, the cut vectors can be expressed as $\beta\mathbf{x} + \alpha\mathbf{1}$, where the elements of $\mathbf{x}$ are all $+1$ or -1. We previously assumed that $\alpha = 0$. Now we see that adding $\alpha\mathbf{1}$ is the same as mixing some of $\mathbf{u}_1$ in with $\mathbf{u}_2$. In terms of the cluster labels, α is the halfway point between them. Changing the distance β from the midpoint to either of the cluster labels has no effect on f, hence has no effect on the minimizer of f. Moving the midpoint α away from zero actually decreases f, since it mixes some of $\mathbf{u}_1 = \mathbf{1}$ in with $\mathbf{u}_2$.

12.3.5 Ratiocut

Simple minimization of cut size has a disadvantage from a practical point of view: it does not take into account the relative size of the two clusters that result from the cut. Although we excluded, by fiat, the trivial cut that puts all nodes in the same cluster, we have not excluded nearly-trivial cuts, such as cuts that have a single node in one cluster and all remaining nodes in the other cluster. Such cuts arise frequently in practice. The smallest nontrivial

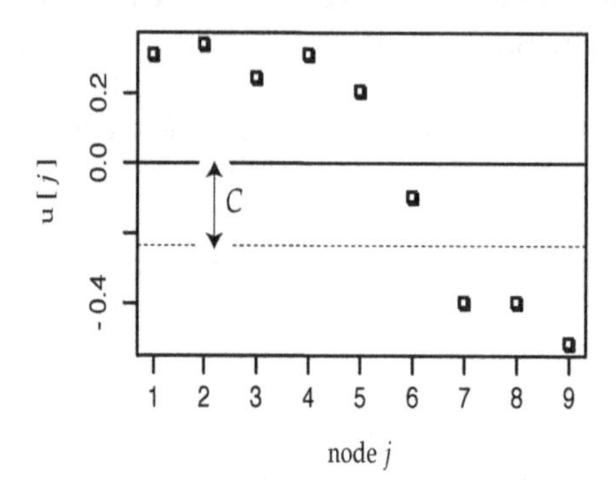

FIGURE 12.11
The effect of adding $\alpha\mathbf{u}_1$ to $\mathbf{u}_2$.

cut severs a single edge, and if such a cut is possible, it is very likely that a single isolated node lies at one endpoint of the edge in question.

An alternative that penalizes highly skewed partitions is the **ratiocut**. It is defined as

$$\text{rcut}(S, \bar{S}) = \frac{\text{size}(\partial S)}{|S| \cdot |\bar{S}|}. \tag{12.37}$$

The denominator of (12.37) is maximized, and hence rcut is minimized, when S and $\bar{S}$ are equally large.

The relaxed clustering representation $\mathbf{y}$ that is most suitable for ratiocut is what we might call a **ratiocut vector**. It has

$$y_i = \begin{cases} |\bar{S}|/n & \text{if } \nu_i \in S \\ -|S|/n & \text{if } \nu_i \in \bar{S} \end{cases}$$

where ν_i represents the i-th node. A ratiocut vector is not the same as a cut vector, since the values for the cluster labels in a ratiocut vector are not constants, but depend on the number of positive elements (the size of S) and the number of negative elements (the size of $\bar{S}$) in the vector. Nonetheless, we can relate $\text{rcut}(S, \bar{S})$ to $f(\mathbf{y})$, where $\mathbf{y}$ is a ratiocut vector.

First, if nodes i and j belong to the same cluster, $y_i - y_j$ is obviously zero. If they belong to different clusters, then either

$$y_i - y_j = |S|/n + |\bar{S}|/n = |S|/n + (1 - |S|/n) = 1$$

or

$$y_i - y_j = -|\bar{S}|/n - |S|/n = (|S|/n - 1) - |S|/n = -1.$$

In either case, $(y_i - y_j)^2 = 1$. It follows that

$$g(\mathbf{y}) = \sum_{ij} \frac{(y_i - y_j)^2}{8} w_{ij} = \frac{1}{8} \sum_{ij} [\![y_i \neq y_j]\!] w_{ij} = \frac{1}{8} \text{cut}(S, \bar{S}).$$

Second, the squared length of $\mathbf{y}$ is equal to $|S| \cdot |\bar{S}|/n$:

$$\mathbf{y}^{\mathrm{T}} \mathbf{y} = |S| \frac{|\bar{S}|^2}{n^2} + |\bar{S}| \frac{|S|^2}{n^2} = |S| \cdot |\bar{S}| \cdot \frac{|\bar{S}| + |S|}{n^2} = |S| \cdot |\bar{S}|/n.$$

Hence

$$\text{rcut}(S, \bar{S}) = \frac{\text{cut}(S, \bar{S})}{|S| \cdot |\bar{S}|} = 8n \frac{g(\mathbf{y})}{\mathbf{y}^{\mathrm{T}} \mathbf{y}} = 8n f(\mathbf{y}).$$

Since $8n$ is a constant, it does not affect minimization. The ratiocut vector that minimizes f represents the clustering $(S, \bar{S})$ that minimizes rcut.

Finally, $\mathbf{y}$ is perpendicular to $\mathbf{1}$, since

$$\mathbf{y} \cdot \mathbf{1} = \sum_i y_i = |S| \frac{|\bar{S}|}{n} - |\bar{S}| \frac{|S|}{n} = 0.$$

As before, the second eigenvector $\mathbf{u}_2$ minimizes f under the constraint that $\mathbf{y}$ be perpendicular to $\mathbf{1}$. Mixing in some of $\mathbf{u}_1 = \mathbf{1}$ reduces f further, and has the effect of adjusting the zero line α (figure 12.11) to create a series of candidate partitions. However, instead of choosing the candidate partition that minimizes cut, we choose the one that minimizes rcut.

12.4 Spectral methods for semisupervised learning

We have just seen how to use spectral methods for clustering. We turn now to the question of how they can be applied to semisupervised learning.

12.4.1 Harmonics and harmonic functions

Taking the second smallest eigenvector cuts the graph into two components along a natural "fault line," where the edges holding the two halves of the graph together are weakest. But in the semisupervised setting, we are additionally given the labels of some of the instances, and there is no guarantee that the division between positive and negative labels lines up neatly with the natural fault line of the graph.

A natural idea is to go back to the matrix (12.15) corresponding to the string with fixed endpoints. One might take the second smallest eigenvector of that matrix, which is the case $k = 2$ in figure 12.6. However, the endpoints are both fixed at the same value, namely zero. We wish to have two different values, one representing the positive class and one representing the negative class. Unfortunately, the value zero is special; we cannot fix one of the endpoints at, say, value one and capture that in a matrix like the one in (12.15). Setting one of the endpoints at value one requires us to put a constant in the place of one of the y_k's in the vector that is being premultiplied by the matrix.

It turns out that we already know what happens when we fix the values of some of the y_k's, though it will take a bit of explaining to see the connection. Let us begin by considering again the quantity $g(\mathbf{y})$ that we defined as a general measure of the cost of a label assignment $\mathbf{y}$. It is equal to $\text{cut}(S, \bar{S})$ whenever $\mathbf{y}$ is a cut vector for S, that is, whenever the components of $\mathbf{y}$ are $+1$ for nodes belonging to S and -1 for nodes belonging to $\bar{S}$. Generalizing to arbitrary values $\alpha+\beta$ for the positive class and $\alpha-\beta$ for the negative class, we saw in (12.34) that the function $g(\mathbf{y})$ could be written in two equivalent forms:

$$g(\mathbf{y}) = \frac{1}{(2\beta)^2}\sum_{ij}(y_i - y_j)^2 w_{ij}$$

$$g(\mathbf{y}) = \frac{1}{(2\beta)^2}\mathbf{y}^{\mathrm{T}} L \mathbf{y}.$$

Choosing $\mathbf{y}$ so as to minimize $g(\mathbf{y})$ is equivalent to minimizing cut size.

If we take a given value y_k to represent the *probability* of having the positive label, then y_k ranges from 0 to 1. Equivalently, we can treat 0 as the "negative" value and 1 as the "positive" value, that is, we take $\alpha = 1/2$ and $\beta = 1/2$. In that case, our two equations for $g(\mathbf{y})$ simplify to:

$$g(\mathbf{y}) = \sum_{ij}(y_i - y_j)^2 w_{ij} \tag{12.38}$$

$$g(\mathbf{y}) = \mathbf{y}^{\mathrm{T}} L \mathbf{y}. \tag{12.39}$$

Taking y_k to represent the probability that the k-th node is positively labeled brings harmonic functions to mind. In the context of harmonic functions, equation (12.38) is familiar: it is twice the "J" energy of $\mathbf{y}$, as defined in (10.12), repeated here with a slight modification in notation:

$$J(\mathbf{y}) = \frac{1}{2}\sum_{ij}(y_i - y_j)^2 w_{ij}. \tag{12.40}$$

Recall that J is defined for any function over nodes of the graph, and a vector $\mathbf{y}$ assigning values to nodes is such a function. In the semisupervised case, we wish to choose $\mathbf{y}$ so as to minimize $g(\mathbf{y})$, subject to the constraint that the values assigned to boundary nodes (that is, labeled nodes) agree with the labeled instances given as input. But the vector that minimizes $g(\mathbf{y})$ is identical to the vector that minimizes $J(\mathbf{y})$, and we know from Thomson's principle (Theorem 10.3) that the vector that minimizes $J(\mathbf{y})$ is the harmonic function that agrees with the given boundary values. In short, the basic spectral methods for semisupervised learning are the now-familiar methods for constructing a harmonic function: the method of relaxations, or the closed form matrix solution given in (10.14).

In short, the objective function (12.38) is equivalent to the energy $J(\mathbf{y})$, which can be written

$$J(\mathbf{y}) = \frac{1}{2}\mathbf{y}^{\mathrm{T}} L\mathbf{y}. \tag{12.41}$$

It is minimized in the unconstrained case by the harmonic $\mathbf{u}_1$ with the smallest eigenvalue, or the harmonic $\mathbf{u}_2$ with the second smallest eigenvalue if we exclude $\mathbf{u}_1$, and it is minimized in the constrained case (the semisupervised case) by the harmonic function determined by the boundary values.

12.4.2 Eigenvalues and energy

Not only is the energy $J(\mathbf{y})$ minimized by the eigenvector with the smallest eigenvalue, but in general, the energy of eigenvector $\mathbf{u}_i$ is exactly equal to one-half the associated eigenvalue λ_i. As we just saw, energy is equal to $(1/2)\mathbf{y}^{\mathrm{T}} L\mathbf{y}$. The Laplacian L is symmetric and hence has the diagonalization

$$L = Q\Lambda Q^{\mathrm{T}}$$

where the columns of Q are eigenvectors $\mathbf{u}_k$, and the entries in the diagonal matrix Λ are the corresponding eigenvalues λ_k. The diagonalization can also be expressed as an outer product:

$$L = \sum_i \lambda_i \mathbf{u}_i \mathbf{u}_i^{\mathrm{T}}.$$

Further, we can assume that the eigenvalues are sorted in order of increasing value.

Let us consider the energy $J(\mathbf{u}_i)$ for an eigenvector $\mathbf{u}_i$:

$$\begin{aligned}
\frac{1}{2}\mathbf{u}_i^{\mathrm{T}} L\mathbf{u}_i &= \frac{1}{2}\mathbf{u}_j^{\mathrm{T}} \left(\sum_j \lambda_j \mathbf{u}_j \mathbf{u}_j^{\mathrm{T}}\right) \mathbf{u}_i \\
&= \frac{1}{2}\left(\sum_j \lambda_j \mathbf{u}_i^{\mathrm{T}} \mathbf{u}_j \mathbf{u}_j^{\mathrm{T}}\right) \mathbf{u}_i \\
&= \frac{1}{2}\lambda_i \mathbf{u}_i^{\mathrm{T}} \mathbf{u}_i \\
&= \frac{1}{2}\lambda_i.
\end{aligned}$$

In both of the last two steps, we have used the facts that $\mathbf{u}_i^{\mathrm{T}}\mathbf{u}_i = 1$ and $\mathbf{u}_i^{\mathrm{T}}\mathbf{u}_j = 0$ for $j \neq i$.

We have just seen that the energy of an eigenvector is equal to half the corresponding eigenvalue. A general value assignment $\mathbf{y}$ can be expressed as a linear combination of eigenvectors:

$$\mathbf{y} = \sum_i b_i \mathbf{u}_i.$$

The energy of $\mathbf{y}$ can be computed from the energies of the eigenvectors. As a preliminary, we require the following fact. If $i \neq j$:

$$\begin{aligned} \mathbf{u}_i^{\mathrm{T}} L \mathbf{u}_j &= \mathbf{u}_i^{\mathrm{T}} \left(\sum_k \lambda_k \mathbf{u}_k \mathbf{u}_k^{\mathrm{T}} \right) \mathbf{u}_j \\ &= \lambda_i \mathbf{u}_i^{\mathrm{T}} \mathbf{u}_j \\ &= 0. \end{aligned}$$

Now we compute $J(\mathbf{y})$:

$$\begin{aligned} \frac{1}{2} \mathbf{y}^{\mathrm{T}} L \mathbf{y} &= \frac{1}{2} \left(\sum_i b_i \mathbf{u}_i \right)^{\mathrm{T}} L \left(\sum_j b_j \mathbf{u}_j \right) \\ &= \frac{1}{2} \sum_{ij} b_i b_j \mathbf{u}_i^{\mathrm{T}} L \mathbf{u}_j \\ &= \frac{1}{2} \sum_i b_i^2 \lambda_i. \end{aligned}$$

Since eigenvalues are sorted by increasing value, it becomes obvious that we minimize energy, for a fixed total $\sum_i b_i^2$, by making b_i large for small values of i and small for large i.

Geometrically, energy corresponds to roughness. This is apparent in figures 12.6 and 12.8. Increasing energy corresponds to increasing eigenvalue, which corresponds to increasing frequency, which corresponds to increasing roughness, in the form of more rapid changes in value in $\mathbf{y}$. By contrast, a value assignment $\mathbf{y}$ that satisfies the cluster hypothesis ("similar instances have similar labels") is locally homogeneous, which is to say smooth, varying slowly across the graph.

12.4.3 The Laplacian and random fields

While we are on the subject of energy, it is perhaps appropriate to revisit random fields. This is something of a digression, but it does serve to draw together some threads that we introduced much earlier, and to underline the commonalities among diverse methods.

In section 9.3 we introduced random fields as providing a direct way of assigning high probability to labelings in which close neighbors have the same label. Specifically, one defines an "energy" or "cost" function $H(\mathbf{y})$ that penalizes label disagreements between neighbors, and the probability of labeling $\mathbf{y}$ is defined as

$$p(\mathbf{y}) = \frac{1}{Z}e^{-H(\mathbf{y})}$$

where Z is a normalizing constant. In section 9.3 we used a simple version of the Ising model, with

$$H(\mathbf{y}) = -\sum_{(i,j)\in E} y_i y_j$$

where it was assumed that the labels y_k are either $+1$ or -1. Let us compare this to our current definition of energy:

$$J(\mathbf{y}) = \frac{1}{2}\sum_{ij}(y_i - y_j)^2 w_{ij}.$$

In the case under consideration, where the graph is unweighted, w_{ij} has value one if there is an edge between i and j, and value zero otherwise.

If neighbors i and j have different labels, then

$$y_i y_j = -1 \qquad 1 - y_i y_j = 2 \qquad \frac{1}{2}(y_i - y_j)^2 = 2.$$

If y_i and y_j have the same label, then

$$y_i y_j = 1 \qquad 1 - y_i y_j = 0 \qquad \frac{1}{2}(y_i - y_j)^2 = 0.$$

In either case, we have

$$\frac{1}{2}(y_i - y_j)^2 = 1 - y_i y_j.$$

Summing over pairs of neighbors yields

$$\begin{aligned} J(\mathbf{y}) &= \sum_{(i,j)\in E} \frac{1}{2}(y_i - y_j)^2 \\ &= \sum_{(i,j)\in E} (1 - y_i y_j) \\ &= |E| + H(\mathbf{y}). \end{aligned}$$

When we consider the resulting probability distribution, the constant $|E|$ gets folded into the normalization:

$$\begin{aligned} \frac{1}{Z}e^{-H(\mathbf{y})} &= \frac{1}{Z}e^{-J(\mathbf{y})+|E|} \\ &= \frac{1}{Z'}e^{-J(\mathbf{y})} \end{aligned}$$

with

$$\frac{1}{Z'} = \frac{1}{Z} e^{|E|}.$$

That is, if we have an unweighted graph and labels in $\{-1, +1\}$, then the Ising model is equivalent to a random field that uses $J(\mathbf{y})$ as its energy function. The functions $H(\mathbf{y})$ and $J(\mathbf{y})$ are *not* equivalent, however, when we generalize to weighted graphs and real vectors $\mathbf{y}$. In the general case, it is common to use $J(\mathbf{y})$ as the energy function, and the resulting random field is called a **Gaussian random field**. Note that the Gaussian random field can be written in terms of the graph Laplacian:

$$\frac{1}{Z'} e^{-\frac{1}{2}\mathbf{y}^{\mathrm{T}} L \mathbf{y}}.$$

12.4.4 Harmonic functions and the Laplacian

We return now to the idea that harmonic function methods are representatives of semisupervised spectral methods. In the next section, we will go beyond the methods that we introduced in chapter 10, but before we do, it is informative to re-examine the derivation of one of those methods, the matrix solution (10.14), in light of the connection between harmonic functions and the graph Laplacian. In chapter 10, we related harmonic functions to random walks, and derived the matrix solution from the fixed-point equation of the random walk:

$$\mathbf{y} = P\mathbf{y}. \tag{12.42}$$

The transition matrix P contains entries p_{ij} defined as

$$p_{ij} = \frac{w_{ij}}{\sum_k w_{ik}}.$$

The denominator is the degree of the i-th node:

$$d_i = \sum_k w_{ik}.$$

Hence, writing D for the diagonal degree matrix, we can express P as

$$P = D^{-1}W$$

and (12.42) becomes

$$\mathbf{y} = D^{-1}W\mathbf{y}$$

which is to say

$$D\mathbf{y} = W\mathbf{y}. \tag{12.43}$$

Given what we know now, instead of taking the roundabout path involving Thomson's principle and random walks, we can derive (12.43) much more directly, using the matrix differentiation methods of section 5.2.3 to minimize energy (12.41). The derivative is

$$\begin{aligned}\mathbf{D_y}J(\mathbf{y}) &= \frac{1}{2}\mathbf{D_y}\mathbf{y}^{\mathrm{T}}L\mathbf{y}\\ &= \mathbf{y}^{\mathrm{T}}L\\ &= L\mathbf{y}.\end{aligned}$$

In both steps we have taken advantage of the symmetry of L. Now we set the derivative to zero, to obtain

$$L\mathbf{y} = \mathbf{0}. \tag{12.44}$$

Given $L = D - W$, this is equivalent to (12.43).

Closer inspection of (12.44) makes the fundamental connection between the Laplacian and harmonic functions clearer. The i-th row of L has the form:

$$[-w_{i,1} \quad \dots \quad -w_{i,i-1} \quad d_i \quad -w_{i,i+1} \quad \dots \quad -w_{i,n}].$$

Hence (12.44) corresponds to a system of equations of form:

$$d_i y_i - \sum_j w_{ij} y_j = 0.$$

Substituting $\sum_k w_{ik}$ for d_i and rearranging yields

$$y_i = \sum_j \frac{w_{ij}}{\sum_k w_{ik}} y_j.$$

In short, the equation (12.44) implies the averaging property for values y_i, which is to say, (12.44) implies that $\mathbf{y}$ is a harmonic function.

Returning now to (12.43) and continuing with the derivation of the solution, we partition $\mathbf{y}$ into two half-vectors, one for the labeled nodes and one for the unlabeled nodes, and we partition D and W similarly:

$$\begin{bmatrix} D_l & 0 \\ 0 & D_u \end{bmatrix} \begin{bmatrix} \mathbf{y}_l \\ \mathbf{y}_u \end{bmatrix} = \begin{bmatrix} W_{ll} & W_{lu} \\ W_{ul} & W_{uu} \end{bmatrix} \begin{bmatrix} \mathbf{y}_l \\ \mathbf{y}_u \end{bmatrix}.$$

This can be written as two equations:

$$D_l\mathbf{y}_l = W_{ll}\mathbf{y}_l + W_{lu}\mathbf{y}_u \tag{12.45}$$

$$D_u\mathbf{y}_u = W_{ul}\mathbf{y}_l + W_{uu}\mathbf{y}_u. \tag{12.46}$$

Since the labels for the labeled nodes are fixed, the equation of interest is (12.46), which determines the label probabilities for the unlabeled nodes. It can be written

$$(D_u - W_{uu})\mathbf{y}_u = W_{ul}\mathbf{y}_l$$

which is

$$\mathbf{y}_u = L_{uu}^{-1} W_{ul}\mathbf{y}_l. \tag{12.47}$$

12.4.5 Using the Laplacian for regularization

Our conclusion so far concerning spectral methods in the semisupervised case is that they reduce to computation of the harmonic function determined by the labeled instances. In fact, there is a more general way to view the problem. In our discussion of the matrix solution for computing the harmonic function, we derived equation (12.45), but put it to the side on the grounds that the values for labeled nodes are fixed. However, by doing so we swept an issue under the carpet. We had asserted that we wanted an assignment $\mathbf{y}$ that satisfies the fixed point equation (12.42). Unless $\mathbf{y}$ satisfies (12.45), it cannot satisfy the fixed point equation. But harmonic functions often fail to satisfy (12.45). Equation (12.45) states that, for each labeled node i:

$$y_i = \sum_j \frac{w_{ij}}{d_i} y_j$$

where j ranges over all nodes, labeled and unlabeled. In other words, (12.45) states that labeled nodes satisfy the averaging property. We have already seen counterexamples: for example, the harmonic functions shown in figures 10.3 and 10.4.

Intuitively speaking, the fixed point equation wants each labeled node to satisfy the averaging property, but the averaging property is often inconsistent with the given labeled data, introducing a "stress" between harmonicity and the fixed labels. The approach we have considered until now chooses a function $\mathbf{y}$ that is maximally congruent with the fixed labels, and ignores the conflicting demands of harmonicity for labeled nodes.

An alternative is to view energy, which is to say, the graph Laplacian, as a regularization term. The idea of regularization was introduced in section 4.2.3, and it was mentioned again in connection with support vector machines in section 6.4.3. A regularization term represents a measure of simplicity that is traded off against fit to the data. For support vector machines, a simple model is one that predicts good separation, that is, large, empty boundary regions, between clusters. For spectral clustering, a simple model is one that has low energy. Geometrically, it is smooth, varying slowly across the graph. Harmonic functions fit the labeled data perfectly, at a cost in smoothness. The more general approach is to trade off fit against simplicity.

We desire a function $\mathbf{y}$ that fits the labeled data well. Writing $\tilde{\mathbf{y}}$ for the values in the labeled data, we can measure (lack of) fit to the data as squared error. The total squared error is

$$\sum_{1 \le i \le \ell} (y_i - \tilde{y}_i)^2 = (\mathbf{y} - \tilde{\mathbf{y}})^{\mathrm{T}} I_\ell (\mathbf{y} - \tilde{\mathbf{y}}) \tag{12.48}$$

where I_ℓ is a diagonal matrix with 1's in the first ℓ positions on the diagonal, and 0's after that. We note that the input label values $\tilde{y}_i$ are discrete: 1 for postively labeled and 0 for negatively labeled. The predicted values y_i are not so constrained.

The function should also be simple, which in this case we interpret as being smooth, as measured by the graph Laplacian. Taking a linear combination of the fit term (12.48) and the simplicity term, we have the objective function

$$\alpha(\mathbf{y} - \tilde{\mathbf{y}})^{\mathrm{T}} I_\ell (\mathbf{y} - \tilde{\mathbf{y}}) + \beta \mathbf{y}^{\mathrm{T}} L \mathbf{y} \tag{12.49}$$

where α and β are free parameters. Since our goal is to minimize the objective function, and multiplication by a constant has no effect on the identity of the minimizer, we can multiply the preceding by $1/\alpha$, and write the objective function more simply as

$$(\mathbf{y} - \tilde{\mathbf{y}})^{\mathrm{T}} I_\ell (\mathbf{y} - \tilde{\mathbf{y}}) + \gamma \mathbf{y}^{\mathrm{T}} L \mathbf{y} \tag{12.50}$$

where we replace the two free parameters α and β with a single free parameter $\gamma = \beta/\alpha$.

We wish to minimize (12.50), so we take its derivative and set it to zero:

$$\begin{aligned}
&\mathbf{D}_{\mathbf{y}}[(\mathbf{y} - \tilde{\mathbf{y}})^{\mathrm{T}} I_\ell (\mathbf{y} - \tilde{\mathbf{y}}) + \gamma \mathbf{y}^{\mathrm{T}} L \mathbf{y}] \\
&= \mathbf{D}_{\mathbf{y}}(\mathbf{y} - \tilde{\mathbf{y}})^{\mathrm{T}} I_\ell (\mathbf{y} - \tilde{\mathbf{y}}) + \gamma \mathbf{D}_{\mathbf{y}} \mathbf{y}^{\mathrm{T}} L \mathbf{y} \\
&= 2(\mathbf{y} - \tilde{\mathbf{y}})^{\mathrm{T}} I_\ell + 2\gamma \mathbf{y}^{\mathrm{T}} L.
\end{aligned}$$

In the last step we have used the fact that both I_ℓ and L are symmetric. The result is a row vector; converting it to a column vector yields

$$2I_\ell^{\mathrm{T}}(\mathbf{y} - \tilde{\mathbf{y}}) + 2\gamma L^{\mathrm{T}} \mathbf{y}.$$

Setting this expression to zero, and using the facts $I_\ell^{\mathrm{T}} = I_\ell$ and $L^{\mathrm{T}} = L$, we have

$$I_\ell \mathbf{y} + \gamma L \mathbf{y} = I_\ell \tilde{\mathbf{y}}.$$

Hence

$$\mathbf{y} = (I_\ell + \gamma L)^{-1} I_\ell \tilde{\mathbf{y}}.$$

If we define the last $n - \ell$ entries in $\tilde{\mathbf{y}}$ to be 0's, this simplifies slightly to

$$\mathbf{y} = (I_\ell + \gamma L)^{-1} \tilde{\mathbf{y}}. \tag{12.51}$$

The only remaining question is how to determine γ. This can be done *a priori*. Let us consider the objective function in the form (12.49). It is natural to average the first term (the squared error term) over labeled instances, meaning that we take $\alpha = 1/\ell$. The second term (the Laplacian) sums over weighted edges, so it is natural to normalize it by total edge weight; that is, we take $\beta = 1/\sum_{ij} w_{ij}$. We then have $\gamma = \beta/\alpha$, which is

$$\gamma = \frac{\ell}{\sum_{ij} w_{ij}}.$$

Alternatively, γ can be estimated from held-out data.

12.4.6 Transduction to induction

We have just considered two semisupervised spectral methods, the harmonic function method (12.47) and the regularized method (12.51). Both of them are transductive: they tell us how to compute labels for the unlabeled instances, but they do not tell us how to label new instances that were not included in the training data. A final question, then, is how to deal with test instances, which is to say, how to do induction using transductive learners.

Given a new instance $\mathbf{x}$ whose label we wish to predict, the simplest idea is to add $\mathbf{x}$ to the graph and re-run the transductive algorithm. That approach is exact, but obviously has the disadvantage of high cost for labeling a single new instance.

An alternative is to optimize the objective function just for the new instance, taking all other instances *and their predicted labels* as fixed. If we are given a new instance $\mathbf{x}$, let us consider the effect on the objective function (12.50). The first term of (12.50), the squared error term, is a sum over labeled instances; since $\mathbf{x}$ is unlabeled it has no effect on that term. The second term, the Laplacian, involves a sum over edges, as we see when it is written in the form:

$$\mathbf{y}^{\mathrm{T}} L \mathbf{y} = \sum_{ij} (y_i - y_j)^2 w_{ij}.$$

Adding a new instance $\mathbf{x}$ adds new terms to the sum, one for each edge between $\mathbf{x}$ and a training instance $\mathbf{x}_i$, but it does not affect the old terms.

In short, let C represent the value of (12.50) computed on the training data. Holding that constant, and adding a single new instance $\mathbf{x}$, the objective function becomes

$$C + \gamma \sum_i (y - y_i)^2 w(\mathbf{x}, \mathbf{x}_i)$$

where y is the value to be computed (the positive-label probability for $\mathbf{x}$) and $w(\mathbf{x}, \mathbf{x}_i)$ represents the similarity between the new instance $\mathbf{x}$ and the old instance $\mathbf{x}_i$. The derivative with respect to y is

$$2\gamma \sum_i (y - y_i) w(\mathbf{x}, \mathbf{x}_i).$$

Setting the derivative to zero yields

$$y \sum_i w(\mathbf{x}, \mathbf{x}_i) = \sum_i y_i w(\mathbf{x}, \mathbf{x}_i)$$

$$y = \sum_i \frac{w(\mathbf{x}, \mathbf{x}_i)}{\sum_j w(\mathbf{x}, \mathbf{x}_j)} y_i.$$

In short, we simply use the averaging property to compute the label for $\mathbf{x}$.

12.5 Bibliographic notes

For further reading on semisupervised learning with spectral methods, the definitive collection is Chapelle et al. [39]. Ding [78] also has a useful tutorial on spectral clustering. The spectral method for approximating ratiocut was proposed by Hagen & Kahng [102]. The method discussed for going from transduction to induction follows Bengio et al. [13].

Bibliography

[1] Abney, Steven. Bootstrapping. *Proceedings of the 40th Annual Meeting of the Association for Computational Linguistics (ACL-2002)*, pp. 360–367. Association for Computational Linguistics. East Stroudsburg, PA. 2002.

[2] Abney, Steven. Understanding the Yarowsky Algorithm. *Computational Linguistics* 30(3):365–395. 2004.

[3] Agichtein, Eugene, Eleazar Eskin, and Luis Gravano. Combining Strategies for Extracting Relations from Text Collections. Columbia University Technical Report CUCS-006-00. 2000.

[4] Agichtein, E. and L. Gravano. Snowball: Extracting Relations from Large Plain-Text Collections. *The 5th International Conference on Digital Libraries.* Association for Computing Machinery. New York. 2000.

[5] Allan, James. Relevance feedback with too much data. *Proceedings of the 18th Annual International Conference on Research and Development in Information Retrieval (SIGIR-1995)*, pp. 337–343. Association for Computing Machinery. New York. 1995.

[6] Ando, Rie Kubota. Semantic lexicon construction: Learning from unlabeled data via spectral analysis. *Proceedings of the Conference on Natural Language Learning (CoNLL-2004).* Association for Computational Linguistics. East Stroudsburg, PA. 2004.

[7] Baluja, S. Probabilistic modeling for face orientation discrimination: Learning from labeled and unlabeled data. *Advances in Neural Information Processing Systems 11 (NIPS-1998).* MIT Press. Cambridge, MA. 1999.

[8] Basu, S., A. Banerjee, and R. Mooney. Semi-supervised clustering by seeding. *Proceedings of the 19th International Conference on Machine Learning (ICML-2002).* Morgan Kaufmann Publishers. San Francisco, CA. 2002.

[9] Bean, David and Ellen Riloff. Unsupervised learning of contextual role knowledge for coreference resolution. *Proceedings of the Conference on Human Language Technology and the Meeting of the North American Chapter of the Association for Computational Linguistics*

(HLT/NAACL-2004). Association for Computational Linguistics. East Stroudsburg, PA. 2004.

[10] Becker, S. and G.E. Hinton. A self-organizing neural network that discovers surfaces in random-dot stereograms. *Nature* 355:161–163. 1992.

[11] Bengio, S. and Y. Bengio. An EM Algorithm for Asynchronous Input/Output Hidden Markov Models. *Proceedings of the 3rd International Conference On Neural Information Processing (ICONIP 1996)*, pp. 328–334. Springer-Verlag. Berlin. 1996.

[12] Bengio, Y. and F. Gingras. Recurrent Neural Networks for Missing or Asynchronous Data. In M. Mozer, D.S. Touretzky, and M. Perrone (eds.), *Advances in Neural Information Processing Systems 8 (NIPS-1995)*. MIT Press. Cambridge, MA. 1996.

[13] Bengio, Yoshua, Olivier Delalleau, and Nicolas Le Roux. Label propagation and quadratic criterion. In Chapelle et al. (eds.), *Semi-Supervised Learning* (Chapter 11). MIT Press. Cambridge, MA. 2006.

[14] Bennett, K.P., and A. Demiriz. Semi-supervised support vector machines. *Advances in Neural Information Processing Systems 10 (NIPS-1997)*, pp. 368–374. MIT Press. Cambridge, MA. 1998.

[15] Bennett, K., A. Demiriz, and R. Maclin. Exploiting unlabeled data in ensemble methods. *Proceedings of 8th International Conference on Knowledge Discovery and Data Mining (KDD-2002)*, pp. 289–296. Association for Computing Machinery. New York. 2002.

[16] Berland, M., and E. Charniak. Finding parts in very large corpora. *Proceedings of the 37th Annual Meeting of the Association for Computational Linguistics (ACL-1999)*, pp. 57–64. Association for Computational Linguistics. East Stroudsburg, PA. 1999.

[17] Bhattacharya, Indrajit, Lise Getoor, and Yoshua Bengio. Unsupervised sense disambiguation using bilingual probabilistic models. *Proceedings of the 42nd Meeting of the Association for Computational Linguistics (ACL-2004)*. Association for Computational Linguistics. East Stroudsburg, PA. 2004.

[18] Bikel, Daniel M., Scott Miller, Richard Schwartz, and Ralph Weischedel. Nymble: A High-Performance Learning Name-finder. *Proceedings of the Fifth Conference on Applied Natural Language Processing*, pp. 194–201. 1997.

[19] Bishop, Christopher M. *Neural Networks for Pattern Recognition.* Clarendon Press. Oxford. 1995.

[20] Blum, Avrim and Shuchi Chawla. Learning from labeled and unlabeled data using graph mincuts. *Proceedings of the 18th International Confer-*

ence on Machine Learning, pp. 19–26. Morgan Kaufmann Publishers. San Francisco, CA. 2001.

[21] Blum, Avrim, and Tom Mitchell. Combining labeled and unlabeled data with co-training. *Proceedings of the 11th Annual Conference on Computational Learning Theory (COLT),* pp. 92–100. Morgan Kaufmann Publishers. San Francisco, CA. 1998.

[22] Blum, Avrim, John Lafferty, Rajashekar Reddy, and Mugizi Robert Rwebangira. Semi-supervised learning using randomized mincuts. *Proceedings of the 21st International Conference (ICML-2004).* Association for Computing Machinery. New York. 2004.

[23] Bodenreider, Olivier, Thomas Rindflesch, and Anita Burgun. Unsupervised, corpus-based method for extending a biomedical terminology. *Proceedings of the Workshop on Natural Language Processing in the Biomedical Domain, Annual Meeting of the Association for Computational Linguistics (ACL-2002).* Association for Computational Linguistics. East Stroudsburg, PA. 2002.

[24] Borthwick, A., J. Sterling, E. Agichtein, and R. Grishman. Exploiting Diverse Knowledge Sources via Maximum Entropy in Named Entity Recognition. *Sixth Workshop on Very Large Corpora (VLC-1998).* Association for Computational Linguistics. East Stroudsburg, PA. 1998.

[25] Brill, Eric. Unsupervised Learning of Disambiguation Rules for Part of Speech Tagging. *Third Workshop on Very Large Corpora (VLC-1995),* pp. 1–13. Association for Computational Linguistics. East Stroudsburg, PA. 1995.

[26] Brin, Sergey. Extracting Patterns and Relations from the World Wide Web. *Proceedings of the WebDB Workshop, 6th International Conference on Extending Database Technology (EDBT-1998).* Springer-Verlag. Berlin. 1998.

[27] Briscoe, Ted and John Carroll. Automatic Extraction of Subcategorization from Corpora. *Proceedings of the Conference on Applied Natural Language Processing (ANLP).* Association for Computational Linguistics. East Stroudsburg, PA. 1997.

[28] Briscoe, Ted and John Carroll. Towards Automatic Extraction of Argument Structure from Corpora. Rank Xerox Research Centre Tech Report MLTT-006. 1994.

[29] Broder, Andrei Z., Robert Krauthgamer, and Michael Mitzenmacher. Improved classification via connectivity information. *Proceedings of the Eleventh Annual Symposium on Discrete Algorithms,* pp. 576–585. Association for Computing Machinery/Society for Industrial and Applied Mathematics. New York. 2000.

[30] Brown, P., V. Della Pietra, P. deSouza, J. Lai, and R. Mercer. Class-Based n-gram Models of Natural Language. *Computational Linguistics* 18(4):467–480. 1992.

[31] Bruce, Rebecca. "A Statistical Method for Word-Sense Disambiguation." PhD diss. New Mexico State University. 1995.

[32] Buckley, C., M. Mitra, J. Walz, and C. Cardie. Using clustering and SuperConcepts within SMART. *Proceedings of the Sixth Text Retrieval Conference (TREC).* National Institute of Standards and Technology. Gaithersburg, MD. 1998.

[33] Buckley, C., G. Salton, and J. Allan. Automatic retrieval with locality information using SMART. *Proceedings of the First Text Retrieval Conference (TREC),* pp. 59–72. National Institute of Standards and Technology. Gaithersburg, MD. 1992.

[34] Cao, Yunbo, Hang Li, and Li Lian. Uncertainty Reduction in Collaborative Bootstrapping: Measure and Algorithm. *Proceedings of the 41st Meeting of the Association for Computational Linguistics (ACL-2003).* Association for Computational Linguistics. East Stroudsburg, PA. 2003.

[35] Caraballo, S.A. Automatic construction of a hypernym-labeled noun hierarchy from text. *Proceedings of the 37th Meeting of the Association for Computational Linguistics (ACL-1999).* Association for Computational Linguistics. East Stroudsburg, PA. 1999.

[36] Castelli, Vittorio and Thomas Cover. On the exponential value of labeled samples. *Pattern Recognition Letters* 16:105–111. 1995.

[37] Castelli, V. and T. Cover. The Relative Value of Labeled and Unlabeled Samples in Pattern Recognition with an Unknown Mixing Parameter. *IEEE Transactions on Information Theory* 42(6):2102–2117. Institute of Electrical and Electronics Engineers. New York. 1996.

[38] Chakrabarti, Soumen. *Mining the Web: Analysis of Hypertext and Semi Structured Data.* Morgan Kaufmann Publishers. San Francisco, CA. 2002.

[39] Chapelle, Olivier, Bernhard Schölkopf, and Alexander Zien (eds). *Semi-Supervised Learning.* MIT Press, Cambridge, MA. 2006.

[40] Chapelle, O., J. Weston, and B. Schölkopf. Cluster kernels for semi-supervised learning. *Advances in Neural Information Processing Systems 15 (NIPS-2002).* MIT Press. Cambridge, MA. 2003.

[41] Charniak, Eugene. Unsupervised learning of name structure from coreference data. *Proceedings of the 2nd Meeting of the North American Chapter of the Association for Computational Linguistics (NAACL-2001),* pp. 48–54. Association for Computational Linguistics. East Stroudsburg, PA. 2001.

[42] Chawla, Nitesh and Grigoris Karakoulas. Learning from labeled and unlabeled data: an empirical study across techniques and domains. *Journal of Artificial Intelligence Research* 23:331–366. 2005.

[43] Chelba, Ciprian and Frederick Jelinek. Exploiting Syntactic Structure for Language Modeling. *Proceedings of the 36th Meeting of the Association for Computational Linguistics (ACL-1998).* Association for Computational Linguistics. East Stroudsburg, PA. 1998.

[44] Church, Kenneth. A Stochastic Parts Program and Noun Phrase Parser for Unrestricted Texts. *Proceedings of the 2nd Conference on Applied Natural Language Processing (ANLP-1988).* Association for Computational Linguistics. East Stroudsburg, PA. 1988.

[45] Church, Kenneth, Patrick Hanks, Donald Hindle, William Gale, and Rosamond Moon. Substitutability. Internal Technical Memorandum. AT&T Bell Laboratories. 1990.

[46] Coates-Stephens, Sam. The Analysis and Acquisition of Proper Names for the Understanding of Free Text. *Computers and the Humanities* 26:441–456. 1993.

[47] Cohen, I., N. Sebe, F.G. Cozman, and T.S. Huang. Semi-supervised Learning for Facial Expression Recognition. *Proceedings of the 5th International Workshop on Multimedia Information Retrieval (MIR-2003),* pp. 17–22. Association for Computing Machinery. New York. 2003.

[48] Cohen, I., N. Sebe, F.G. Cozman, M.C. Cirelo, and T.S. Huang. Semi-supervised Learning of Classifiers: Theory and Algorithms for Bayesian Network Classifiers and Applications to Human-Computer Interaction. *IEEE Transactions on Pattern Analysis and Machine Intelligence.* Institute of Electrical and Electronics Engineers. New York. 2004.

[49] Cohn, D.A., Z. Ghahramani, and M.I. Jordan. Active learning with statistical models. *Journal of Artificial Intelligence Research* 4:129–145. 1996.

[50] Collins, Michael and Yoram Singer. Unsupervised Models for Named Entity Classification. *Proceedings of the Conference on Empirical Methods in Natural Language Processing (EMNLP-1999),* pp. 100–110. Association for Computational Linguistics. East Stroudsburg, PA. 1999.

[51] Cormen, T., C. Leiserson, and R. Rivest. *Introduction to Algorithms.* MIT Press. Cambridge, MA. 1990.

[52] Cover, T. and J. Thomas. *Elements of Information Theory.* John Wiley & Sons. New York. 1991.

[53] Cozman, Fabio G. and Ira Cohen. Unlabeled data can degrade classification performance of generative classifiers. *Proceedings of the 15th*

International Florida Artificial Intelligence Research Society Conference (FLAIRS-2002), pp. 327–331. AAAI Press/MIT Press. Cambridge, MA. 2002.

[54] Cozman, Fabio G., I. Cohen, and M.C. Cirelo. Semi-supervised learning of mixture models. *Proceedings of the 20th International Conference on Machine Learning (ICML-2003).* AAAI Press/MIT Press. Cambridge, MA. 2003.

[55] Craven, M. and S. Slattery. Relational Learning with Statistical Predicate Invention: Better Models for Hypertext. *Machine Learning* 43:97–119. 2001.

[56] Craven, Mark, Dan DiPasquo, Dayne Freitag, Andrew McCallum, Tom Mitchell, Kamal Nigam, and Sean Slattery. Learning to Construct Knowledge Bases form the World Wide Web. *Artificial Intelligence* 118:69–113. 2000.

[57] Craven, M., D. DiPasquo, D. Freitag, A. McCallum, T. Mitchell, K. Nigam, and S. Slattery. Learning to extract symbolic knowledge from the World Wide Web. *Proceedings of the 15th National Conference on Artificial Intelligence (AAAI-1998).* AAAI Press/MIT Press. Cambridge, MA. 1998.

[58] Cucerzan, Silviu and David Yarowsky. Language independent minimally supervised induction of lexical probabilities. *Proceedings of the 38th Meeting of the Association for Computational Linguistics (ACL-2000)*, pp. 270–277. Association for Computational Linguistics. East Stroudsburg, PA. 2000.

[59] Cucerzan, Silviu and David Yarowsky. Language Independent Named Entity Recognition Combining Morphological and Contextual Evidence. In *Proceedings of the Joint SIGDAT Conference on Empirical Methods in Natural Language Processing and Very Large Corpora (EMNLP/VLC-1999)*, pp. 90–99. Association for Computational Linguistics. East Stroudsburg, PA. 1999.

[60] Cucerzan, Silviu and David Yarowsky. Minimally supervised induction of grammatical gender. *Proceedings of the Conference on Human Language Technology and the Meeting of the North American Chapter of the Association for Computational Linguistics (HLT/NAACL-2003).* Association for Computational Linguistics. East Stroudsburg, PA. 2003.

[61] Curran, James R. Supersense tagging of unknown nouns using semantic similarity. *Proceedings of the 43rd Meeting of the Association for Computational Linguistics (ACL-2005).* Association for Computational Linguistics. East Stroudsburg, PA. 2005.

[62] Cutting, Doug, Julian Kupiec, Jan Pedersen, and Penelope Sibun. A Practical Part-of-Speech Tagger. *Third Conference on Applied Natural*

Language Processing (ANLP), pp. 133–140. Association for Computational Linguistics. East Stroudsburg, PA. 1992.

[63] Dagan, Ido and Alon Itai. Word Sense Disambiguation Using a Second Language Monolingual Corpus. *Computational Linguistics* 20:563–596. 1994.

[64] Dagan, Ido, Alon Itai, and Ulrike Schwall. Two Languages are more Informative than One. *Proceedings of the 29th Annual Meeting of the Association for Computational Linguistics (ACL-1991),* pp. 130–137. Association for Computational Linguistics. East Stroudsburg, PA. 1991.

[65] Dagan, Ido, Lillian Lee, and Fernando Pereira. Similarity-based models of word coocurrence probabilities. *Machine Learning* 34(1–3):43–69. 1999.

[66] Dasgupta, Sanjoy. Learning Mixtures of Gaussians. Berkeley Tech Report CSD-99-1047. University of California Berkeley. 1999.

[67] Dasgupta, Sanjoy, Michael Littman, and David McAllester. PAC generalization bounds for co-training. *Advances in Neural Information Processing Systems 14 (NIPS-2001).* MIT Press. Cambridge, MA. 2002.

[68] De Comité, F., F. Denis, R. Gilleron, and F. Letouzey. Positive and unlabeled data help learning. *Proceedings of the Conference on Computational Learning Theory (COLT-1999).* Association for Computing Machinery. New York. 1999.

[69] De Marcken, Carl. The Unsupervised Acquisition of a Lexicon from Continuous Speech. A.I. Memo 1558. MIT AI Lab and Department of Brain and Cognitive Sciences. Massachusetts Institute of Technology. 1995.

[70] De Marcken, Carl. "Unsupervised Language Acquisition." PhD diss. Massachusetts Institute of Technology. 1996.

[71] De Sa, Virginia. Learning classification with unlabeled data. In Cowan, Tesauro, and Alspector (eds), *Advances in Neural Information Processing Systems 6.* Morgan Kaufmann Publishers. San Francisco, CA. 1993.

[72] De Sa, Virginia. "Unsupervised Classification Learning from Cross-Modal Environmental Structure." PhD diss. University of Rochester. 1994.

[73] Deerwester, Scott, Susan T. Dumais, George W. Furnas, Thomas K. Landauer, and Richard Harshman. Indexing by latent semantic indexing. *Journal of the American Society for Information Science* 41(6):391–407. 1990.

[74] Dempster, A.P., N.M. Laird, and D.B. Rubin. Maximum likelihood from incomplete data via the EM algorithm. *Journal of the Royal Statistical Society B* 39:1–38. 1977.

[75] DeRose, S. Grammatical Category Disambiguation by Statistical Optimization. *Computational Linguistics* 14(1). 1988.

[76] Dietterich, Thomas G., and Ghulum Bakiri. Solving multiclass learning problems via Error-Correcting Output Codes. *Journal of Artificial Intelligence Research* 2:263–286. 1995.

[77] Dietterich, Thomas G., Richard H. Lathrop, and Tomas Lozano-Perez. Solving the multiple-instance problem with axis-parallel rectangles. *Artificial Intelligence* 89:1–2, 31–71. 1997.

[78] Ding, Chris. A tutorial on spectral clustering. *Proceedings of the 21st International Conference on Machine Learning (ICML-2004)*. Association for Computing Machinery. New York. 2004.

[79] Doyle, P., and J. Snell. *Random walks and electric networks.* Mathematical Association of America. Washington, D.C. 1984.

[80] Du Toit, S.H.C., A.G.W. Steyn, and R.H. Stumpf. *Graphical Exploratory Data Analysis.* Springer-Verlag. Berlin. 1986.

[81] Duda, Richard O., Peter E. Hart, and David G. Stork. *Pattern Classification* (second edition). John Wiley & Sons. New York. 2001.

[82] Efthimiadis, N.E. Query expansion. *Annual Review of Information Systems and Technology* 31:121–187. 1996.

[83] Elworthy, David. Does Baum-Welch Re-estimation Help Taggers? *4th Conference on Applied Natural Language Processing (ANLP),* pp. 53–58. Association for Computational Linguistics. East Stroudsburg, PA. 1994.

[84] Etzioni, Oren, Michael Cafarella, Doug Downey, Stanley Kok, Ana-Maria Popescu, Tal Shaked, Stephen Soderland, Daniel S. Weld, and Alexander Yates. Web-scale information extraction in KnowItAll. *Proceedings of the 13th International Conference on World Wide Web.* Association for Computing Machinery. New York. 2004.

[85] Everitt, B.S. *An Introduction to Latent Variable Models.* Chapman and Hall/Kluwer Academic Publishers. Dordrecht. 1984.

[86] Fawcett, Tom. ROC Graphs: Notes and practical considerations for researchers. HP Labs Tech Report HPL-2003-4. 2003.

[87] Fellbaum, Christiane (ed). *WordNet: An Electronic Lexical Database.* MIT Press. Cambridge, MA. 1998.

[88] Finch, Steven Paul. *Finding Structure in Language.* University of Edinburgh. Edinburgh. 1993.

[89] Flake, Gary W., Kostas Tsioutsiouliklis, and Robert E. Tarjan. Graph clustering techniques based on minimum cut trees. Technical Report 2002-06. NEC Laboratories America. Princeton, NJ. 2002.

[90] Freund, Yoav, H. Sebastian Seung, Eli Shamir, and Naftali Tishby. Selective Sampling Using the Query by Committee Algorithm. *Machine Learning* 28:133–168. 1997.

[91] Gale, W., K. Church, and D. Yarowsky. One sense per discourse. *Proceedings of the Fourth Speech and Natural Language Workshop,* pp. 233–237. Defense Advanced Projects Research Agency (DARPA). Morgan Kaufmann Publishers. San Francisco, CA. 1992.

[92] Ganesalingam, S. and G. McLachlan. The efficiency of a linear discriminant function based on unclassified initial samples. *Biometrika* 65:658–662. 1978.

[93] Ganesalingam, S. and G. McLachlan. Small sample results for a linear discriminant function estimated from a mixture of normal populations. *Journal of Statistical Computation and Simulation* 9:151–158. 1979.

[94] Ghahramani, Zoubin and Michael Jordan. Supervised learning from incomplete data via an EM approach. *Advances in Neural Information Processing Systems 6 (NIPS-1993).* MIT Press. Cambridge, MA. 1994.

[95] Goldman, Sally and Yan Zhou. Enhancing supervised learning with unlabeled data. *Proceedings of the 17th International Conference on Machine Learning (ICML-2000).* Morgan Kaufmann Publishers. San Francisco, CA. 2000.

[96] Goldsmith, John. Unsupervised Learning of the Morphology of a Natural Language. *Computational Linguistics* 27(2):153–198. 2001.

[97] Golub, G.H. and C.F.V. Loan. *Matrix Computations* (3rd edition). Johns Hopkins University Press. Baltimore, MD. 1996.

[98] Greenspan, H., R. Goodman, and R. Chellappa. Texture analysis via unsupervised and supervised learning. *International Joint Conference on Neural Networks (IJCNN-1991),* vol. 1, pp. 639–644. Institute of Electrical and Electronics Engineers. New York. 1991.

[99] Grefenstette, G. A new knowledge-poor technique for knowledge extraction from large corpora. *Proceedings of the 15th Annual International Conference on Research and Development in Information Retrieval (SIGIR-1992).* Association for Computing Machinery. New York. 1992.

[100] Grefenstette, G. Sextant: Extracting semantics from raw text implementation details. Technical Report CS92-05. University of Pittsburgh, Computer Science Dept. 1992.

[101] Grefenstette, G. and M. Hearst. A Knowledge-Poor Method for Refining Automatically-Discovered Lexical Relations: Combining Weak Techniques for Stronger Results. *AAAI Workshop on Statistically-Based*

NLP Techniques, 10th National Conference on Artificial Intelligence (AAAI-1992). AAAI Press/MIT Press. Cambridge, MA. 1992.

[102] Hagen, L. and A. Kahng. New spectral methods for ratio cut partitioning and clustering. *IEEE Transactions on CAD* 11:1074–1085. Institute of Electrical and Electronics Engineers. New York. 1992.

[103] Hartigan, J.A. *Clustering Algorithms.* John Wiley & Sons. New York. 1975.

[104] Hartley, H.O. and J.N.K. Rao. Classification and estimation in analysis of variance problems. *Review of International Statistical Institute* 36:141–147. August 1968.

[105] Hastie, T. and W. Stuetzle. Principal curves. *Journal of the American Statistical Association* 84:502–516. 1989.

[106] Hastie, Trevor, Robert Tibshirani, and Jerome Friedman. *The Elements of Statistical Learning.* Springer-Verlag. Berlin. 2001.

[107] Hearst, Marti A. Automated discovery of WordNet relations. In Christiane Fellbaum (ed.), *WordNet: An Electronic Lexical Database and Some of its Applications.* MIT Press. Cambridge, MA. 1998.

[108] Hearst, M. Automatic acquisition of hyponyms from large text corpora. *Proceedings of the 14th International Conference on Computational Linguistics (COLING),* pp. 539–545. Morgan Kaufmann Publishers. San Francisco, CA. 1992.

[109] Hearst, M. Noun homonym disambiguation using local context in large text corpora. *Proceedings, Seventh Annual Conference of the UW Centre for the New OED and Text Research,* pp. 1–22. University of Waterloo. Waterloo, Ontario. 1991.

[110] Hendrickson, Bruce and Robert Leland. A multi-level algorithm for partitioning graphs. *Proceedings of the International Conference for High Performance Computing, Networking, Storage and Analysis (Supercomputing-1995).* Association for Computing Machinery. New York. 1995.

[111] Hindle, Don and Mats Rooth. Structural ambiguity and lexical relations. *Proceedings of the 29th Annual Meeting of the Association for Computational Linguistics (ACL-1991).* Association for Computational Linguistics. East Stroudsburg, PA. 1991.

[112] Hofmann, Thomas. Unsupervised learning by probabilistic latent semantic analysis. *Machine Learning* 42(1–2):177–196. 2001.

[113] Horn, R. and C.R. Johnson. *Matrix Analysis.* Cambridge University Press. Cambridge. 1985.

[114] Inoue, N. Automatic noun classification by using Japanese-English word pairs. *Proceedings of the 29th Annual Meeting of the Association for Computational Linguistics (ACL-1991),* pp. 201–208. Association for Computational Linguistics. East Stroudsburg, PA. 1991.

[115] Jaakkola, T., M. Meila, and T. Jebara. Maximum entropy discrimination. *Advances in Neural Information Processing Systems 12 (NIPS-1999).* MIT Press. Cambridge, MA. 2000.

[116] Joachims, Thorsten. A probabilistic analysis of the Rocchio algorithm with TFIDF for text categorization. *Proceedings of the 14th International Conference on Machine Learning (ICML-1997).* Morgan Kaufmann Publishers. San Francisco, CA. 1997.

[117] Joachims, T. Transductive inference for text classification using support vector machines. *Proceedings of the 16th International Conference on Machine Learning (ICML-1999).* Morgan Kaufmann Publishers. San Francisco, CA. 1999.

[118] Jones, Rosie, Andrew McCallum, Kamal Nigam, and Ellen Riloff. Bootstrapping for Text Learning Tasks. *Workshop on Text Mining: Foundations, Techniques, and Applications, International Joint Conference on Artificial Intelligence (IJCAI-1999).* Morgan Kaufmann Publishers. San Francisco, CA. 1999.

[119] Jurafsky, D., and J. Martin. *Speech and Language Processing.* Prentice Hall. Upper Saddle River, NJ. 2000.

[120] Karger, David. Random sampling in cut, flow, and network design problems. *Proceedings of the ACM Symposium on Theory of Computing (STOC-1994),* pp. 648–657. Association for Computing Machinery. New York. 1994.

[121] Karger, David R. "Random Sampling in Graph Optimization Problems." PhD diss. Stanford University. 1995.

[122] Karov, Yael and Shimon Edelman. Learning similarity-based word sense disambiguation from sparse data. Technical Report CS-TR 96-05. The Weizmann Institute of Science. 1996.

[123] Kaski, Samuel. Discriminative clustering. *Bulletin of the International Statistical Institute, Invited Paper Proceedings of the 54th Session,* vol. 2, pp. 270–273. International Statistical Institute. Voorburg, The Netherlands. 2003.

[124] Kaufmann, L. and P.J. Rousseeuw. *Finding Groups in Data.* John Wiley & Sons. New York. 1990.

[125] Kishida, Kazuaki. Pseudo relevance feedback method based on Taylor expansion of retrieval function in NTCIR-3 patent retrieval task.

Proceedings of the Workshop on Patent Corpus Processing, Conference on Human Language Technologies and the Meeting of the North American Chapter of the Association for Computational Linguistics (HLT/NAACL-2003). Association for Computational Linguistics. East Stroudsburg, PA. 2003.

[126] Klein, Dan and Christopher D. Manning. Corpus-based induction of syntactic structure: models of dependency and constituency. *Proceedings of the 42nd Meeting of the Association for Computational Linguistics (ACL-2004).* Association for Computational Linguistics. East Stroudsburg, PA. 2004.

[127] Kleinberg, Jon M. and Eva Tardos. Approximation algorithms for classification problems with pairwise relationships: Metric labeling and Markov random fields. *IEEE Symposium on Foundations of Computer Science,* pp. 14–23. Institute of Electrical and Electronics Engineers. New York. 1999.

[128] Ko, Youngjoong and Jungyun Seo. Learning with unlabeled data for text categorization using bootstrapping and feature projection techniques. *Proceedings of the 42nd Meeting of the Association for Computational Linguistics (ACL-2004).* Association for Computational Linguistics. East Stroudsburg, PA. 2004.

[129] Kohonen, Teuvo. Improved versions of Learning Vector Quantization. *Proceedings of the International Joint Conference on Neural Networks (IJCNN-1990),* vol. I, pp. 545–550. Institute of Electrical and Electronics Engineers. New York. 1990.

[130] Kuhn, Jonas. Experiments in parallel-text based grammar induction. *Proceedings of the 42nd Meeting of the Association for Computational Linguistics (ACL-2004).* Association for Computational Linguistics. East Stroudsburg, PA. 2004.

[131] Kupiec, Julian. Robust part-of-speech tagging using a Hidden Markov Model. *Computer Speech and Language* 6. 1992.

[132] Lafferty, J., A. McCallum, and F. Pereira. Conditional random fields: Probabilistic models for segmenting and labeling sequence data. *Proceedings of the 18th International Conference on Machine Learning (ICML-2001).* Morgan Kaufmann Publishers. San Francisco, CA. 2001.

[133] Lapata, Mirella and Frank Keller. The Web as a baseline: Evaluating the performance of unsupervised Web-based models for a range of NLP tasks. *Proceedings of the Conference on Human Language Technology and the Meeting of the North American Chapter of the Association for Computational Linguistics (HLT/NAACL-2004).* Association for Computational Linguistics. East Stroudsburg, PA. 2004.

[134] Lawrence, Neil D., and Michael I. Jordan. Semi-supervised learning via Gaussian processes. *Advances in Neural Information Processing Systems 17 (NIPS-2004)*. MIT Press. Cambridge, MA. 2005.

[135] Lee, Lillian Jane. "Similarity-Based Approaches to Natural Language Processing." PhD diss. Harvard University. 1997.

[136] Lewis, D. and J. Catlett. Heterogenous uncertainty sampling for supervised learning. *Proceedings of the 11th International Conference on Machine Learning (ICML-1994)*. Morgan Kaufmann Publishers. San Francisco, CA. 1994.

[137] Lewis, David D. and William A. Gale. A Sequential Algorithm for Training Text Classifiers. *Proceedings of the 17th Annual International Conference on Research and Development in Information Retrieval (SIGIR-1994)*. Association for Computing Machinery. New York. 1994.

[138] Li, Cong and Hang Li. Word translation disambiguation using bilingual bootstrapping. *Proceedings of the 40th Meeting of the Association for Computational Linguistics (ACL-2002)*. Association for Computational Linguistics. East Stroudsburg, PA. 2002.

[139] Li, Hang and Cong Li. Word translation disambiguation using bilingual bootstrapping. *Computational Linguistics* 30(1):1–22. 2004.

[140] Li, Hang and Naoki Abe. Clustering Words with the MDL Principle. *Proceedings of the 16th International Conference on Computational Linguistics (COLING-1996)*. Morgan Kaufmann Publishers. San Francisco, CA. 1996.

[141] Manning, Chris and Hinrich Schütze. *Foundations of Statistical Natural Language Processing*. MIT Press. Cambridge, MA. 1999.

[142] Marcus, Mitchell, Beatrice Santorini, and Mary Ann Marcinkiewicz. Building a large annotated corpus of English: the Penn Treebank. *Computational Linguistics* 19.2:313–330. 1993.

[143] Mark, Kevin, Michael Miller, and Ulf Grenander. Markov Random Field Models for Natural Languages. *Proceedings of the International Symposium on Information Theory*. Institute of Electrical and Electronics Engineers. New York. 1995.

[144] Mark, Kevin, Michael Miller, Ulf Grenander, and Steve Abney. Parameter Estimation for Constrained Context-Free Language Models. *Proceedings of the Fifth Darpa Workshop on Speech and Natural Language*. Morgan Kaufman Publishers. San Mateo, CA. 1992.

[145] Massart, D.L. and L. Kaufman. *The Interpretation of Analytical Chemical Data by the Use of Cluster Analysis*. John Wiley & Sons. New York. 1983.

[146] McCallum, Andrew Kachites and Kamal Nigam. Employing EM and Pool-Based Active Learning for Text Classification. *Proceedings of the 15th International Conference on Machine Learning (ICML-1998)*, pp. 350–358. Morgan Kaufmann Publishers. San Francisco, CA. 1998.

[147] McCallum, Andrew Kachites and Kamal Nigam. Text Classification by Bootstrapping with Keywords, EM and Shrinkage. *Proceedings of the Workshop on Unsupervised Learning, Meeting of the Association for Computational Linguistics (ACL-1999)*. Association for Computational Linguistics. East Stroudsburg, PA. 1999.

[148] McCallum, Andrew Kachites, Kamal Nigam, Jason Rennie, and Kristie Seymore. Automating the Construction of Internet Portals with Machine Learning. *Information Retrieval Journal* 3:127–163. 2000.

[149] McLachlan, G. Iterative reclassification procedure for constructing an asymptotically optimal rule of allocation in discriminant analysis. *Journal of the American Statistical Association* 70:365–369. 1975.

[150] McLachlan, G.J. Estimating the linear discriminant function from initial samples containing a small number of unclassified observations. *Journal of the American Statistical Association* 72:403–406. 1977.

[151] McLachlan, G.J. and K.E. Basford. *Mixture Models.* Marcel Dekker. New York. 1988.

[152] McLachlan, G.J. and S. Ganesalingam. Updating the discriminant function on the basis of unclassified data. *Communications Statistics – Simulation* 11(6):753–767. 1982.

[153] Merialdo, Bernard. Tagging English Text with a Probabilistic Model. *Computational Linguistics* 20(2):155–172. 1994.

[154] Mihalcea, Rada. Co-training and self-training for word sense disambiguation. *Proceedings of the Conference on Natural Language Learning (CoNLL-2004)*. Association for Computational Linguistics. East Stroudsburg, PA. 2004.

[155] Mikheev, Andrei. Unsupervised Learning of Word-Category Guessing Rules. *Proceedings of the 34th Meeting of the Association for Computational Linguistics (ACL-1996)*. Association for Computational Linguistics. East Stroudsburg, PA. 1996.

[156] Miller, David and Hasan Uyar. A mixture of experts classifier with learning based on both labelled and unlabelled data. *Advances in Neural Information Processing Systems 9 (NIPS-1996)* pp. 571–577. MIT Press. Cambridge, MA. 1997.

[157] Milun, David. Generating Markov Random Field Image Analysis Systems from Examples. CS Tech Report 95-23. State University of New York/Buffalo. 1995.

[158] Mitchell, T.M. The role of unlabeled data in supervised learning. *Proceedings of the 6th International Colloquium on Cognitive Science.* Kluwer Academic Publishers. Dordrecht. 1999.

[159] Morris, Stephen. Contagion. *Review of Economic Studies* 67:57–58. 2000.

[160] Muslea, Ion, Steven Minton, and Craig A. Knoblock. Active + Semi-Supervised Learning = Robust Multi-View Learning. *Proceedings of the 19th International Conference on Machine Learning (ICML-2002).* Morgan Kaufmann Publishers. San Francisco, CA. 2002.

[161] Muslea, Ion, Steven Minton, and Craig A. Knoblock. Selective Sampling With Redundant Views. *Proceedings of the 17th National Conference on Artificial Intelligence (AAAI-2000).* AAAI Press/MIT Press. Cambridge, MA. 2000.

[162] Navigli, Roberto and Paola Velardi. Learning domain ontologies from document warehouses and dedicated web sites. *Computational Linguistics* 30(2). 2004.

[163] Ng, Vincent and Claire Cardie. Weakly supervised natural language learning without redundant views. *Proceedings of the Conference on Human Language Technology and the Meeting of the North American Chapter of the Association for Computational Linguistics (HLT/NAACL-2003).* Association for Computational Linguistics. East Stroudsburg, PA. 2003.

[164] Nigam, Kamal. "Using Unlabeled Data to Improve Text Classification." PhD diss. Tech Report CMU-CS-01-126. Computer Science, Carnegie Mellon University. 2001.

[165] Nigam, Kamal and Rayid Ghani. Analyzing the effectiveness and applicability of co-training. *Proceedings of the 9th International Conference on Information and Knowledge Management.* Association for Computing Machinery. New York. 2000.

[166] Nigam, Kamal and Rayid Ghani. Understanding the Behavior of Co-training. *Proceedings of the Workshop on Text Mining, 6th International Conference on Knowledge Discovery and Databases (KDD-2000).* Association for Computing Machinery. New York. 2000.

[167] Nigam, Kamal, Andrew Kachites McCallum, Sebastian Thrun, and Tom Mitchell. Text Classification from Labeled and Unlabeled Documents using EM. *Machine Learning* 39:103–134. 2000.

[168] Nigam, Kamal, Andrew Kachites McCallum, Sebastian Thrun, and Tom Mitchell. Learning to Classify Text from Labeled and Unlabeled Documents. *Proceedings of the 15th National Conference on Artificial*

Intelligence (AAAI-1998). AAAI Press/MIT Press. Cambridge, MA. 1998.

[169] Niu, Cheng, Wei Li, Jihong Ding, and Rohini K. Srihari. A bootstrapping approach to named entity classification using successive learners. *Proceedings of the 41st Meeting of the Association for Computational Linguistics (ACL-2003)*. Association for Computational Linguistics. East Stroudsburg, PA. 2003.

[170] Niu, Cheng, Wei Li, and Rohini K. Srihari. Weakly supervised learning for cross-document person name disambiguation supported by information extraction. *Proceedings of the 42nd Meeting of the Association for Computational Linguistics (ACL-2004)*. Association for Computational Linguistics. East Stroudsburg, PA. 2004.

[171] Oflazer, Kemal, Marjorie McShane, and Sergei Nirenburg. Bootstrapping morphological analyzers by combining human elicitation and machine learning. *Computational Linguistics* 27(1). 2001.

[172] O'Neill, T.J. Normal discrimination with unclassified observations. *Journal of the American Statistical Association* 73(364):821–826. 1978.

[173] Osborne, Miles and Jason Baldridge. Ensemble-based active learning for parse selection. *Proceedings of the Conference on Human Language Technology and the Meeting of the North American Chapter of the Association for Computational Linguistics (HLT/NAACL-2004)*. Association for Computational Linguistics. East Stroudsburg, PA. 2004.

[174] Pakhomov, Serguei. Semi-supervised maximum entropy based approach to acronym and abbreviation normalization in medical texts. *Proceedings of the 40th Meeting of the Association for Computational Linguistics (ACL-2002)*. Association for Computational Linguistics. East Stroudsburg, PA. 2002.

[175] Pantel, Patrick and Deepak Ravichandran. Automatically labeling semantic classes. *Proceedings of the Conference on Human Language Technology and the Meeting of the North American Chapter of the Association for Computational Linguistics (HLT/NAACL-2004)*. Association for Computational Linguistics. East Stroudsburg, PA. 2004.

[176] Pantel, Patrick and Dekang Lin. An unsupervised approach to prepositional phrase attachment using contextually similar words. *Proceedings of the 38th Meeting of the Association for Computational Linguistics (ACL-2000)*. Association for Computational Linguistics. East Stroudsburg, PA. 2000.

[177] Pao, Y.-H. and D.J. Sobajic. Combined use of supervised and unsupervised learning for dynamic security assessment. *IEEE Transactions on Power Systems* 7(2):878–884. Institute of Electrical and Electronics Engineers. New York. 1992.

[178] Papadimitriou, Christos H., Prabhakar Raghavan, Hisao Tamaki, and Santosh Vempala. Latent semantic indexing: A probabilistic analysis. *Journal of Computer and System Sciences* 61:217–235. 2000.

[179] Peng, Fuchun and Andrew McCallum. Accurate information extraction from research papers using conditional random fields. *Proceedings of the Conference on Human Language Technology and the Meeting of the North American Chapter of the Association for Computational Linguistics (HLT/NAACL-2004).* Association for Computational Linguistics. East Stroudsburg, PA. 2004.

[180] Pereira, Fernando and Yves Schabes. Inside-outside reestimation from partially bracketed corpora. *Proceedings of the 30th Annual Meeting of the Association for Computational Linguistics (ACL-1992),* pp. 128–135. Association for Computational Linguistics. East Stroudsburg, PA. 1992.

[181] Pereira, Fernando, Naftali Tishby, and Lillian Lee. Distributional Clustering of English Words. *Proceedings of the 31st Annual Meeting of the Association for Computational Linguistics (ACL-1993),* pp. 183–190. Association for Computational Linguistics. East Stroudsburg, PA. 1993.

[182] Phillips, William, and Ellen Riloff. Exploiting strong syntactic heuristics and co-training to learn semantic lexicons. *Proceedings of the Conference on Empirical Methods in Natural Language Processing (EMNLP-2002).* Association for Computational Linguistics. East Stroudsburg, PA. 2002.

[183] Purandare, Amruta and Ted Pedersen. Word sense discrimination by clustering contexts in vector and similarity spaces. *Proceedings of the Conference on Natural Language Learning (CoNLL-2004).* Association for Computational Linguistics. East Stroudsburg, PA. 2004.

[184] Quillian, M. Semantic memory. In M. Minsky (ed.), *Semantic Information Processing,* 227–270. MIT Press. Cambridge, MA. 1968.

[185] Ratsaby, Joel. "The Complexity of Learning from a Mixture of Labeled and Unlabeled Examples." PhD diss. University of Pennsylvania. 1994.

[186] Ratsaby, Joel and Santosh S. Vankatesh. Learning from a mixture of labeled and unlabeled examples with parametric side information. *Proceedings of 8th Annual Conference on Computational Learning Theory (COLT-1995),* pp. 412–417. Association for Computing Machinery. New York. 1995.

[187] Rattray, Magnus. A model-based distance for clustering. *Proceedings of the International Joint Conference on Neural Networks (IJCNN-2000).* Institute of Electrical and Electronics Engineers. New York. 2000.

[188] Riloff, Ellen. An Empirical Study of Automated Dictionary Construction for Information Extraction in Three Domains. *AI Journal* 85. August 1996.

[189] Riloff, Ellen. Automatically Constructing a Dictionary for Information Extraction Tasks. *Proceedings of the 11th National Conference on Artificial Intelligence (AAAI-1993).* AAAI Press/MIT Press. Cambridge, MA. 1993.

[190] Riloff, Ellen. Automatically Generating Extraction Patterns from Untagged Text. *Proceedings of the 13th National Conference on Artificial Intelligence (AAAI-1996).* AAAI Press/MIT Press. Cambridge, MA. 1996.

[191] Riloff, E. and M. Schmelzenbach. An Empirical Approach to Conceptual Case Frame Acquisition. *Proceedings of the Sixth Workshop on Very Large Corpora (VLC-1998).* Association for Computational Linguistics. East Stroudsburg, PA. 1998.

[192] Riloff, E. and J. Shepherd. A Corpus-Based Approach for Building Semantic Lexicons. *Proceedings of the Conference on Empirical Methods in Natural Language Processing (EMNLP-1997),* pp. 127–132. Association for Computational Linguistics. East Stroudsburg, PA. 1997.

[193] Riloff, E. and J. Shepherd. A Corpus-Based Bootstrapping Algorithm for Semi-Automated Semantic Lexicon Construction. *Journal of Natural Language Engineering* 5(2):147–156. 1999.

[194] Riloff, Ellen and Rosie Jones. Learning Dictionaries for Information Extraction by Multi-Level Bootstrapping. *Proceedings of the 16th National Conference on Artificial Intelligence (AAAI-1999).* AAAI Press/MIT Press. Cambridge, MA. 1999.

[195] Roark, Brian and Eugene Charniak. Noun-phrase co-occurrence statistics for semi-automatic semantic lexicon construction. *Proceedings of the 36th Annual Meeting of the Association for Computational Linguistics and 17th International Conference on Computational Linguistics.* Association for Computational Linguistics. East Stroudsburg, PA. 1998.

[196] Rocchio, J.J., Jr. Relevance feedback in information retrieval. In G. Salton (ed.), *The SMART Retrieval System: Experiments in Automatic Document Processing.* Prentice-Hall. Englewood Cliffs, NJ. 1971.

[197] Roget, Peter Mark. *Roget's Thesaurus of English Words and Phrases.* 1911 edition available from Project Gutenberg (`http://www.gutenberg.org/`). 2004.

[198] Roth, Dan and Dmitry Zelenko. Toward a Theory of Learning Coherent Concepts. *Proceedings of the 17th National Conference on Artificial*

Intelligence (AAAI-2000). AAAI Press/MIT Press. Cambridge, MA. 2000.

[199] Russell, Stuart J. and Peter Norvig. *Artificial Intelligence: A Modern Approach*, 2nd Edition. Prentice Hall. Upper Saddle River, NJ. 2002.

[200] Samuel, A. Some studies in machine learning using the game of checkers. *IBM Journal of Research and Development* 3:210–229. 1959. (Reprinted in vol. 44:206–227, 2000.)

[201] Sapir, Edward. *Language: An Introduction to the Study of Speech.* Harcourt, Brace and Company. New York. 1921.

[202] Sarkar, Anoop. "Combining Labeled and Unlabeled Data in Statistical Natural Language Parsing." PhD diss. University of Pennsylvania. 2001.

[203] Schapire, Robert, Yoram Singer, and Amit Singhal. Boosting and Rocchio applied to text filtering. *Proceedings of the 21st Annual International Conference on Research and Development in Information Retrieval (SIGIR-1998).* Association for Computing Machinery. New York. 1998.

[204] Schütze, Hinrich. Word Space. *Advances in Neural Information Processing Systems* 5. Morgan Kaufmann Publishers. San Mateo, CA. 1993.

[205] Sedgewick, Robert. *Algorithms.* Second edition. Addison-Wesley. Reading, MA. 1988.

[206] Seeger, M. Learning with labeled and unlabeled data. Technical Report, University of Edinburgh. Edinburgh. 2001.

[207] Shahshahani, B.M. and D.A. Landgrebe. On the asymptotic improvement of supervised learning by utilizing additional unlabeled samples; normal mixture density case. *Proceedings of the International Conference on Neural and Stochastic Methods in Image and Signal Processing* 1766:143–155. Society of Photographic Instrumentation Engineers (SPIE). Bellingham, WA. 1992.

[208] Shahshahani, B.M. and D.A. Landgrebe. The effect of unlabeled samples in reducing the small sample size problem and mitigating the Hughes phenomenon. *IEEE Transactions on Geoscience and Remote Sensing* 32(5):1087–1095. Institute of Electrical and Electronics Engineers. New York. 1994.

[209] Shen, Dan, Jie Zhang, Jian Su, Guodong Zhou, and Chew-Lim Tan. Multi-criteria-based active learning for named entity recognition. *Proceedings of the 42nd Meeting of the Association for Computational Linguistics (ACL-2004).* Association for Computational Linguistics. East Stroudsburg, PA. 2004.

[210] Sinkkonen, Janne and Samuel Kaski. Semisupervised clustering based on conditional distributions in an auxiliary space. Technical Report A60. Helsinki University of Technology, Publications in Computer and Information Science. Espoo, Finland. 2000.

[211] Smith, Noah A. and Jason Eisner. Annealing techniques for unsupervised statistical language learning. *Proceedings of the 42nd Meeting of the Association for Computational Linguistics (ACL-2004)*. Association for Computational Linguistics. East Stroudsburg, PA. 2004.

[212] Steedman, Mark, Rebecca Hwa, Stephen Clark, Miles Osborne, Anoop Sarkar, Julia Hockenmaier, Paul Ruhlen, Steven Baker, and Jeremiah Crim. Example selection for bootstrapping statistical parsers. *Proceedings of the Conference on Human Language Technology and the Meeting of the North American Chapter of the Association for Computational Linguistics (HLT/NAACL-2003)*. Association for Computational Linguistics. East Stroudsburg, PA. 2003.

[213] Strang, Gilbert. *Introduction to Linear Algebra*. Third edition. Wellesley-Cambridge Press. Wellesley, MA. 2003.

[214] Sudo, Kiyoshi, Satoshi Sekine, and Ralph Grishman. An improved extraction pattern representation model for automatic IE pattern acquisition. *Proceedings of the 41st Meeting of the Association for Computational Linguistics (ACL-2003)*. Association for Computational Linguistics. East Stroudsburg, PA. 2003.

[215] Szummer, M. and T. Jaakkola. Kernel expansions with unlabeled examples. *Advances in Neural Information Processing Systems 13 (NIPS-2000)*. MIT Press. Cambridge, MA. 2001.

[216] Szummer, M. and T. Jaakkola. Partially labeled classification with Markov random walks. *Advances in Neural Information Processing Systems 14 (NIPS-2001)*. MIT Press. Cambridge, MA. 2002.

[217] Tang, Min, Xiaoqiang Luo, and Salim Roukos. Active learning for statistical natural language parsing. *Proceedings of the 40th Annual Meeting of the Association for Computational Linguistics (ACL-2002)*, pp. 120–127. Association for Computational Linguistics. East Stroudsburg, PA. 2002.

[218] Tesauro, Gerald. TD-Gammon, a self-teaching backgammon program, achieves master-level play. *Neural Computation* 6.2:215–219. 1994.

[219] Thelen, Michael, and Ellen Riloff. A bootstrapping method for learning semantic lexicons using extracting pattern contexts. *Proceedings of the Conference on Empirical Methods in Natural Language Processing (EMNLP-2002)*. Association for Computational Linguistics. East Stroudsburg, PA. 2002.

[220] Thrun, S. Exploration in Active Learning. In M. Arbib (ed.), *Handbook of Brain and Cognitive Science.* MIT Press. Cambridge, MA. 1995.

[221] Tipping, M. Deriving cluster analytic distance functions from Gaussian mixture models. *Proceedings of the 9th International Conference on Artificial Neural Networks.* Institute of Electrical and Electronics Engineers. New York. 1999.

[222] Tishby, Naftali Z., Fernando Pereira, and William Bialek. The information bottleneck method. In Bruce Hajek and R.S. Sreenivas, eds., *Proceedings of the 37th Allerton Conference on Communication, Control and Computing.* University of Illinois/Urbana-Champaign. Urbana, Illinois. 1999.

[223] Titterington, D.M., A.F.M. Smith, and U.E. Makov. *Statistical Analysis of Finite Mixture Distributions.* John Wiley & Sons. New York. 1985.

[224] Tolat, V.V. and A.M. Peterson. Nonlinear mapping with minimal supervised learning. *Proceedings of the Hawaii International Conference on System Science* 1. Jan 1990.

[225] Tolstov, Georgi P. *Fourier Series.* Prentice-Hall. Englewood Cliffs, NJ. 1962.

[226] Tong, S. and D. Koller. Support vector machine active learning with applications to text classification. *Proceedings of the 17th International Conference on Machine Learning (ICML-2000),* pp. 999–1006. Morgan Kaufmann Publishers. San Francisco, CA. 2000.

[227] Vapnik, Vladimir N. *Statistical Learning Theory.* John Wiley & Sons. New York. 1998.

[228] Weiss, G.M. and F. Provost. Learning When Training Data are Costly: The Effect of Class Distribution on Tree Induction. *Journal of Artificial Intelligence Research* 19:315–354. 2003.

[229] Weston, Jason et al. Protein ranking: From local to global structure in the protein similarity network. *Proceedings of the National Academy of Sciences of the United States of America* 101(17):6559–6563. 2004.

[230] Widdows, Dominic. Unsupervised methods for developing taxonomies by combining syntactic and statistical information. *Proceedings of the Conference on Human Language Technology and the Meeting of the North American Chapter of the Association for Computational Linguistics (HLT/NAACL-2003).* Association for Computational Linguistics. East Stroudsburg, PA. 2003.

[231] Widdows, Dominic, and Beate Dorow. A graph model for unsupervised lexical acquisition. *19th International Conference on Computational Linguistics (COLING-2002),* pp. 1093–1099. Morgan Kaufmann Publishers. San Francisco, CA. 2002.

[232] Widdows, Dominic, Stanley Peters, Scott Cederberg, Chiu-Ki Chan, Diana Steffen, and Paul Buitelaar. Unsupervised Monolingual and Bilingual Word-Sense Disambiguation of Medical Documents using UMLS. *Workshop on Natural Language Processing in Biomedicine, 41st Meeting of the Association for Computational Linguistics (ACL-2003)*, pp. 9–16. Association for Computational Linguistics. East Stroudsburg, PA. 2003.

[233] Winkler, Gerhard. *Image Analysis, Random Fields and Dynamic Monte Carlo Methods.* Springer-Verlag. Berlin. 1995.

[234] *Wordnet Reference Manual.* `http://wordnet.princeton.edu/doc`.

[235] Wu, Dekai, Weifeng Su, and Marine Carpuat. A kernel PCA method for superior word sense disambiguation. *Proceedings of the 42nd Meeting of the Association for Computational Linguistics (ACL-2004).* Association for Computational Linguistics. East Stroudsburg, PA. 2004.

[236] Yangarber, Roman. Counter-training in discovery of semantic patterns. *Proceedings of the 41st Meeting of the Association for Computational Linguistics (ACL-2003).* Association for Computational Linguistics. East Stroudsburg, PA. 2003.

[237] Yangarber, Roman, Ralph Grishman, Pasi Tapanainen, and Silja Huttunen. Unsupervised Discovery of Scenario-Level Patterns for Information Extraction. *Proceedings of the Conference on Applied Natural Language Processing and the Meeting of the North American Chapter of the Association for Computational Linguistics (ANLP/NAACL-2000)*, pp. 282–289. Association for Computational Linguistics. East Stroudsburg, PA. 2000.

[238] Yarowsky, David. One sense per collocation. *Proceedings of the Advanced Research Projects Agency (ARPA) Workshop on Human Language Technology.* Morgan Kaufmann Publishers. San Mateo, CA. 1993.

[239] Yarowsky, David. Unsupervised word sense disambiguation rivaling supervised methods. *Proceedings of the 33rd Annual Meeting of the Association for Computational Linguistics (ACL-1995)*, pp. 189–196. Association for Computational Linguistics. East Stroudsburg, PA. 1995.

[240] Yarowsky, D. and R. Wicentowski. Minimally supervised morphological analysis by multimodal alignment. *Proceedings of the 38th Meeting of the Association for Computational Linguistics (ACL-2000)*, pp. 207–216. Association for Computational Linguistics. East Stroudsburg, PA. 2000.

[241] Zhang, Q., and Sally Goldman. EM-DD: An Improved Multiple-Instance Learning Technique. *Advances in Neural Information Processing Systems 14 (NIPS-2001).* MIT Press. Cambridge, MA. 2002.

[242] Zhang, Tong. The value of unlabeled data for classification problems. *Proceedings of the 17th International Conference on Machine Learning (ICML-2000).* Morgan Kaufmann Publishers. San Francisco, CA. 2000.

[243] Zhang, Tong and Frank J. Oles. A probability analysis on the value of unlabeled data for classification problems. *Proceedings of the 17th International Conference on Machine Learning (ICML-2000),* pp. 1191–1198. Morgan Kaufmann Publishers. San Francisco, CA. 2000.

[244] Zhu, Xiaojin. "Semi-Supervised Learning with Graphs." PhD diss. Carnegie Mellon University. 2005.

[245] Zhu, Xiaojin, John Lafferty, and Zoubin Ghahramani. Combining active learning and semi-supervised learning using Gaussian fields and harmonic functions. *Proceedings of the 20th International Conference on Machine Learning (ICML-2003).* AAAI Press/MIT Press. Cambridge, MA. 2003.

[246] Zhu, Xiaojin, John Laffery, and Zoubin Ghahramani. Semi-supervised learning: from Gaussian fields to Gaussian processes. CMU Tech Report CMU-CS-03-175. Carnegie Mellon University. 2003.

[247] Zhu, Xiaojin, Zoubin Ghahramani, and John Lafferty. Semi-supervised learning using Gaussian fields and harmonic functions. *Proceedings of the 20th International Conference on Machine Learning (ICML-2003).* AAAI Press/MIT Press. Cambridge, MA. 2003.

Index

Nonalphabetic.

A.

B.

C.

D.

M.

N.

O.

P.

Q.

R.

S.

T.

U.

V.

W.

X.Y.Z.